Metaverse Communication and Computing Networks

IEEE Press
445 Hoes Lane
Piscataway, NJ 08854

IEEE Press Editorial Board
Sarah Spurgeon, *Editor in Chief*

Jón Atli Benediktsson
Anjan Bose
James Duncan
Amin Moeness
Desineni Subbaram Naidu

Behzad Razavi
Jim Lyke
Hai Li
Brian Johnson

Jeffrey Reed
Diomidis Spinellis
Adam Drobot
Tom Robertazzi
Ahmet Murat Tekalp

Metaverse Communication and Computing Networks

Applications, Technologies, and Approaches

Edited by

Dinh Thai Hoang
University of Technology Sydney, Australia

Diep N. Nguyen
University of Technology Sydney, Australia

Cong T. Nguyen
Duy Tan University, Vietnam

Ekram Hossain
University of Manitoba, Canada

Dusit Niyato
Nanyang Technological University, Singapore

Copyright © 2024 by The Institute of Electrical and Electronics Engineers, Inc.
All rights reserved.

Published by John Wiley & Sons, Inc., Hoboken, New Jersey.
Published simultaneously in Canada.

No part of this publication may be reproduced, stored in a retrieval system, or transmitted in any form or by any means, electronic, mechanical, photocopying, recording, scanning, or otherwise, except as permitted under Section 107 or 108 of the 1976 United States Copyright Act, without either the prior written permission of the Publisher, or authorization through payment of the appropriate per-copy fee to the Copyright Clearance Center, Inc., 222 Rosewood Drive, Danvers, MA 01923, (978) 750-8400, fax (978) 750-4470, or on the web at www.copyright.com. Requests to the Publisher for permission should be addressed to the Permissions Department, John Wiley & Sons, Inc., 111 River Street, Hoboken, NJ 07030, (201) 748-6011, fax (201) 748-6008, or online at http://www.wiley.com/go/permission.

Trademarks: Wiley and the Wiley logo are trademarks or registered trademarks of John Wiley & Sons, Inc. and/or its affiliates in the United States and other countries and may not be used without written permission. All other trademarks are the property of their respective owners. John Wiley & Sons, Inc. is not associated with any product or vendor mentioned in this book.

Limit of Liability/Disclaimer of Warranty
While the publisher and author have used their best efforts in preparing this book, they make no representations or warranties with respect to the accuracy or completeness of the contents of this book and specifically disclaim any implied warranties of merchantability or fitness for a particular purpose. No warranty may be created or extended by sales representatives or written sales materials. The advice and strategies contained herein may not be suitable for your situation. You should consult with a professional where appropriate. Further, readers should be aware that websites listed in this work may have changed or disappeared between when this work was written and when it is read. Neither the publisher nor authors shall be liable for any loss of profit or any other commercial damages, including but not limited to special, incidental, consequential, or other damages.

For general information on our other products and services or for technical support, please contact our Customer Care Department within the United States at (800) 762-2974, outside the United States at (317) 572-3993 or fax (317) 572-4002.

Wiley also publishes its books in a variety of electronic formats. Some content that appears in print may not be available in electronic formats. For more information about Wiley products, visit our web site at www.wiley.com.

Library of Congress Cataloging-in-Publication Data applied for:

Hardback ISBN: 9781394159987

Cover Design: Wiley
Cover Image: © Olga Siletskaya/Getty Images

Set in 9.5/12.5pt STIXTwoText by Straive, Chennai, India

To my family
— Dinh Thai Hoang

To my family
— Diep N. Nguyen

To my family
— Cong T. Nguyen

To my parents
— Ekram Hossain

To my family
— Dusit Niyato

Contents

Editors' Biography

Dinh Thai Hoang received his PhD degree from the School of Computer Science and Engineering, Nanyang Technological University, Singapore, in 2016. He is currently a faculty member at the University of Technology Sydney (UTS), Australia. Over the past 10 years, he has significantly contributed to advanced wireless communications and networking systems. This is evidenced by his excellent record with one patent filed by Apple Inc., two authored books, one edited book, four book chapters, more than 80 IEEE Q1 journals, and 60 flagship IEEE conference papers in the areas of communications and networking. Most of his journal papers have been published in top IEEE journals, including IEEE JSAC, IEEE TWC, IEEE COMST, and IEEE TCOM. Furthermore, his research papers have had a high impact, as evidenced by nearly 14,000 citations with an h-index of 44 (according to Google Scholar) over the past 10 years. Since joining UTS in 2018, he has received more than AUD 3 million in external funding and several precious awards, including the Australian Research Council Discovery Early Career Researcher Award for his project "Intelligent Backscatter Communications for Green and Secure IoT Networks" and IEEE TCSC Award for Excellence in Scalable Computing for Contributions on "Intelligent Mobile Edge Computing Systems" (Early Career Researcher). Alternatively, he is the lead author of two authored books, "*Ambient Backscatter Communication Networks,*" published by Cambridge Publisher in 2020 and "Deep Reinforcement Learning for Wireless Communications and Networking," published by IEEE-Wiley Publisher in 2022. He is currently an Editor of IEEE TMC, IEEE TWC, IEEE TCCN, IEEE TVT, and IEEE COMST.

Diep N. Nguyen is a faculty member of the Faculty of Engineering and Information Technology, University of Technology Sydney (UTS). He received ME and PhD in Electrical and Computer Engineering from the University of California San Diego (UCSD) and the University of Arizona (UA), respectively. Before joining UTS, he was a DECRA Research Fellow at Macquarie University, a member of technical staff at Broadcom (California), ARCON Corporation (Boston), consulting the Federal Administration of Aviation on turning detection of UAVs and aircraft, US Air Force Research Lab on anti-jamming. He has received several awards from LG Electronics, the University of California San Diego, the University of Arizona, US National Science Foundation, and Australian Research Council, including nominations for the outstanding RA (2013) awards, the best paper award at the WiOpt conference (2014), Discovery Early Career Researcher Award (DECRA, 2015), and outstanding Early Career Researcher award (SEDE, University of Technology Sydney, 2018). His recent research interests are in the areas of computer networking, wireless communications, and machine learning application, with an emphasis on systems' performance and security/privacy. Dr. Nguyen is a senior member of IEEE and an editor/associate editor of the *IEEE Transactions on Mobile Computing, IEEE Access, Sensors* journal, and *IEEE Open Journal of the Communications Society* (OJ-COMS).

Cong T. Nguyen received his BE degree in Electrical Engineering and Information Technology from Frankfurt University of Applied Sciences in 2014, his MSc in Global Production Engineering and Management from the Technical University Berlin in 2016, and his PhD in Information Technology from University of Technology Sydney in 2023. He is currently with Duy Tan University, Vietnam. His research interests include blockchain technology, operation research, game theory, and optimization.

Ekram Hossain is a professor and an associate head (Graduate Studies) at the Department of Electrical and Computer Engineering, University of Manitoba, Canada. He is a member (Class of 2016) of the College of the Royal Society of Canada. His current research interests include design, analysis, and optimization of 6G cellular wireless networks. He was listed as a Clarivate Analytics Highly Cited Researcher in Computer Science for six years in a row from 2017 to 2022. He received the 2017 IEEE ComSoc Technical Committee on Green Communications and Computing Distinguished Technical Achievement Recognition Award "for outstanding technical leadership and achievement in green wireless communications and networking." He has won several research awards, including the 2017 IEEE Communications Society Best Survey Paper Award and the 2011 IEEE Communications Society Fred Ellersick Prize Paper Award. He served as the editor-in-chief for the IEEE Communications Surveys and Tutorials from 2012 to 2016 and the editor-in-chief for IEEE Press. He was a distinguished lecturer of the IEEE Communications Society and the IEEE Vehicular Technology Society. He was an elected member of the board of governors of the IEEE Communications Society for the term from 2018 to 2020. He was elevated to an IEEE fellow "for contributions to spectrum management and resource allocation in cognitive and cellular radio networks." He is a fellow of the Canadian Academy of Engineering and a fellow of the Engineering Institute of Canada.

Dusit Niyato is a professor at the School of Computer Science and Engineering, Nanyang Technological University, Singapore. He received BE from King Mongkuk's Institute of Technology Ladkrabang (KMITL), Thailand, in 1999 and a PhD in Electrical and Computer Engineering from the University of Manitoba, Canada, in 2008. Dusit's research interests are in the areas of distributed collaborative machine learning, the Internet of Things (IoT), edge intelligent metaverse, mobile and distributed computing, and wireless networks. Dusit won the Best Young Researcher Award of IEEE Communications Society (ComSoc) Asia Pacific and the 2011 IEEE Communications Society Fred W. Ellersick Prize Paper Award and the IEEE Computer Society Middle Career Researcher Award for Excellence in Scalable Computing in 2021 and Distinguished Technical Achievement Recognition Award of IEEE ComSoc Technical Committee on Green Communications and Computing 2022. Dusit also won a number of best paper awards, including IEEE Wireless Communications and Networking Conference (WCNC), IEEE International Conference on Communications (ICC), IEEE ComSoc Communication Systems Integration and Modelling Technical Committee, and IEEE ComSoc Signal Processing and Computing for Communications Technical Committee 2021. Currently, Dusit is serving as editor-in-chief of IEEE Communications Surveys and Tutorials, an area editor of IEEE Transactions on Vehicular Technology, editor of IEEE Transactions on Wireless Communications, associate editor of IEEE Internet of Things Journal, IEEE Transactions on Mobile Computing, IEEE Wireless Communications, IEEE Network, and ACM Computing Surveys. He was a guest editor of the IEEE Journal on Selected Areas on Communications. He was a distinguished lecturer of the IEEE Communications Society for 2016–2017. He was named the 2017–2021 highly cited researcher in computer science. He is a fellow of IEEE and a fellow of IET.

List of Contributors

Mshari Aljumaie
School of Electrical and Data Engineering
University of Technology Sydney
Ultimo, NSW
Australia

and

Department of Information Technology
Taif University
Taif
Saudi Arabia

Carlos Bermejo
Department of Computer Science and Engineering, School of Engineering
The Hong Kong University of Science and Technology
Hong Kong SAR
China

Sweta Bhattacharya
School of Information Technology and Engineering
Vellore Institute of Technology
Tamil Nadu
India

Tristan Braud
Division of Integrative Systems and Design
Hong Kong University of Science and Technology
Hong Kong SAR
China

Wei Cai
School of Science and Engineering
The Chinese University of Hong Kong
Shenzhen, Shenzhen
China

Dimitris Chatzopoulos
School of Computer Science
University College Dublin
Dublin
Ireland

Mahdi Chehimi
Wireless@VT
Bradley Department of Electrical and Computer Engineering, Virginia Tech
Arlington, VA
USA

Rajeswari Chengoden
School of Information Technology and Engineering
Vellore Institute of Technology
Tamil Nadu
India

Hieu Chi Nguyen
School of Electrical and Data Engineering
University of Technology Sydney
Ultimo, NSW
Australia

Nam H. Chu
School of Electrical and Data Engineering
University of Technology Sydney
Ultimo, NSW
Australia

Tan Do-Duy
Department of Computer and Communication Engineering
Ho Chi Minh City University of Technology and Education
Ho Chi Minh City
Vietnam

Haihan Duan
School of Science and Engineering
The Chinese University of Hong Kong
Shenzhen, Shenzhen
China

Eryk Dutkiewicz
School of Electrical and Data Engineering
University of Technology Sydney
Ultimo, NSW
Australia

Thippa Reddy Gadekallu
School of Information Technology and Engineering
Vellore Institute of Technology
Tamil Nadu
India

and

Department of Electrical and Computer Engineering
Lebanese American University
Byblos
Lebanon

Zhu Han
Department of Electrical and Computer Engineering
University of Houston
Houston, TX
USA

Yue Han
Alibaba-NTU Singapore Joint Research Institute
Nanyang Technological University
Singapore

Pawan Kumar Hegde
School of Information Technology and Engineering
Vellore Institute of Technology
Tamil Nadu
India

Dinh Thai Hoang
School of Electrical and Data Engineering
University of Technology Sydney
Ultimo, NSW
Australia

Pan Hui
Department of Computer Science and Engineering, School of Engineering
The Hong Kong University of Science and Technology
Hong Kong SAR
China

Dong In Kim
Department of Electrical and Computer Engineering
Sungkyunkwan University
Suwon, Gyeonggi-do
Korea

Cyril Leung
Department of Electrical and Computer Engineering
University of British Columbia
Vancouver, BC
Canada

Lik-Hang Lee
Department of Industrial and Systems Engineering (ISE)
The Hong Kong Polytechnic University
Hong Kong SAR
China

Chin-Teng Lin
Computational Intelligence and Brain-Computer Interface, Australian Artificial Intelligence Institute
University of Technology Sydney
Ultimo, NSW
Australia

Dongxiao Liu
Department of Electrical and Computer Engineering
University of Waterloo
Waterloo, ON
Canada

Tom H. Luan
School of Cyber Science and Engineering
Xi'an Jiaotong University
Xi'an, Shaanxi
China

Praveen Kumar Reddy Maddikunta
School of Information Technology and Engineering
Vellore Institute of Technology
Tamil Nadu
India

Shiwen Mao
Department of Electrical and Computer Engineering
Auburn University
Auburn, AL
USA

Dusit Niyato
School of Computer Science and Engineering
Nanyang Technological University
Singapore

Cong T. Nguyen
School of Electrical and Data Engineering
University of Technology Sydney
Ultimo, NSW
Australia

Diep N. Nguyen
School of Electrical and Data Engineering
University of Technology Sydney
Ultimo, NSW
Australia

Quoc-Viet Pham
School of Computer Science and Statistics
Trinity College Dublin
Dublin
Ireland

Xuan-Qui Pham
ICT Convergence Research Center
Kumoh National Institute of Technology
Gumi
Korea

Walid Saad
Wireless@VT
Bradley Department of Electrical and Computer Engineering, Virginia Tech
Arlington, VA
USA

Xuemin Shen
Department of Electrical and Computer Engineering
University of Waterloo
Waterloo, ON
Canada

Zhou Su
School of Cyber Science and Engineering
Xi'an Jiaotong University
Xi'an, Shaanxi
China

Thien-Huynh The
Department of Computer and Communication Engineering
Ho Chi Minh City University of Technology and Education
Ho Chi Minh City
Vietnam

Nancy Victor
School of Information Technology and Engineering
Vellore Institute of Technology
Tamil Nadu
India

Yuntao Wang
School of Cyber Science and Engineering
Xi'an Jiaotong University
Xi'an, Shaanxi
China

Rui Xing
School of Cyber Science and Engineering
Xi'an Jiaotong University
Xi'an, Shaanxi
China

Minrui Xu
School of Computer Science and Engineering
Nanyang Technological University
Singapore

Hongliang Zhang
School of Electronics
Peking University
Beijing
China

Ning Zhang
Department of Electrical and
Computer Engineering
University of Windsor
Windsor, ON
Canada

Pengyuan Zhou
School of Cyber Science and
Technology
University of Science and Technology
of China
Hefei
China

Howe Yuan Zhu
Computational Intelligence and
Brain-Computer Interface
Australian Artificial Intelligence
Institute
University of Technology Sydney
Ultimo, NSW
Australia

Preface

Recently, Metaverse has gained paramount interest and huge investment from the tech industry. Microsoft acquired Activision Blizzard for $70 billion in 2022 to set its first footsteps in the Metaverse game development race. Along with its huge investment in AR, one of the core technologies of Metaverse, Google has invested $39.5 million in a private equity fund for all Metaverse projects. Nvidia has created Omniverse, a developing tool for Metaverse applications. Besides huge investments from big tech companies, the economic activities of virtual worlds are also significant, with transactions that exceed the magnitude of millions of dollars. As a result, there is no doubt that the Metaverse will become one of the most prominent directions of development in both industry and academia. However, the development of the Metaverse, especially in academia, is still in a nascent stage. Currently, researchers are striving to judge the shape and boundary of the future Metaverse. They are only able to envision some of its possible characteristics, such as open space, decentralization, human–computer interaction experience, digital assets, and digital economy. Moreover, Metaverse applications are expected to face various challenges such as massive resource demands, ultralow latency requirements, interoperability among applications, and security and privacy concerns. Given the above, this book aims to provide a comprehensive overview of Metaverse and discuss its enabling technologies and how these technologies can be utilized to develop Metaverse applications.

Sydney, Australia

Dinh Thai Hoang
Diep N. Nguyen
Cong T. Nguyen
Ekram Hossain
Dusit Niyato

Acknowledgments

The contribution made by Dr. Dinh Thai Hoang was supported in part by the Australian Research Council's Discovery Projects funding scheme (project DE210100651).

The contribution done by Prof. Dusit Niyato was supported in part by the National Research Foundation (NRF), Singapore, and Infocomm Media Development Authority under the Future Communications Research Development Programme (FCP); DSO National Laboratories under the AI Singapore Programme (AISG Award No: AISG2-RP-2020-019); and under DesCartes and the Campus for Research Excellence and Technological Enterprise (CREATE) programme.

Introduction

Edited by: Dinh Thai Hoang, Diep N. Nguyen, Cong T. Nguyen, Ekram Hossain, and Dusit Niyato

The term "Metaverse" refers to next-generation Internet applications that aim to create virtual 3D environments where humans can interact with each other and the applications' functionalities via digital avatars. Although the original concept dates back to 1992, Metaverse has recently attracted paramount attention due to the huge potential to rival, or even replace, conventional Internet applications in the near future.

However, the development of the Metaverse, especially in academia, is still in a nascent stage. Currently, researchers are striving to judge the shape and boundary of the future Metaverse. They can only envision some of its possible characteristics, such as open space, decentralization, human–computer interaction experience, digital assets, and digital economy. Moreover, Metaverse applications are expected to face various challenges, such as massive resource demands, ultralow latency requirements, application interoperability, and security and privacy concerns.

Given the above, this book aims first to introduce the emerging paradigm of Metaverse, which is expected to pave the way for the evolution of the future Internet. The book also provides a comprehensive review of the state-of-the-art research and development covering different aspects of Metaverse for a wide range of readers, from general readers to experts. Advanced knowledge including innovative models, techniques, and approaches to overcome the limitations and challenges in developing Metaverse are then discussed. Finally, emerging applications of Metaverse are presented, along with the related challenges and open issues.

1

Metaverse: An Introduction

Lik-Hang Lee[1], Dimitris Chatzopoulos[2], Pengyuan Zhou[3], and Tristan Braud[4#]

[1] *Department of Industrial and Systems Engineering (ISE), The Hong Kong Polytechnic University, Hong Kong SAR, China*
[2] *School of Computer Science, University College Dublin, Dublin, Ireland*
[3] *School of Cyber Science and Technology, University of Science and Technology of China, Hefei, China*
[4] *Division of Integrative Systems and Design, Hong Kong University of Science and Technology, Hong Kong SAR, China*

After reading this chapter you should be able to:

- Understand the current trends and challenges that building such a virtual environment will face.
- Focus on three major pillars to guide the development of the Metaverse: privacy, governance, and ethical design and to guide the sustainable yet acceptable development of the Metaverse.
- Illustrate a preliminary modular-based framework for an ethical design of the Metaverse.

1.1 Introduction

The term "Metaverse" was first introduced to the public in 1992 by Neal Stephenson in his work of science fiction, "Snow Crash." The main characters of the book are shown to coexist with their avatars in a world that is an integration of the virtual and the real, and it is populated by persistent virtual entities that are superimposed on our actual surroundings. People are able to execute a wide variety of immersive activities in this integrated reality.

# All authors equally contributed to this chapter.

Metaverse Communication and Computing Networks: Applications, Technologies, and Approaches, First Edition.
Edited by Dinh Thai Hoang, Diep N. Nguyen, Cong T. Nguyen, Ekram Hossain, and Dusit Niyato.
© 2024 The Institute of Electrical and Electronics Engineers, Inc. Published 2024 by John Wiley & Sons, Inc.

Several noticeable instances of this trend include people getting together with their friends in a different location, working jointly with their coworkers, and participating in shared virtual experiences (e.g. dating and virtual fitting). In other words, diverse digital or virtual contents originating from cyberspace will eventually go beyond the boundary of 2D displays in the Internet that we are now using and gradually make their way into three-dimensional (3D) settings.

As was said before, coincidentally, the projected environment is congruent with Mark Weiser's vision of ubiquitous computing in 1991: computer services would be integrated into a multitude of facets of our lives, and users will have access to virtual information whenever and wherever they choose. With such a compelling vision, the landscape of ubiquitous computing has been advanced throughout the course of the previous three decades by the proliferation of computing devices. These computing devices include laptop computers, smartphones, the Internet of Things (IoTs), and intelligent wearables.

According to Milgram and Kishino's Reality-Virtuality Continuum [16], the current cyberspace has undergone significant development in recent years, and recent attempts have been made to provide human users with services and digital experiences by means of virtual environments such as augmented reality (AR) and virtual reality (VR). Although no one can say for certain what the Metaverse will bring about once it is fully realized, recent pre-metaverse apps have most likely identified AR and VR on smartphones as the major testbed for immersive user experiences. Pokémon Go, for example, has become the most popular AR program on ubiquitous smartphones, astoundingly with 1 billion downloads, while Google Cardboard is bringing VR content to mainstream audiences (for example, YouTube VR) [4].

As such, the term "Metaverse" refers to a blended space at the intersection between physical and digital in which multiple users can concurrently interact with a persistent and unified computer-generated environment, and other users. This space has the potential to become the next important milestone in the development of cyberspace as it exists today.

It is worth noting that modern devices that enable entrance to Metaverse get access to multiple types of users' data. Also services based on artificial intelligence (AI) use derivatives of data generated by users in their function, making data the new commodity that spawns a lucrative, fast-growing industry.

This introductory chapter focuses primarily on discussing the evolution of the Metaverse as well as the difficulties that have been encountered. First, we will provide a concise overview of the evolution of cyberspace as well as the importance of technological enablers. As a result, our bottom-up methodology places an emphasis on the following three crucial technological enablers for the Metaverse: networks, systems, and users. In addition, we emphasize a number of essential challenges, both from a technical and an ecosystemic point of view, that are necessary for the construction and maintenance of the Metaverse.

1.2 The Metaverse: Fantasy, Text, 3D Worlds

It should come as no surprise that technological advancements always play the role of an essential constituent in the construction of the subsequent phase of cyberspace's evolution into the Metaverse. Before we get to the status quo, also known as the pre-Metaverse (V), we will have passed through four distinct phases in rapid succession. The stages are as follows: (i) Literature: imaginary worlds described in written works; (ii) Text-based virtual worlds: computer-mediated cyberspace solely driven by text-based interaction; (iii) 3D Virtual Worlds: enriched cyberspace with graphics; and (vi) AR and VR on Ubiquitous Devices: virtual-physical blended worlds appearing on mobile phones, tablets, and smartglasses. In other words, the above four phases have been triggered by technological advancement, and they are interconnected.

At the beginning of the process, Literature evolves into the limitless area of imagination that is required for technological development. Even though the technologies were not quite developed enough at the time, some of the most representative novels are "The Master Key," "Dungeons & Dragons," "Neuromancer," and "Snow Crash." These novels depict what it might be like to have immersive experiences. Multi-User Dungeon games, often known as MUD games, have been popular since the introduction of personal computers (II), and they enable user interaction via the use of text. Next, in 1995, the very first 3D virtual environment, known as Active World, was released. This occurred when personal computers became capable of running computer graphics programs (III). Multiplayer 3D online games like Second Life and Minecraft have become more popular as a result of the proliferation of online users. In addition, the proliferation of smartphones in the latter part of the 2000s (VI) laid the groundwork for the mobile AR and VR applications that were discussed before.

After 2015 (V), the commercialization of AR and VR headsets, such as Google Glass and Meta Oculus, will enable users to be enclosed by computer-mediated environments and to see digital overlays in virtual-physical blended environments, such as VR Chat, Horizon Workrooms, and HoloMeeting, respectively. It is important to note that the pandemic that has been going on since 2020 can be regarded as one of the "experiments" that has been carried out on the largest scale in the history of the world: are people willing to accept the continued movement of various life functions into virtual environments? Obviously, we should. Certain visionary multinational companies in the technology sector, such as Microsoft and Meta, are promoting virtual collaboration spaces for everyone via inexpensive AR and VR ubiquitous devices. This includes both end users and businesses. As a result, we anticipate that the further progressions of cyberspace that incorporate immersiveness, also known as immersive cyberspace, will need to take into consideration various technological enablers.

The subsequent paragraphs in this chapter provide a reality check on the various technological enablers and the issues they provide. We take a bottom-up approach to discuss the three most important aspects, which have been inherited from the viewpoints of ubiquitous computing in particular: the network and mobile edge, system interoperability, and user-centric immersive design, which are representing the network, the system, and human users. We are aware that the breadth of the technological enablers for the Metaverse [14]) (such as artificial intelligence, hyperledger, computer vision, and robotics) may transcend beyond the boundaries of our current conversation.

1.3 The Rise of Edge Computing

In order to deliver the same degree of experience as reality, it is vital in the Metaverse to ensure that the users have the sense of being immersed in the environment. Latency, for example the motion to photon (MTP) latency,[1] is one of the most critical factors to consider. According to what the researchers have discovered, the MTP latency requirement has to be below the human perceptible limit in order for users to be able to engage with holographic scenes and objects in a smooth and natural manner [15].

For example, the registration procedure for AR may create nausea and dizziness because of excessive latency, which typically results in virtual items trailing behind the desired location. As a result, lowering latency is essential for the Metaverse, particularly in settings where real-time data processing is required, for example, real-time AR overlapping with the actual environment, such as in AR surgeries. On the other hand, the Metaverse often necessitates a computation that is too demanding for mobile devices, which severely restricts its usefulness.

1.3.1 Offloading

The process of offloading is often utilized as a means of relieving the burden of compute and memory in this context, despite the fact that doing so incurs extra delay in the networking process. As a result, achieving a satisfactory compromise is essential in order to make the offloading process visible to the end users, which is regrettably not a simple task. For instance, while presenting a locally navigable 1 MTP, latency refers to the amount of time that passes between an action taken by the user and the moment when that action's related consequence is reflected on the screen.

1 MTP latency is the amount of time between the user's action and its corresponding effect to be reflected on the display screen.

It is required to have an egocentric view that is two times wider than the Field-of-View (FOV) of the headset in order to compensate for the networking delay that occurs during offloading. However, there is a conflict between the needed user's egocentric view(s) and the networking latency. Longer networking latency can be caused by a bigger size of display and resolution, and thus the streaming of more digital content, both of which result in even longer networking delay. As a result, a solution that improves physical deployment may be a more practical option than one that focuses just on resource orchestration. Because of the fluctuating and unexpected nature of the high latency, cloud offloading is unable to constantly achieve the optimal balance. This results in long-tail latency performance, which has a negative influence on the user experience [5].

Therefore, a supplementary solution is required in order to ensure that users will have a consistent and immersive experience inside the Metaverse. Edge computing, which computes, stores, and transmits the data physically closer to end-users and their devices, can reduce the user-experienced latency when compared with cloud offloading. Edge computing works by computing, storing, and transmitting the data physically closer to end-users and their devices. As early as 2009, Satyanarayana et al. [20] noticed that installing powerful cloud-like infrastructure only one wireless hop away from the end users' mobile devices, namely Cloudlet, might completely change the course of the game.

The most recent iterations of mobile AR frameworks have begun to use edge-based solutions in an effort to boost the overall performance of Metaverse applications. For instance, EdgeXAR is a mobile AR framework that uses the advantages of edge offloading to achieve lightweight tracking with six degree-of-freedom (DOF) while reducing the offloading delay from the user's view [25]. Additionally, edge-facilitated augmented vision in vehicle-to-everything (EAVVE) provides a new system of cooperative AR vehicular perception between multiple drivers, which are remarkably assisted by edge servers. This system helps to lower the total offloading latency and compensates for the inadequate processing capability that is present in vehicles [26].

1.3.2 Scale to Outdoor

Users are able to have a more engaging and immersive experience at higher frame rates without compromising the level of details present in immersive settings by offloading processing to an edge server (for example, a high-end personal computer). On the other hand, such systems can only be used in interior settings where the user's movement is restricted. It is essential to have seamless movement outside in order to enable for a Metaverse experience that is both really and completely ubiquitous. With the development of 5G and 6G, it is anticipated that multiaccess edge computing, also known as MEC, will enhance the Metaverse user experience.

MEC can deliver standard and universal edge offloading services, which is just one hop away from cellular-connected user devices, such as AR smartglasses.

Not only does it have the potential to cut down on the round-trip time (RTT) of packet delivery, but it also paves the way for near real-time orchestration of multiuser interactions. MEC is essential for outdoor Metaverse services to have in order to perceive the specifics of the local situation and organize intimate collaborations between users or devices in close proximity to one another. For instance, 5G MEC servers may manage the AR content of adjacent users with just one hop of packet transfer. This enables real-time user engagement for social AR apps like "InGress[2]." Utilizing MEC to improve the experience of the Metaverse has gained interest from the academic community. Also, MEC has been employed by several Metaverse enterprises to enhance the experience of their customers.[3]

1.4 Universality, Interoperability, and Openness

When this chapter was being prepared, the majority of the plans and initiatives pertaining to the Metaverse were separate projects that were entirely controlled by a single organization. In spite of the fact that such fragmentation makes it possible to experiment with many thoughts and ideas during an experimental age, this approach cannot be maintained over the long run. Even more so, we contend that interoperability and openness will be the two key factors that will determine the success or the failure of the Metaverse's push for global acceptance.

1.4.1 Using the World Wide Web as an Illustration

In multitudinous ways, the Metaverse may be seen as the next stage of development after the World Wide Web in terms of a new type of content generation and dissemination. The World Wide Web was conceptualized from the very beginning with decentralization and interoperability [2] in mind. A website may be created by almost anybody or any organization provided they have enough technical knowledge, e.g. a computer engineer or programmer, and access to the Internet. The operation of the World Wide Web is not dependent on any centralized body, and the more centralized services (such as Domain Name Registration) are only handy extensions to a wholly decentralized technological base. In point of fact, Tim Berners-personal Lee's computer at the European Organization for Nuclear Research (CERN) was the host of the very first website that was ever made available to the public. The fact that there is a very low barrier to entry, in addition to the

2 https://ingress.com
3 https://nianticlabs.com/blog/niantic-planet-scale-ar-5g-urban-legends/

fact that the user has full control over what is published, gave people ownership of the medium, which contributed to the success of the Web.

Interoperability and openness were critical factors in the development of the Web into the pervasive technology that it is happening. On the other hand, this resulted in a number of different actors inventing their own versions of the HTML language and rendering engines. Tim Berners-Lee established the World Wide Web Consortium (W3C)[4] in 1994 with the intention of promoting interoperability throughout the industry and, more crucially, maintaining the vendor-neutrality of the Internet. Nowadays, the W3C serves as an organization for standardization. Under the direction of its members, it writes and sets the standards that govern the World Wide Web.

1.4.2 Realizing the Potential of Interoperability in the Metaverse

The Metaverse needs to be based on a paradigm that is analogous to that of the World Wide Web, in which users may access a wide variety of virtual worlds that are hosted by a variety of entities by means of devices and browsers that have been developed by a large number of businesses and organizations. On the other hand, the Metaverse has properties that have a substantial impact on this operation. Before it was made available to the general public, the World Wide Web was a technology that was only developed and administered by a single organization.

Comparatively, there have been several efforts to construct the Metaverse since the late 1990s. These attempts began with the French virtual world "Deuxi'eme Monde"[5] (1997) (1997) and continued with the more well-known "Second Life[6]" (2003). In contrast to the World Wide Web, the majority of these initiatives were developed using the concept of massively multiplayer online games. In this approach, a single corporation retains complete control over the material that is published on their platform.

More recent hyperledger-based concepts, such as Decentraland,[7] take things one step further by storing material on distributed servers (IPFS) and tracking ownership using a public hyperledger. However, the online platform that the user uses to access the Metaverse world continues to be reliant on the organization that is its parent, and there is very little interchange between the web platform and other platforms. However, we need to highlight that deleting data from decentralized storage mechanisms may be challenging but necessary to comply with the "right to be forgotten" that has been advocated for by several organizations all over the globe. In light of the growing number of Metaverse initiatives that are

4 World Wide Web Consortium – https://www.w3.org/
5 Mémoires du Deuxième Monde – www.bimondiens.com/
6 Second Life – https://secondlife.com/
7 Welcome to Decentraland – https://decentraland.org/

incompatible with one another, it is more important than ever to build standardizing organisms that are able to solve the following aspects:

1. The opportunity for anybody to build a server and run a virtual world that is linked to the rest of the Metaverse.
2. The ability to access the Metaverse using any device and browser as long as they adhere to a predetermined specification (e.g. client-based rendering).
3. Keeping a record of who owns each piece of digital property across all of the servers and the clients.
4. Allowing avatars to provide access to their data in exchange for credit in Metaverse spaces.
5. Making it possible for a single avatar to communicate with other avatars located on other servers.
6. Providing users with the ability to produce, display, buy, sell, and remove digital assets inside the Metaverse.
7. Integrating inter-Metaverse (in equivalence to interplanetary) learning mechanisms [19, 23] that will allow users, via their avatars, to update and use personalized models that are adapted to different virtual worlds within the Metaverse.
8. It is essential for users to have these traits in order to take ownership of the Metaverse and to encourage ubiquitous adoption, which will pave the road for a Metaverse that is omnipresent.

1.4.3 The Argument in Favor of Immersive Technologies

Both ubiquitous computing and the Metaverse rely heavily on technologies that provide users with an immersive experience, such as AR and VR. When it comes to ubiquitous computing, they make it possible for greater contextualization of data to be visualized, and when it comes to the Metaverse, they help users feel more immersed in the virtual environment they are in.

A common way to think about these technologies is as a spectrum spanning from realism to complete virtuality, with applications integrating digital material with a physical setting to varying degrees. This spectrum may be thought of as a continuum. Because the Metaverse is still in its infancy, there is no concrete definition of the extent to which these technologies should interact with the real world. This ambiguity will undoubtedly have an effect on interoperability, since certain Metaverse programs may be firmly rooted at a specific physical place, while others may be fully virtual [6] and accessible through a conventional desktop computer. According to the findings of a research on the creation of an AR Metaverse campus [3], an entirely virtual experience should be replicated of the real-world setting, via the use of digital twins, which make it possible for geolocalized experiences to be accessed by people located elsewhere. On the other hand, in a manner that is analogous to the problems that have been discussed so far, the only entity that can definitively specify how to produce such duality is a standardization organism.

1.5 Steps Toward Mobile User Interaction Within the Metaverse

Users should have their own channels to convert their intents into actions in virtual or immersive scenarios whenever and wherever they choose [11]. This is especially important in light of the fact that the Metaverse will ultimately permeate every facet of everyday activities. The already available AR and VR apps for smartphones provide light on the widespread acceptance of virtual material that is layered on the real environment. Smartphones cannot compare to the level of immersion provided by the latest generation of Metaverse devices, such as AR and VR headsets. Nevertheless, modern headsets do not provide users with effective options for user input. Under the premise that interfaces are increasing smaller in size, user interaction with virtual contents in the Metaverse becomes more complex and laborious as a result.

A great number of research prototypes are working toward the goal of enhancing unique input channels that incorporate user mobility. For example, Google's Jacquard[8] incorporates integrated sensors into clothing, which augments our everyday clothes as a huge touch interface for user interaction that is similar to that of a touchscreen. Wearable addendums are another sort of mobile input. These addendums expand users' bodies as the focus of user engagement with virtual contents. Examples of wearable addendums include wristbands, gloves, and finger-worn devices. There is still a considerable difference in input capabilities between sedentary solutions like the keyboard and mouse duos and the mobile input solutions that were discussed before, despite the fact that an increasing number of mobile input solutions have been offered in recent years [12].

In addition, the information that can be found in physical environments might be deemed limitless; however, the augmentation of the information that can be seen inside the relatively limited field of vision of headsets can be difficult [13]. As a result, we are going to have to perfect the way that virtual material is presented. In the event that this does not occur, users of the Metaverse who take a simplistic approach to the administration of virtual material may experience an information overflow, which will result in a considerable increase in the amount of time spent choosing augmentation. Context awareness is a notable method that

8 https://atap.google.com/jacquard/

involves taking into consideration the users, settings, and social dynamics [10]. Edge AI, such as recommendation systems, are able to analyze user context and provide the most relevant augmentation.

In addition, the current study on user contexts focuses largely on the users' five senses, which are presented in a plain and truthful manner throughout the research. Instead, further efforts should be taken to allow the user-centric Metaverse to grasp the abstract but difficult-to-quantify sensations on top of the five senses. For example, a home with a dark and purple backdrop as opposed to a daunted house may elicit different feelings in the user (abstract feeling). In the Metaverse, a higher granularity of context-awareness enables the provision of services that are both more exact and more personally tailored. Even if edge computing has the potential to enhance not just user experiences but also user privacy, Metaverse users will nevertheless leave behind a large number of user interaction traces in a cyberspace that blends the virtual and the real.

User privacy and design space are not fully explored when seen from the edge infrastructure. As a result of the fact that it is anticipated that users would participate in a variety of virtual activities, it follows that every incorrect pop-ups of AR/VR entities might lead to new privacy issues. For instance, users may choose to leave their "augmented" discussions in the public realm of the immersive environment. There is a potential risk to users' privacy if every line of such discussion with numerous users is visually augmented [9]. The proprietors of virtual spaces need to work to strengthen information flows, maybe by having them be driven by contextual integrity. User and data contexts are impacted by a wide variety of elements, including the receiver of the data, the locations of the data, the sensitivity of the data, and so on. On the other hand, our knowledge of the design of information flows for highly diverse augmentation is quite limited, and that's not even taking into account the undiscovered types of augmentation that will emerge in the Metaverse.

1.6 Bringing Users' Profiles and Assets on the Metaverse

Notably, all the immersive technologies that will deliver entry points to Metaverse via multiple types of devices [22] will have access to potentially private users' data. These data are useful for

1. Producing personalised artificial intelligence (AI) models that will be enhance users' quality of experience (QoE) while being on the Metaverse.
2. Analyzing users' behavior and producing data of financial value (e.g. for advertising).

Representative Example A marathon athlete is trained for the next competition and is wearing a smart watch, smart shoes, and a VR headset while being remotely supervised by her coach in a virtual world that looks like the next marathon. The shoes are collecting data related to the athlete's steps and running technique, the watch gathers data associated with the current condition of her body, and the VR headset simulates the conditions of a marathon trail and virtually places the athlete together with other athletes in the trail. The athlete is training in virtual worlds, placing past versions of herself on the trail as different avatars to compare her performance with previously recorded attempts. All these data are used as input to a coaching service for athletes that simulate running environments, predict injuries, and assesses the health of the athlete. Unfortunately, the company that offers this service has access to the generated data.

The athlete should be able to decide which types of her data she is willing to share with the coaching company and the coaching company should employ privacy-preserving techniques to process sensitive data [8]. Additionally, the athlete's devices should support on-device privacy-aware training techniques that guarantee that her private data are not shared with any centralized computing entity [18]. Moreover, in the scenario where the athlete is invited by fellow athlete to train together in a different virtual world that is hosted by a different company, the athlete should not only be able to join the virtual world, but she should also be able to "bring together her past selves" as avatars who are running in the new virtual world [7, 21]. Note that since the athlete is running for the first time in this new virtual world, the hosting company should be able to predict the performance of her past selves based on the produced AI models the athlete is bringing with her.

Last, considering that advertisements will be a major source income at Metaverse, the hosting company of the new virtual world should be able to get access to derivatives of the analytics produced by the previous world in order to present meaningful advertisements. Such information can be considered as a form of credit exchange between the worlds (i.e. similarly to the roaming service provided by mobile operators).

1.6.1 The Role of Distributed Ledgers, Smart Contracts, and Decentralized Storage

Distributed ledgers and smart contracts are developed from the Byzantine Generals' dilemma of achieving data consensus in a decentralized and trustless way. Remarkably, the first yet well-known example is Bitcoin [17]. The development of the technical details of such technology is described in Chapter 7. Accordingly, all the aforementioned challenges can be tackled with decentralized Metaverse-ready solutions that are based on distributed ledgers (e.g. the Ethereum blockchain [24]) and decentralized storage (e.g. IPFS [1]). AI models can be stored on IPFS and produced by IPLS [19] while being assisted by compute nodes [23]. Although multiple virtual worlds can share a distributed ledger for retrieving instantly information related to the users via specifically designed smart contracts, adding information to such ledger will be time demanding, depending on the employed consensus protocol and the size of the consensus network. Furthermore, an envisioned solution should require a set of smart contracts per user that will provide the flexibility to each user to control the information they are willing to share.

1.7 Conclusions and Future Research Directions

Because the Metaverse places an emphasis on user experiences that are at an entirely new level, spanning both the real and the virtual worlds, it necessitates concurrent efforts from networking, system, and user-centric components. As a closing remark, we draw attention to a number of pressing concerns about three different facets of technology, each of which has the potential to become the stumbling block in the process of implementing the "ubiquitous" Metaverse on a large scale. The offloading of Metaverse applications in a way that is delay-insensitive is the first obstacle to overcome.

Real-time offloading is essential for delay-sensitive applications in the Metaverse since it is required to compensate for the restricted capacity of mobile devices. Cloud has been an industry leader in offloading for many years; however, recent research has shown that its latency does not satisfy the MTP criteria described above. A key topic of worry is the rise of privacy concerns brought on by the unreliability of deep learning, which is frequently employed in cloud computing. The rapidly evolving 5G and 6G network technologies are demonstrating tremendous potential for a very low last-mile latency, which will allow delay-transparent edge offloading. As a result, the Metaverse is able to envision a future in which ubiquitous service supply would enable users to freely enjoy an immersive experience while adventuring in the outdoors.

The second challenge is the social and technical roadblocks that stand in the way of an interoperable Metaverse. It is now difficult to envisage a standard arising since the technologies that will enable the Metaverse are not yet well defined. This makes it difficult to understand how a standard may emerge. The Metaverse should, in a manner similar to that of the Environment Wide Web, make it possible for any entity to develop and host its own virtual world, which users would then be able to explore using avatars of their choosing.

The key technical difficulty in achieving such intercompatibility will be the development of digital material and ownership of that content across different virtual worlds. Concerning the sociocultural dimensions, we anticipate seeing a pattern that is analogous to that of open-source software, in which groups will strive to achieve interoperability for either humanistic (free access to the medium) or pragmatic reasons (openness drives innovation). Concurrently, other entities will attempt to create monopolies over the Metaverse in an effort to increase their power.

The last part concerns the throughput rates of users as well as the users' perceptions of virtual environments. The throughput rates of mobile input devices are much lower than those of standard interfaces, which are designed to be used when sedentary. In addition, content management and presentation on AR/VR headsets, in addition to a better understanding of users' abstract sensations, are still areas that need further research.

The construction of the Metaverse, which is often referred to as the immersive Internet, will take many decades, much as the creation of the Internet did. More significantly, once virtual environments are extensively deployed in our surrounds, this will take the Metaverse a significant amount of time to iterate and refine its ecosystems. This will also take place over a lengthy period of time, i.e. a decade. As a result, it is necessary for us to take into consideration many technological issues and ecosystem problems that have potential effects on the long-term viability of the Metaverse [14]. The main obstacle is the effect of the Metaverse on social human life, as the avatar's life might be different from what the individual is experiencing in real life. So the effect of the Metaverse on social human life needs to be further addressed. Avatar behaviors in the wild, trust in multiple avatar identities, content creations, cross-generational contents, cancel culture, decentralized design of virtual economy and governance, digital humanity, user diversity and fairness, online addiction, user data ownership and ethics, user safety with AR and VR, stewardship, and accountability would be also some of the potential issues.

Bibliography

1 Juan Benet. IPFS-content addressed, versioned, P2P file system. *arXiv preprint arXiv:1407.3561*, 2014.

2 Tim Berners-Lee. History of the web. *World Wide Web Foundation. Retrieved from https://webfoundation.org/about/vision/history-of-the-web*, 2015.

3 Tristan Braud, Carlos Bermejo Fernández, and Pan Hui. Scaling-up AR: University campus as a physical-digital Metaverse. In *2022 IEEE Conference*

on Virtual Reality and 3D User Interfaces Abstracts and Workshops (VRW), pages 169–175. IEEE, 2022.

4 Dimitris Chatzopoulos, Carlos Bermejo, Zhanpeng Huang, and Pan Hui. Mobile augmented reality survey: From where we are to where we go. *IEEE Access*, 5:6917–6950, 2017. doi: 10.1109/ACCESS.2017.2698164.

5 Jeffrey Dean and Luiz André Barroso. The tail at scale. *Communications of the ACM*, 56:74–80, 2013. URL http://cacm.acm.org/magazines/2013/2/160173-the-tail-at-scale/fulltext.

6 Haihan Duan, Jiaye Li, Sizheng Fan, Zhonghao Lin, Xiao Wu, and Wei Cai. Metaverse for social good: A university campus prototype. In *Proceedings of the 29th ACM International Conference on Multimedia*, pages 153–161, 2021.

7 Yao-Chieh Hu, Ting-Ting Lee, Dimitris Chatzopoulos, and Pan Hui. Analyzing smart contract interactions and contract level state consensus. *Concurrency and Computation: Practice and Experience*, 32(12):e5228, 2020. doi: https://doi.org/10.1002/cpe.5228.

8 Vlasis Koutsos, Dimitrios Papadopoulos, Dimitris Chatzopoulos, Sasu Tarkoma, and Pan Hui. Agora: A privacy-aware data marketplace. *IEEE Transactions on Dependable and Secure Computing*, 19(6):3728–3740, 2021. doi: 10.1109/TDSC.2021.3105099.

9 Abhishek Kumar, Tristan Braud, Lik-Hang Lee, and Pan Hui. Theophany: Multimodal speech augmentation in instantaneous privacy channels. In *Proceedings of the 29th ACM International Conference on Multimedia (MM '21), October 20–24, 2021, Virtual Event, China*. Association for Computing Machinery (ACM), 2021. ISBN 978-1-4503-8651-7/21/10. doi: 10.1145/3474085.3475507.

10 Kit Yung Lam, Lik-Hang Lee, and Pan Hui. A2W: Context-aware recommendation system for mobile augmented reality web browser. In *ACM International Conference on Multimedia*, United States, October 2021. Association for Computing Machinery (ACM). doi: https://doi.org/10.1145/3474085.3475413.

11 Lik-Hang Lee, Tristan Braud, Farshid Hassani Bijarbooneh, and Pan Hui. UbiPoint: Towards non-intrusive mid-air interaction for hardware constrained smart glasses. *Proceedings of the 11th ACM Multimedia Systems Conference*, 2020.

12 Lik-Hang Lee, Tristan Braud, S. Hosio, and Pan Hui. Towards augmented reality-driven human-city interaction: Current research and future challenges. *ArXiv*, abs/2007.09207, 2020.

13 Lik-Hang Lee, Tristan Braud, Kit-Yung Lam, Yui-Pan Yau, and Pan Hui. From seen to unseen: Designing keyboard-less interfaces for text entry on the constrained screen real estate of augmented reality headsets. *Pervasive and Mobile Computing*, 64:101148, 2020.

14 Lik-Hang Lee, Tristan Braud, Pengyuan Zhou, Lin Wang, Dianlei Xu, Zijun Lin, Abhishek Kumar, Carlos Bermejo, and Pan Hui. All one needs to know about Metaverse: A complete survey on technological singularity, virtual ecosystem, and research agenda. *ArXiv*, abs/2110.05352, 2021.

15 Katerina Mania, Bernard D. Adelstein, Stephen R. Ellis, and Michael I. Hill. Perceptual sensitivity to head tracking latency in virtual environments with varying degrees of scene complexity. In *Proceedings of the 1st Symposium on Applied Perception in Graphics and Visualization*, APGV '04, pages 39–47, New York, NY, USA, 2004. Association for Computing Machinery. ISBN 1581139144. doi: 10.1145/1012551.1012559.

16 P. Milgram and F. Kishino. A taxonomy of mixed reality visual displays. *IEICE Transactions on Information and Systems*, 77:1321–1329, 1994.

17 Satoshi Nakamoto. Bitcoin: A peer-to-peer electronic cash system. 2008.

18 Kamalesh Palanisamy, Vivek Khimani, Moin Hussain Moti, and Dimitris Chatzopoulos. SplitEasy: A practical approach for training ML models on mobile devices. In *Proceedings of the 22nd International Workshop on Mobile Computing Systems and Applications*, HotMobile '21, pages 37–43, New York, NY, USA, 2021. Association for Computing Machinery. ISBN 9781450383233. doi: 10.1145/3446382.3448362.

19 Christodoulos Pappas, Dimitris Chatzopoulos, Spyros Lalis, and Manolis Vavalis. IPLS: A framework for decentralized federated learning. In *2021 IFIP Networking Conference (IFIP Networking)*, pages 1–6, 2021. doi: 10.23919/IFIPNetworking52078.2021.9472790.

20 Mahadev Satyanarayanan, Paramvir Bahl, Ramon Caceres, and Nigel Davies. The case for VM-based cloudlets in mobile computing. *IEEE Pervasive Computing*, 8(4):14–23, 2009. doi: 10.1109/MPRV.2009.82.

21 Pierre Schutz, Stanislas Gal, Dimitris Chatzopoulos, and Pan Hui. Decentralizing indexing and bootstrapping for online applications. *IET Blockchain*, 1(1):3–15, 2021. doi: https://doi.org/10.1049/blc2.12001. URL https://ietresearch.onlinelibrary.wiley.com/doi/abs/10.1049/blc2.12001.

22 Kirill A. Shatilov, Dimitris Chatzopoulos, Lik-Hang Lee, and Pan Hui. Emerging ExG-based NUI inputs in extended realities: A bottom-up survey. *ACM Transactions on Interactive Intelligent Systems*, 11(2):1–49, 2021. doi: 10.1145/3457950.

23 Joana Tirana, Christodoulos Pappas, Dimitris Chatzopoulos, Spyros Lalis, and Manolis Vavalis. The role of compute nodes in privacy-aware decentralized AI. In *Proceedings of the 6th International Workshop on Embedded and Mobile Deep Learning*, EMDL '22, pages 19–24. Association for Computing Machinery, 2022. doi: 10.1145/3539491.3539594.

24 Gavin Wood. Ethereum: A secure decentralised generalised transaction ledger. *Ethereum Project Yellow Paper*, 151(2014):1–32, 2014.

25 Wenxiao Zhang, Sikun Lin, Farshid Bijarbooneh, Hao-Fei Cheng, Tristan Braud, Pengyuan Zhou, Lik-Hang Lee, and Pan Hui. EdgeXAR: A 6-DoF camera multi-target interactionframework for MAR with user-friendly latencycompensation using edge computing. In *Proceedings of the ACM on HCI (Engineering Interactive Computing Systems)*, 2022.

26 Pengyuan Zhou, Tristan Braud, Aleksandr Zavodovski, Zhi Liu, Xianfu Chen, Pan Hui, and Jussi Kangasharju. Edge-facilitated augmented vision in vehicle-to-everything networks. *IEEE Transactions on Vehicular Technology*, 69(10):12187–12201, 2020. doi: 10.1109/TVT.2020.3015127.

2

Potential Applications and Benefits of Metaverse

Mshari Aljumaie[1,2], Hieu Chi Nguyen[1], Nam H. Chu[1], Cong T. Nguyen[1], Diep N. Nguyen[1], Dinh Thai Hoang[1], and Eryk Dutkiewicz[1]

[1] *School of Electrical and Data Engineering, University of Technology Sydney, Ultimo, NSW, Australia*
[2] *Department of Information Technology, Taif University, Taif, Saudi Arabia*

After reading this chapter, you should be able to:

- Aware of potential Metaverse applications in notable areas, including entertainment, virtual office, education, healthcare, autonomous vehicles, and virtual tourism.
- Understand the benefits of Metaverse for each area of applications as well as the specific challenges that these applications are facing.

2.1 Metaverse Applications for Entertainment

2.1.1 Introduction to Entertainment

Undoubtedly, the entertainment industry is one of the most important pillars to pushing Metaverse's implementation forward. Briefly, entertainment can be defined as an activity or event that is designed to attract and bring delight or pleasure to attendees [23]. As such, there are many forms of entertainment such as music, game, and performance. Content, user experience, and community are among the most important factors that keep users returning to an entertainment application. Specifically, even though excellent content can impress users at the beginning, a bad user experience (e.g. too many bugs, lags, and intermittent connections) can lead to a "dead" application (i.e. no one cares about or uses it). Moreover, people join entertainment applications to relax, so they do not want to

Metaverse Communication and Computing Networks: Applications, Technologies, and Approaches, First Edition.
Edited by Dinh Thai Hoang, Diep N. Nguyen, Cong T. Nguyen, Ekram Hossain, and Dusit Niyato.
© 2024 The Institute of Electrical and Electronics Engineers, Inc. Published 2024 by John Wiley & Sons, Inc.

return to a toxic community full of foul languages, cheaters, and bullies. As such, maintaining a good community is vital for entertainment providers to attract users.

2.1.2 Existing Entertainment Activities

Before the rise of technology, people enjoy entertainment by "physically" participating in activities, e.g. concerts, performances, and games. Thanks to the evolution of technology and computer-based applications, the way we immerse ourselves in entertainment activities has rapidly changed over the last few decades, e.g. from passively watching entertainment programs at home through television to actively joining activities, such as creating content and interacting with other users. Currently, several fundamental concepts of the Metaverse (i.e. a "virtual" community, a blend of digital and physical worlds, and a universal virtual world) are partially observed in current online-entertainment applications. For example, we can not only watch videos on YouTube, a social platform, but also create and upload our videos to share with others on this platform. In addition, users can interact with others via a comment section for each video. However, users may not feel that this is a "real" community since it lacks the "physical visualization."

In contrast, in three-dimensional (3D)-based applications, e.g. Massive Multiplayer Online Games (MMPOGs), users feel more connected with their communities in the games due to characters/avatars' appearances and behaviors. In addition, as 3D avatars only appear when users log in, users can perceive the existence of surrounding avatars. Furthermore, with recent advances in Extended-Reality (XR) technologies, full integration of the digital and physical world is no longer a futuristic notion, as shown in Figure 2.1. This results in new ways of interaction in the digital world rather than using traditional equipment such as a mouse, keyboard, and gamepad. For instance, there are many XR-based applications (e.g. Pokemon Go, Beat Saber, and Firewall Zero Hour) allowing us to interact with virtual objects (e.g. saber and shield) in the same way that we experience in the physical world (e.g. grabbing, gestures, and body movements). By employing XR, the physical feel of the virtual world becomes more realistic. For example, in Pokemon Go, a camera on a mobile device (e.g. a phone or tablet) is used to capture real-world scenes (e.g. street, park, and building). Then, augmented reality (AR) is used to blend the 3D virtual Pokemons (i.e. fictional creatures) on this scene in real-time. On the device's screen, Pokemons look like existing creatures in real-world scenes. However, most of the current MMPOGs do not allow users to create new objects freely. For example, users in World of Warcraft can level up an item (e.g. a sword and armor), but the new item's appearance and specification must follow the upgrade mechanism of this game.

Recently, several entertainment types have been integrated into a single virtual environment. Specifically, some virtual concerts attracting a massive number of

Figure 2.1 Pokemon Go [5].

online users have been organized in game environments. For instance, about 12.3 million concurrent users attended Travis Scott's concert that was organized on the Fortnite platform in April 2020. Fortnite was also chosen by the BTS, a K-pop group, to debut their new single in September 2020 [6]. Similarly, other artists (e.g. David Guetta, Lil Nas X, and Charli XCX) organized their concerts in Roblox [27]. In these events, both artists and audiences are represented by their avatars. Thus, it will bring a new user experience that we may have never seen before.

Given the above, it can be observed that each current design of entertainment applications only partially fulfills user experience demands. Specifically, social media platforms (e.g. Facebook and YouTube) lack a physical feel of communication, while users' creativity is limited in MMPOGs. In addition, current 3D virtual entertainment events are still unable to leverage recent advances in virtual reality (VR) technology (e.g. XR). Moreover, since each entertainment is independent, users need to create and manage multiple avatars, each for one application. In Section 2.1.3, we will discuss how the Metaverse can address these gaps in the entertainment sector.

2.1.3 Entertainment Activities in Metaverse

The Metaverse introduces a seamless integration of multiple XR-based virtual worlds (i.e. applications) and limitless creativity for users. First, a Metaverse user only needs one account to move between entertainment applications while keeping his/her assets intact, e.g. virtual objects, electronic wallet, and avatar's appearance. Thus, the Metaverse makes account management easier than that of traditional approaches. After logging in, Metaverse users may start at their virtual homes that are analogous to home pages on social network platforms such as Facebook and YouTube. In Metaverse, users' homes can be their virtual properties, e.g. a house, flat, or even an island. From there, users select entertainment applications as they desire and can also invite their friends to join an entertainment application. Note that similar to what we experience in the physical world, Metaverse users may be free or have to pay (e.g. buying tickets) to enter an entertainment application, depending on the owner of it. In addition, the full integration of XR in Metaverse blends the physical and digital worlds to create a more realistic communication in entertainment applications. Second, limitless creativity is a standout characteristic of Metaverse, which brings new experiences to the entertainment industry. Specifically, it allows users to create virtual objects in the Metaverse by using tools provided by this platform, e.g. users can customize their appearances as they like. Moreover, users also can bring their creations in the physical world to this platform via digitizing techniques, e.g. Digital Twin [44]. Based on these principles, the Metaverse can also allow users to create their own virtual worlds. As such, users can design their entertainment world (i.e. application) and then easily share/sell them with/to others in the Metaverse. By doing so, entertainment applications in the Metaverse come not only from companies but also from Metaverse users, leading to a wide variety of entertainment applications. It is worth mentioning that users can keep their world for themself, similar as we enjoy our private spaces in real life.

Next, we will discuss some examples of entertainment activities in Metaverse. For years, the global revenue of video game industry has been being one of the largest values in comparison with others in the entertainment industry, e.g. box office and music [7]. In the Metaverse, fully integrating XR technologies will introduce new gaming experiences to users. First, XR technologies change the way users interact in gaming environments. For example, XR devices, e.g. Hololen, can recognize hand gestures and movements to replace mouse, keyboard, and gamepad usage. Other gaming systems, such as Omni One VR Platform, also leverage body movements (e.g. jump and run) to control game characters. As such, these new control systems help us feel more natural and comfortable during gaming activities. On the other hand, they not only allow us to burn more calories due to the demands of body movements but also avoid some health issues because of sitting in front of a computer/console and using a keyboard/mouse for a long time. Second, XR technologies provide an environment that is a blend of digital and physical worlds. As such, instead of using only a virtual environment, gaming in the Metaverse can take advantage of XR, which allows users to interact with virtual objects in a real-world environment. For example, Pokemon Go, a massive phenomenon in mobile gaming, can be seen as a primitive form of this genre.

In the entertainment industry, concert is another activity expected to be significantly transformed in the Metaverse era. Currently, we can physically attend concerts or virtually go to virtual concerts organized on online platforms. In the Metaverse, the ways that we experience concerts can be both virtual and physical [33, 35] at once, as shown in Figure 2.2. In particular, a concert is held in the real world, and people can attend it physically. At the same time, the digital version of this concert is opened in the Metaverse for people who cannot attend it physically or choose to do so. Thanks to XR technologies, two types of audiences (i.e. who physically and virtually attend the concert) can interact with other as they are all in the same scene. In addition, some visual effects only are experienced via XR technologies, so physically attended people may still need to use XR-based devices. As the Metaverse development is still at its early stage, many new experiences to every online-entertainment activity will eventually be introduced. In Section 2.1.4, we will discuss current challenges that are impeding the implementation of Metaverse in entertainment.

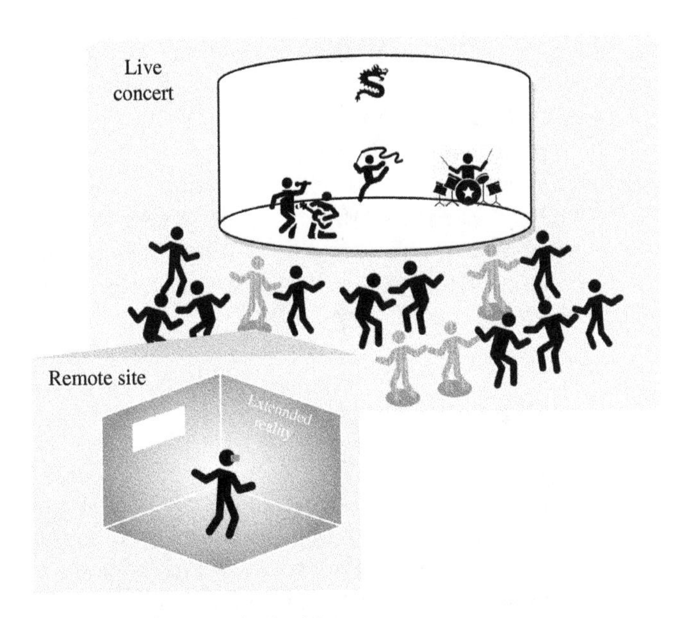

Figure 2.2 A concert in the Metaverse.

2.1.4 Challenges of Entertainment in the Metaverse

As discussed above, the Metaverse platform is expected to bring the next level of user experience for entertainment activities. On the other hand, to achieve that goal, there are many problems with the underlying infrastructure that need to be effectively handled to guarantee the operation of Metaverse. Among common challenges in creating and managing Metaverse, two problems are especially important for entertainment applications in the Metaverse.

First, an entertainment activity (e.g. game and concert) may consist of millions of concurrent attendees. Although current platforms (e.g. Roblox and Fortnite) can support events with millions of concurrent users, they are only traditional massive online systems without the XR technologies. Recall that XR-based applications demand strict requirements of high computing resources and connections with low-latency (e.g. 7 ms) and high-speed (e.g. 250 Mbps) [45]. Thus, two or more events occurring at the same time in the Metaverse will put tremendous stress on the underlying infrastructure and management system.

Second, since user experience is a standout feature of entertainment in Metaverse compared to conventional activities, the Quality of Experiences (QoE) is among the most important metrics for Metaverse's entertainment. However, guaranteeing a high QoE is a very challenging task due to the massive number of ongoing users in each entertainment application. Moreover, a Metaverse entertainment event can be a hybrid, i.e. both real-world and digital at once. Thus, a large number of people are likely gathering and entering the Metaverse simultaneously in an area, making it more challenging to maintain QoE.

To address these challenges, recent technological advances need to be employed. For instance, the multitier and over-the-air computing approaches can be leveraged to alleviate the computing stress due to the Metaverse. In addition, the novel next-generation multiple access (NGMA) for 6G is being studied to provide high-speed, low-latency, and reliable connections. Thus, this multiple access scheme can be used to support unprecedented massive user access and holographic telepresence in the Metaverse.

2.2 Virtual Office in Metaverse

2.2.1 Introduction to Virtual Office

Historically, the term "telework," or telecommuting has been used since 1975 to describe a working condition where employees perform their work remotely outside a traditional office space (e.g. working from home) by using information and communications technologies [38]. Virtual office, or virtual workplace, is expected to be the next generation of telework [34] that can enhance connections and enable virtual collaboration between multiple coworkers. A virtual office platform can provide a broad category of applications and services, including teleconferencing and group meetings, team chatting apps, e-mail, digital whiteboard, and remote working access.

The COVID-19 pandemic made numerous companies to adopt telework solutions to sustain business operations and protect their employees. Even after the epidemic, many big companies including Facebook, Twitter, Shopify, Slack, and Square, planned to establish new policies to allow their staff to select a fully remote or hybrid working mode in the long term [16]. As a result, the global virtual office market is predicted to increase from US$34.77 billion in 2020 to US$101.39 billion in 2027, that is, at a compound annual growth rate (CAGR) of 16.5% [17].

A virtual office allows employees to work from anywhere at any time. Thus, it encourages autonomy and work flexibility for employees as well as organizational flexibility for employers. From the employee's perspective, working in a virtual office can offer a healthier work–life balance and make the workers feel more comfortable. For example, one can keep wearing his favorite sweatpants while working online. From another viewpoint, running a virtual office can allow an employer to reduce operational and renting expenses for real estate, open up opportunities for hiring global talents, and remove barriers to business scaling up. Regarding the environmental impact, a virtual office produces a lower amount of greenhouse gas emissions than the traditional one because of the reduction in employee commuting [43] and electricity usage, as well as producing less waste.

2.2.2 Current Virtual Office Platforms Toward Metaverse

In 2021, Facebook (now called Meta) launched the beta version of Horizon Workrooms, a virtual collaborative platform for remote working [22]. Workrooms allow coworkers to join the same virtual meeting room from anywhere and work together efficiently. Users can use a VR headset (i.e. the Oculus Quest 2) or a web browser to connect to the virtual environment. Workrooms bring several engaging experiences for collaborative working such as brainstorming with other team members and sketching new ideas on the virtual whiteboard, having more natural conversations, or working together on a document. By using the VR device and mixed reality (MR) technology, users can bring their desk, computer, and keyboard into the virtual workplace with all functions as in the real world, including the ability to share a screen with other colleagues. Users can also use a handheld controller as a pen to write and draw on the whiteboard, stick it around in Workrooms, or export it as an image if needed.[1] To create an immersive

---

1 See the video demonstration at https://www.youtube.com/watch?v=lgj50IxRrKQ.

and conversation experience, Workrooms applies spatial audio for automatic volume adjusting and hand tracking technology to allow users to point or give a thumbs-up with their fingers.

VIVE Sync is an all-in-one VR solution introduced by HTC for meetings and collaboration [18]. The solution provides powerful avatar creation tools, where users can create personal avatars with a high degree of realism. Users can also customize the avatars with a vast selection of various clothing, eyeglasses, and hairstyles. Additionally, tools such as spatial audio, gestures, and facial and eye tracking are available to help in improving interhuman interactions, for example one can give high-fives and shake hands with other colleagues. Another attractive feature of VIVE Sync is that it offers several spectacular conference halls and meeting spaces from a lovely garden, a tranquil beach villa, an innovative metropolis, and many others as shown in Figure 2.3. VIVE Sync is accessible from a wide range of devices, including smartphones, tablets, PCs, and VR headsets, and users can even join a meeting from other platforms such as Zoom and Microsoft Teams.

Microsoft's latest MR technology, called Mesh, can bring individuals from many places together and give them the sense that they are sitting in the same room [19]. With the use of projected virtual holograms, Mesh can be used to create a virtual team meeting and allow people to collaborate on 3D projects remotely. Initially, Microsoft Mesh only supports animated representations of humans in an MR environment with virtual avatars taken from the prior social VR platform AltspaceVR. However, future designs of Mesh include the ability to transport an entirely virtual representation of a person, replete with motions and expressions that Microsoft calls "holoportation" technology as illustrated in Figure 2.4. The entire platform is

Figure 2.3 A meeting space in VIVE Sync. Source: HTC Corporation.

Figure 2.4 Virtual collaboration in Microsoft Mesh by Microsoft. Licensed under a Creative Commons Attribution 4.0 International. Downloaded from https://github.com/MicrosoftDocs/mesh-docs/blob/main/mesh/media/holographic-collab.png on 25th October 2022.

built on top of Microsoft Azure cloud infrastructure, and it supports high-quality 3D graphical model streaming, thus allowing designers, engineers, and scientists to collaborate on 3D projects at a distance.

2.2.3 Benefits and Potential Use Cases of Metaverse Workplace

The COVID-19 pandemic has revealed several drawbacks of traditional remote meeting platforms such as Zoom and Microsoft Teams, where everything is carried through a computer screen. People who use those platforms may often feel a lack of dynamism, overwhelmed with information, and stressed from having too much close-up eye contact. A recent study from Stanford University on Zoom fatigue shows that 13.8% and 5.5% of participants said they were "very" and "extremely" exhausted after the video meetings [20]. In the Metaverse, we can have much more interesting virtual meetings thanks to highly realistic 3D avatars and human-to-human interaction as in the real world. As people interact and move around freely in a virtual workplace that mimics the real one, they will enjoy genuine connections and feel less distant from coworkers.

In addition, VR and AR technologies can bring a seamless experience for Metaverse users as they can work from anywhere with their optimum desktop setup. Users can have multiple virtual monitors in the arrangement as they want,

Figure 2.5 Future collaborative work in the Metaverse by Microsoft. Licensed under a Creative Commons Attribution 4.0 International. Downloaded from https://github.com/MicrosoftDocs/mesh-docs/blob/main/mesh/media/scene-1.png on 25th October 2022.

and a virtual keyboard if needed. This feature is substantial for designers, artists, software developers, and many other professionals that work with multiscreens and usually jump around between different software and tools. A Metaverse workplace can also provide an infinite whiteboard for better information relaying and team collaboration. Users can stick any documents and windows on the whiteboard directly from their PCs, such as an informative dashboard, and sketch out new ideas together. [2]

Interaction with 3D objects and models is another benefit of the Metaverse, especially in industries such as manufacturing, architecture, healthcare, and design. Digital twin technology allows replicating real-world objects, thus enabling more accurate mock-ups, which improve comprehension of the objects' materials or physical structure. For example, a group of automotive engineers can discuss a 3D architecture of a novel braking system (as illustrated in Figure 2.5), or an interior designer can share a new design at a true scale with a customer and quickly get feedback on it just like an in-person meeting [37].

2.2.4 Challenges of Virtual Office in Metaverse

Thus far, the Metaverse has the potential to provide an immersive 3D space that takes the virtual office and remote working experience to the next level. However, current Metaverse workplace platforms are still at their early development stages, and several technology-related issues must be resolved to realize their full potential. The first thing to be considered is the infrastructure capacity, especially

2 See the video demonstration at https://www.youtube.com/watch?v=lgj50IxRrKQ.

network resources, due to the massive workload for immersive technologies (i.e. VR, AR, and MR) and high-quality 3D content streaming. Also, more computation resources are required to concurrently serve a large number of users and satisfy the high demands of the cloud computing and cloud rendering concept used by the Metaverse application. Guaranteeing data security and privacy is another challenge while working in the Metaverse. Extensive volumes of personal data of workers and customers will be generated by their activity on Metaverse platforms, such as behaviors, interactions with other people, and even health-related based data, including body movements, eye-gaze tracking, and facial expressions. A data breach or loss might have severe consequences for companies and employees.

To address the challenges facing the development of the Metaverse workplace, several enabling network architecture and technologies can be considered. First, effective schemes for jointly computing and allocating wireless resources are necessary to ensure the smooth functioning of a wireless system for Metaverse applications. Such schemes should be designed proficiently to provide users with reliable and uninterrupted access to the Metaverse by optimizing the wireless system's performance and enhancing the overall user experience. Additionally, a collaborative computing paradigm between cloud, edge, and end devices can be leveraged to minimize latency for mobile users by utilizing computing capabilities present in the mobile edge networks. This is critical as mobile users demand seamless access to the Metaverse from any location and at any time. Furthermore, the use of advanced security solutions, such as privacy-preserving distributed learning and blockchain-based data governance system, is crucial to preventing privacy issues related to the extensive volumes of personal data generated by users' activity on Metaverse workplace.

2.3 Education

2.3.1 The Development of Online Learning

Over the last three decades, thanks to the rapid growth of technology, we have witnessed the strong development of education, especially in online learning. Online learning (or e-learning) refers to a new form of learning in which students and teachers can communicate in a virtual online class from their computers via the Internet network. In this way, we can create a flexible and effective learning environment for both students and teachers as they can learn and teach from their conventional places (as long as with Internet connections). Furthermore, e-learning can save a lot of costs and human resources because we do not need to build facilities (e.g. schools, classrooms, tables, chairs) and a team of staff to manage and maintain the school. As a result, many online courses are offered globally and have rapidly grown over the last ten years. For example, edX provides over 2500 free online courses from top universities, such as Yale, Columbia, Stanford, and so on [15]. Coursera, developed by Andrew Ng, offers degrees from top universities, e.g. the University of London, the University of Michigan. Moreover, Coursera offers programs from leading companies such as Google and IBM [15]. Recently, due to the impact of the COVID-19 pandemic, we have seen tremendous growth in online learning. As shown in [13], within just two years (from 2020 to 2021), the total number of registered learners and a total number of enrolments increased more than double, i.e. from 44 to 92 M and from 76 to 189 M, respectively. All this information shows the great potential and benefits that e-learning can bring to human beings.

2.3.2 Current Online Learning Platforms and Challenges

Currently, there are many online learning platforms. However, only a few of them occupy the market. There are two typical online platforms now, i.e. Zoom and Microsoft Teams. Each of them has advantages and disadvantages, as shown in [26]. However, we can observe in [26] that both current platforms have a lot of modern functions which enable more effective teacher–student interaction. Let's take Zoom [14] as an example. The teacher and students can interact and communicate in class via the cameras, voices, and chatbox. The student can share their lectures (e.g. slides) via the share screen function. Furthermore, the teacher also can make break rooms for students to discuss in groups. Alternatively, the teachers can record their lectures and share them with the students after class. All these functions are excellent conditions to make teaching and learning more efficient (Figure 2.6).

Current online learning platforms can partly solve people's teaching and learning needs, mainly due to the serious effects of the global pandemic that has left millions of students worldwide unable to learn and cannot go to school. It also exposes many limitations that are difficult to overcome, especially for specialized subjects such as engineering, chemistry, and medicine. For example, teaching steps of doing a surgery via Zoom is unclear because the students could not practice. As a result, most subjects could not achieve high performance when taught through online learning platforms because it is difficult for students to grasp the teacher's actions, and at the same time, the students have no chance to practice the knowledge they learn. Therefore, online learning platforms such as Zoom and Microsoft Teams must overcome the abovementioned challenges to enable an effective online learning environment for students.

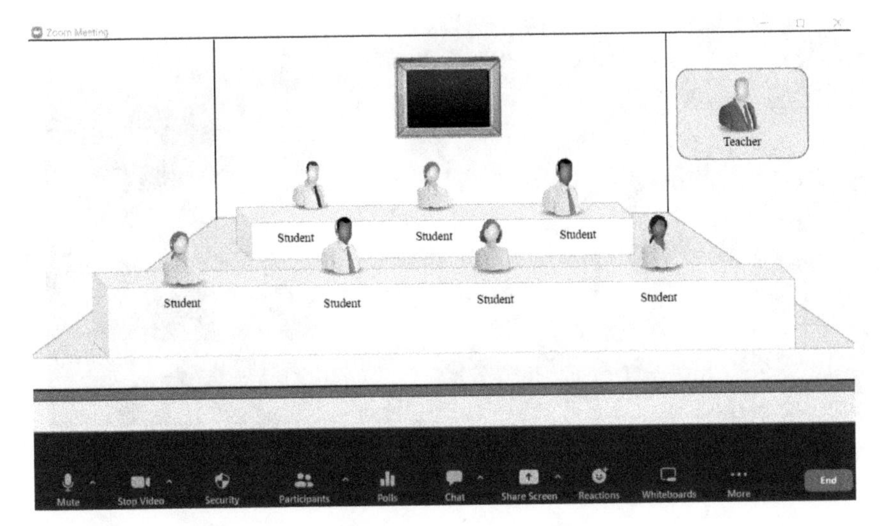

Figure 2.6 An illustration of a Zoom session. Source: Zoom Video Communications, Inc.

2.3.3 Education in Metaverse

Metaverse development brings entirely new experiences to users because it provides intuitive approaches and lively learning environments for students. Specifically, Metaverse can connect both the physical world and the virtual world. This will let the students from different locations have the same learning environment [39]. Moreover, Metaverse can provide an enjoyable environment [41]. For example, the users can create digital representatives in the virtual world and enjoy learning while interacting. In addition, Metaverse can offer a safe environment, e.g. we can avoid school bully and violations.

Additionally, Metaverse can enable the students to freely move within the Metaverse environments, thereby overcoming the space restriction. This could help to significantly enhance the students' learning experience in certain subjects. For example, in the astronomy subject, students can virtually travel to different planets within our solar system, as illustrated in Figure 2.7 [36]. The attractive environment in Metaverse could benefit different subjects such as astronomy. Besides astronomy, history is another subject that can greatly benefit from Metaverse. For example, students can virtually be at historical places, seeing how people work in the past, hearing sound effects, or even witnessing historical events. Finally, new technology used in Metaverse can contribute to the education with its entertaining feature, for example, the mix of gaming and education can bring joy and let students be motivated.

Figure 2.7 The Solar System by Meta licensed under CC-BY-SA 3.0. Source: Meta / https://www.youtube.com/watch?v=KLOcj5qvOio / last accessed: 2022-09-16.

2.3.4 Challenges of Education in Metaverse

This section shows how important it is to adapt new technology such as Metaverse to future e-learning. Metaverse can be used effectively in education as it can take advantage of many technologies such as AR, VR, digital twin, and life logging via Metaverse [42]. The virtual environment can address many issues in our physical learning systems, e.g. limited space and impacts of pandemics and diseases, as well as those of the online platforms by providing more interaction with more illustrative learning environments. However, the development of the Metaverse is still in a nascent stage, and it is important to recognize that there are potential risks and challenges associated with its implementation. One of the primary challenges of the Metaverse is the potential disorientation and discomfort that students may experience when participating in a Metaverse classroom based on immersive interactive technologies such as VR, AR, and MR. While these technologies offer a high degree of interactivity, they may also lack visual and dynamic interactive realism, which can lead to cybersickness and other negative effects, particularly when the students are confined to a limited physical space. Therefore, optimizing Quality of Experience (QoE) metrics, such as motion-to-photon latency, Field of View (FOV), graphic quality and lighting, and other navigation parameters, is necessary to mitigate the impact of cybersickness. Additionally, developing a Metaverse classroom requires addressing synchronization and stringent latency challenges, where real-time synchronization of users' actions is crucial for seamless interaction, and the allowable latency is typically less than 100 ms. As a result, utilizing cloud and edge-based solutions for pre-rendering digital content streams may be necessary. Moreover, the Metaverse is in needs to develop metaverse-based educational models because, until now, there is no particular framework for education [30].

2.4 Metaverse for Healthcare Services

In the past few years, due to the outbreak of the COVID pandemic, we have witnessed an explosion in the healthcare sector, especially smart online healthcare systems. Unlike traditional healthcare systems that require patients to visit clinics or hospitals for doctors to diagnose and treat. Online health screening systems allow patients to meet and interact with doctors directly through online medical platforms provided by companies, hospitals, or the government, e.g. Sesame Care, PlushCare, and Teladoc [8]. Although telemedicine has achieved a lot of success and brought huge benefits to people, especially during the COVID period between 2019 and 2022 (e.g. reducing medical examination and treatment costs, reducing the burden on diseases, avoiding the spread of viruses), they also revealed problems that need to be improved and overcome in the future. First, the current online medical examination and treatment systems are mostly performed through online calls in 2D space. This makes many difficulties for patients to accurately describe their conditions and sometimes lead to misdiagnosis by doctors. Likewise, due to the limitations of communication between doctors and patients in current online medical examination and treatment systems, patients may misunderstand doctors' instructions and cause serious consequences for patients.

Metaverse has recently emerged as an effective solution with many useful applications in the real world, e.g. in education, entertainment, and social networks [32]. In the field of virtual healthcare, many recent studies have shown that Metaverse can bring huge benefits, not only in terms of public health but also in many other social benefits. e.g. reducing stress for the medical team, improving service quality, opening up new solutions in medical examination and treatment. [9–11, 28]. Different from traditional healthcare systems, healthcare systems in Metaverse can enable patients and doctors to discuss in a vivid and very detailed visual way in a 3D virtual reality environment [12]. In addition, thanks to the development of many advanced technologies, such as AR, VR, artificial intelligence (AI), and digital twins, doctors can provide visualizations to patients and interact directly with them [12]. In addition, digital twin technology can create simulations of the impact of the surrounding environment on the patient's health status, thereby allowing doctors to offer effective treatment solutions appropriate to the patient's condition.

While the Metaverse holds great potential for online healthcare systems, it also presents several new challenges in terms of technology and regulation. One significant concern is the need to ensure the privacy and security of personal health-related data in the Metaverse. This includes advanced solutions for data acquisition, communication, and management. For example, tokenization and conversion of patient data into NFTs can improve privacy and safeguard patient data integrity and confidentiality [40]. Additionally, legislations and regulations must also be strengthened or formulated to protect users from criminal or abusive behaviors, especially when insurance companies, pharmaceutical firms, and governments become involved.

2.5 Metaverse for Autonomous Vehicles

In the past 10 years, we have witnessed a strong boom in the auto industry, especially in the autonomous vehicle industry. Thanks to the outstanding development of science and technology, especially AI and IoT technology, autonomous vehicles are continuously being suitable with many outstanding features, not only providing a complete driving experience and completely new for humans but also significantly improving driving safety for drivers. Specifically, by using modern AI applications and advanced IoT control systems (e.g. LiDAR sensors, image processing, and LTE-V2X), autonomous vehicles allow autonomous vehicles to operate precisely with minimal human control. In this way, drivers can still do their daily work right in the car without worrying and losing time on driving. As a result, the development of autonomous vehicles is bringing forth outstanding benefits and is gradually reshaping the car industry in the near future [1, 2, 25].

However, the development of autonomous vehicles is still facing certain difficulties and challenges. In particular, for autonomous vehicles to be used in practice, they need to be tested very thoroughly both in simulation and in real-world environments. However, simulated environment and real environments are practically very different, and thus transferring driving experiences from the simulated environment to the real environment is often not very effective.

In this case, Metaverse can offer a great virtual environment for testing autonomous vehicles [3]. In particular, Metaverse allows to the creation of diverse near-real environments (e.g. streets, buildings, traffic lights, obstacles, weather, etc.) for testing autonomous cars. In [4], a new concept called "Metaverse on Wheels" is introduced that allows drivers to experience Metaverse environment in a real-time manner when they are driving their cars. Furthermore, Metaverse can obtain a huge amount of data from many autonomous cars and drivers worldwide and extract useful information that could be very useful for testing and training autonomous vehicles. This information might be also very useful to help the developers and manufacturers better design autonomous vehicles in the future.

However, since Metaverse is still in its infancy, various challenges need to be addressed to enable Metaverse for autonomous vehicles. First, the Metaverse system is highly complex, and many of its fundamental technologies are still immature (e.g. digital twin, virtual reality, and ultralow latency and high-reliable connections). For example, creating the digital twin of vehicles and maintaining the required connections for Metaverse are still challenging tasks with current technologies. Second, autonomous vehicles only observe a limited vision of the surrounding environment (e.g. road conditions), so an ecosystem (e.g. a network of roadside units) is needed to support Metaverse autonomous vehicles. In the near future, these can be addressed by the rise of quantum computing and the 6G network or beyond. Third, safety is another concern since Metaverse may distract drivers. Note that even in an autonomous vehicle, drivers still need to monitor the vehicle to avoid collisions if the system has issues. Finally, the privacy of users is also questionable since autonomous vehicles need to exchange information (e.g. driver's behaviors and characteristics) with Metaverse. Thus, a comprehensive solution from both regulation and technology perspectives is needed to alleviate these challenges.

2.6 Metaverse for Virtual Travelling

Metaverse will never replace traditional physical travel, but it is expected to bring significant changes to the future travel industry. First, due to the severe impact of the global COVID pandemic, a new type of tourism ecosystem has been introduced recently, namely virtual travelling [21, 46]. This type of travel allows travel enthusiasts to explore new places in a very intuitive way without taking flights to travel. For example, the South Korean city of Seoul, one of the top destinations for international visitors, has recently introduced an ambitious project, namely "Metaverse Seoul" [29]. This project allows for the creation of a virtual environment of Seoul city in Metaverse [24] and allows visitors around the world to experience Seoul without the need of flying to Korea. Such projects are expected to bring huge benefits to tourists as well as to the city government by providing cheap virtualization packages for tourists, avoiding the spread of disease, reducing environmental pollution, and promoting and introducing Korean culture, services, and products to the world [24, 29, 31].

As discussed above, Metaverse can bring such new experiences and benefit for virtual traveling user; however, several challenges are still impeding users' fully immersed experiences. Specifically, the Metaverse-based virtual traveling should offer a "real feeling" of the environment (e.g. weather) at the visited location since, with some people, this is an important factor in deciding their visiting places. It may need a dedicated room or cabin equipped with various device types to provide the weather experience. However, it is still challenging with the current technologies since it requires synchronizing huge amounts of data between a real location and the user location. In addition, reproducing the weather is also a big challenge due to the overcomplexity of weather. Thus, Metaverse-based virtual traveling may not replace traditional travel in the near future, but it can enhance the experience of current travel.

2.7 Conclusions and Future Research Directions

In this chapter, we have presented and discussed Metaverse applications in major areas including entertainment, virtual office, education, healthcare, autonomous vehicle, and virtual tourism. In some areas such as entertainment, virtual office, and autonomous vehicle, Metaverse applications have been adequately developed and even commercialized, whereas the Metaverse applications in the others areas are still at the early stages of development. Nevertheless, we believe that they possess a huge potential and attract massive attention from various industries in the near future.

Besides the applications that are discussed in this chapter, there are other promising applications areas. For example, Metaverse can be used to create more immersive and low-cost training via simulations in military. Moreover, Metaverse can provide a wide range of investment opportunities for banks and financial institutions, e.g. virtual real estates, currencies, and artistic items. In manufacturing, Metaverse can assist in multiple stages of a product's life cycle, from design (prototype development) to production (worker training).

The development of Metaverse applications is still at a nascent stage, and there exist various challenges that need to be addressed for future Metaverse applications. The massive resource demand is the most significant challenge, which necessitates intelligent resource allocation and user resource contribution encouragement. Moreover, advanced wireless technologies such as next-generation multiple access (NGMA) and 6G should be studied to address the ultralow latency requirements of Metaverse applications. Furthermore, for such complex environments with millions of users, privacy is of the main concern for users, to which blockchain technology can be a promising solution.

Bibliography

1 7 benefits of autonomous cars.
2 Automotive revolution - perspective towards 2030 | McKinsey.
3 Metaverse simulates rare scenarios in autonomous car testing.
4 Holograktor | Metaverse on Wheels.
5 Pika-Who? How Pokémon Go Confused the Canadian Military - The New York Times.
6 BTS 'Dynamite' New Music Video to Premiere in 'Fortnite' - Variety.
7 Infographic: Gaming: The Most Lucrative Entertainment Industry By Far.
8 Best Telemedicine Companies in 2022.
9 Metaverse is Revolutionizing Healthcare: Are you Ready for Change?.
10 Importance of Metaverse and Virtual Reality In the Healthcare Industry.

11 Metaverse In Healthcare Will Transform The Industry - eLearning Industry.
12 Biomet Zimmer. OptiVu Mixed Reality. 2022 https://www.zimmerbiometer.com/en/products-and-solutions/zb-edge/optivu.html.
13 These 3 charts show how online learning is growing globally | World Economic Forum.
14 Educator Guide.
15 10 Surprising Benefits of Online Learning | Coursera.
16 These companies plan to make working from home the new normal. As in forever | CNN Business.
17 Virtual Office Market: Size, Dynamics, Regional Insights and Market Segment Analysis (by Type, Services, and End-Users).
18 What does VIVE Sync offer?.
19 Microsoft Mesh (Preview) overview.
20 Nonverbal Overload: A Theoretical Argument for the Causes of Zoom Fatigue. Volume 2, Issue 1.
21 11 Virtual Travel Experiences You Can Enjoy From Home.
22 Horizon Workrooms for VR Remote Collaboration, 2021. URL https://about.fb.com/news/2021/08/introducing-horizon-workrooms-remote-collaboration-reimagined/.
23 Entertainment, 2022. https://en.wikipedia.org/w/index.php?title=Entertainment&oldid=1107533843.
24 BizIn. Seoul Became The First Metaverse City in The World, 2022. URL https://thebizin.com/international/seoul-became-the-first-metaverse-city-in-the-world/.
25 Jeremy Carter. Driverless car benefits | Automated Transport | Self-driving Vehicles.
26 Julia Deien. Microsoft Teams vs. Zoom: Which Tool Should I Use and When? 2020. URL https://anderscpa.com/microsoft-teams-vs-zoom/.
27 Grace Doyle. All artists who have performed a concert in Roblox, 2022. URL https://progameguides.com/roblox/all-artists-who-have-performed-a-concert-in-roblox/.
28 Vishal Gondal. Benefits of metaverse in the healthcare industry. *The Times of India*. ISSN 0971-8257. URL https://timesofindia.indiatimes.com/blogs/voices/benefits-of-metaverse-in-the-healthcare-industry/.
29 Ashleigh Hollowell. How Seoul is creating a metaverse for a smarter city, 2022. URL https://venturebeat.com/ai/how-seoul-is-creating-a-metaverse-for-a-smarter-city/.
30 Gwo-Jen Hwang and Shu-Yun Chien. Definition, roles, and potential research issues of the metaverse in education: An artificial intelligence perspective. *Computers and Education: Artificial Intelligence*, 3:100082, 2022.

31 Jonathan Keane. South Korea is betting on the metaverse – and it could provide a blueprint for others.

32 Lik-Hang Lee, Tristan Braud, Pengyuan Zhou, Lin Wang, Dianlei Xu, Zijun Lin, Abhishek Kumar, Carlos Bermejo, and Pan Hui. All one needs to know about metaverse: A complete survey on technological singularity, virtual ecosystem, and research agenda. *arXiv preprint arXiv:2110.05352*, 2021.

33 MaziMatic. MaziMatic: Building World's First Entertainment Metaverse, 2022. URL https://www.youtube.com/watch?v=3MgqYz90GCM.

34 Jon C. Messenger and Lutz Gschwind. Three generations of Telework: New ICTs and the (R)evolution from Home Office to Virtual Office. *New Technology, Work and Employment*, 31(3):195–208, 2016. doi: https://doi.org/10.1111/ntwe.12073. URL https://onlinelibrary.wiley.com/doi/abs/10.1111/ntwe.12073.

35 Meta. Entertainment in the metaverse, 2021. URL https://www.youtube.com/watch?v=kKPqNd9zfnk.

36 Meta. Education in the metavers, 2021. URL https://www.youtube.com/watch?v=KLOcj5qvOio.

37 Meta. Work in the metaverse, 2021. URL https://www.youtube.com/watch?v=uVEALvpoiMQ.

38 J. Nilles. Telecommunications and organizational decentralization. *IEEE Transactions on Communications*, 23 (10):1142–1147, 1975. doi: 10.1109/TCOM.1975.1092687.

39 Ryu SeonSuk. An exploratory study on the possibility of metaverse-based korean language subject design. *Korean Journal of General Education*, 16(2):289–305, 2022. doi: 10.46392/kjge.2022.16.2.289. URL https://j-kagedu.or.kr/journal/view.php?number=1095.

40 Ioannis Skalidis, Olivier Muller, and Stephane Fournier. CardioVerse: The cardiovascular medicine in the era of metaverse. *Trends in Cardiovascular Medicine*, 7(36):1–6, 2022.

41 Woong Suh and Seongjin Ahn. Utilizing the metaverse for learner-centered constructivist education in the post-pandemic era: An analysis of elementary school students. *Journal of Intelligence*, 10(1):17, 2022. ISSN 2079-3200. URL https://www.mdpi.com/2079-3200/10/1/17.

42 Ahmed Tlili, Ronghuai Huang, Boulus Shehata, Dejian Liu, Jialu Zhao, Ahmed Hosny Saleh Metwally, Huanhuan Wang, Mouna Denden, Aras Bozkurt, Lik-Hang Lee, et al. Is metaverse in education a blessing or a curse: A combined content and bibliometric analysis. *Smart Learning Environments*, 9(1):1–31, 2022.

43 Margaret Walls and Peter Nelson. Telecommuting and emissions reductions: Evaluating results from the ecommute program. page 28, 2004. doi: https://doi.org/10.22004/ag.econ.10628. URL http://ageconsearch.umn.edu/record/10628.

44 Yiwen Wu, Ke Zhang, and Yan Zhang. Digital twin networks: A survey. *IEEE Internet of Things Journal*, 8(18): 13789–13804, 2021. doi: 10.1109/JIOT.2021.3079510.

45 Minrui Xu, Wei Chong Ng, Wei Yang Bryan Lim, Jiawen Kang, Zehui Xiong, Dusit Niyato, Qiang Yang, Xuemin Sherman Shen, and Chunyan Miao. A full dive into realizing the edge-enabled metaverse: Visions, enabling technologies, and challenges. *arXiv preprint arXiv:2203.05471*, 2022.

46 Shu-Ning Zhang. Would you enjoy virtual travel? The characteristics and causes of virtual tourists' sentiment under the influence of the COVID-19 pandemic. *Tourism Management*, 88:104429, 2021. doi: 10.1016/j.tourman.2021.104429.

3

Metaverse Prototype: A Case Study

Haihan Duan and Wei Cai

School of Science and Engineering, The Chinese University of Hong Kong, Shenzhen, Shenzhen, China

After reading this chapter you should be able to:

- Follow the iteration procedure from *Newbie at CUHKSZ*, a single-player campus orientation game, to *The Chinese University of Hong Kong, Shenzhen Metaverse* (abbreviated as *CUHKSZ Metaverse*), a university campus Metaverse prototype.
- Overview the university campus Metaverse prototype *CUHKSZ Metaverse* and understand the three-layer design logic of *CUHKSZ Metaverse*, including the infrastructure layer, interaction layer, and ecosystem layer.
- Explore the detailed features, applications, functions, and components of *CUHKSZ Metaverse* from the infrastructure layer, interaction layer, and ecosystem layer, respectively.

3.1 Overview

In this chapter, we will illustrate a Metaverse prototype for readers to better understand the Metaverse. The selected case is a university campus Metaverse prototype designed for the Chinese University of Hong Kong, Shenzhen (CUHKSZ), namely *CUHKSZ Metaverse* [4].

3.1.1 Related Work

Before introducing the prototype *CUHKSZ Metaverse*, we have a brief overview of the existing works of the university campus Metaverse:

Metaverse Communication and Computing Networks: Applications, Technologies, and Approaches, First Edition.
Edited by Dinh Thai Hoang, Diep N. Nguyen, Cong T. Nguyen, Ekram Hossain, and Dusit Niyato.
© 2024 The Institute of Electrical and Electronics Engineers, Inc. Published 2024 by John Wiley & Sons, Inc.

2D Map: McMaster University built a 2D virtual campus map to introduce its buildings, infrastructures, roads, etc., for newcomers to explore.[1] Also, Huron University[2] provides a virtual campus guide about the locations and affiliated introduction text, image, or video corresponding to each location. This campus guide is supported by *CampusTours*,[3] a provider of virtual and video tour and interactive map services to the academic, nonprofit, and government markets.

Guided Panoramic Photo: Galveston College takes a 360-degree tour of each of the campuses, either on desktop, laptop, phone, or virtual reality VR headset.[4] The users can view the highlight reel of the walk-through or scan one of the links below to jump right in. Through the virtual guide, the users can easily become familiar with the inner space of Galveston College.

3D Campus Model: Hiroshima Institute of Technology [18] proposed a virtual campus guidance system using VR, which could support the visualization on smartphone VR for campus guidance and demonstration. The Communication University of China [12] built a 3D campus model by means of *3ds Max*[5] and imported it into the *Unity*[6] interaction platform. This prototype provided multiple functions for their users, including the first and third person-roaming mode, automatic roaming mode, building information display system, background music switch, etc.

Overall, the existing university campus prototypes mainly focused on the campus guide function without social attributes, which could not be regarded as Metaverse projects. However, the initial start point of the virtual campus guide is similar to our first version before building *CUHKSZ Metaverse*. In Section 3.1.2, we will discuss the motivation and implementation of *CUHKSZ Metaverse* and the iteration process from a virtual campus guide game to a Metaverse project.

3.1.2 Motivation and Implementation

The Chinese University of Hong Kong, Shenzhen, is a campus of the Chinese University of Hong Kong (CUHK), which is set in Shenzhen, Guangdong, China. In 2020, the Coronavirus disease 2019 (COVID-19) pandemic broke out, making the incoming university students unable to attend campus orientation events due to the COVID-19 epidemic prevention policy. Under this situation, we intended to design a virtual university campus orientation game for new students to comprehensively understand their university and build a friendship with their cohorts earlier.

1 https://discover.mcmaster.ca/map/
2 https://huronatwestern.ca/contact/huron-campus-map/
3 https://campustours.com/
4 https://gc.edu/virtualtour/
5 https://www.autodesk.com/products/3ds-max/overview
6 https://unity.com/

Figure 3.1 Campus map of the Chinese University of Hong Kong, Shenzhen. Source: CUHK(SZ).

Specifically, the campus of CUHKSZ is located in the northeastern part of Shenzhen, surrounded by mountainous areas, so the university campus is divided into two parts by the topography, as shown in Figure 3.1. Thereinto, the lower campus occupies a more prominent space used for education, administration, meeting, sports, etc., while the upper campus is mainly for living. Thus, we consider making digital twins of the lower campus in the virtual orientation because most students would spend more time on the lower campus.

In fact, the scale of CUHKSZ is relatively small compared with other universities with more than 40,000 students and faculties, making it feasible to construct the university campus by part-time work of current students. Therefore, we recruited more than ten undergraduate students to develop the virtual campus orientation game using *Unity*. The *Unity* is a cross-platform development engine, which can be used to create three-dimensional (3D) and two-dimensional (2D) games, as well as interactive simulations and other experiences. Also, the engine has been adopted by industries outside video gaming, such as film, automotive, architecture, engineering, construction.[7] The development procedure followed a crowd-sourcing manner, where the team leader published some tasks, and other students could accept and finish the assignments in their spare time.

Cost about half a year, we developed the first version as a virtual university campus orientation application, named *Newbie at CUHKSZ*, which is a single-player puzzle game for the incoming students to explore the university. However, a single-player game cannot provide a social experience for the students, a crucial feature of a university. Therefore, we redeveloped an entire university campus

7 https://en.wikipedia.org/wiki/Unity_(game_engine)

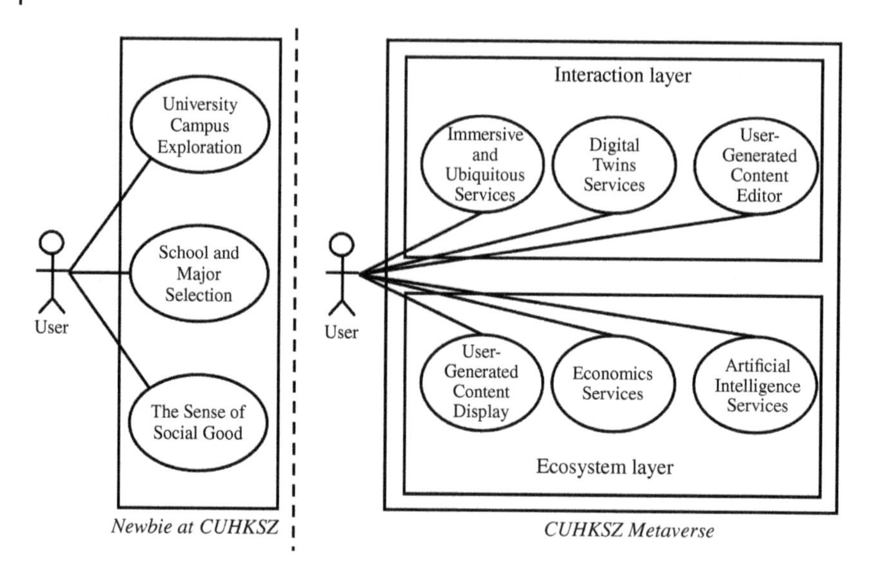

Figure 3.2 Simple use cases of *Newbie at CUHKSZ* and *CUHKSZ Metaverse.*

Metaverse, *CUHKSZ Metaverse*, and the target users were also extended from incoming students to all current students and faculties. In the remaining of this chapter, the readers can obtain more detailed features, applications, functions, and components of *Newbie at CUHKSZ* and *CUHKSZ Metaverse*, where we provide a simple use case figure of *Newbie at CUHKSZ* and *CUHKSZ Metaverse* in Figure 3.2 to help the readers understand the differences. Through the iteration procedure and comparison of the two prototypes, we intend to provide the readers with an intuitive cognition of a Metaverse.

3.2 Newbie at CUHKSZ

Newbie at CUHKSZ is a 3D game featuring a voxelized version of CUHKSZ campus. Figure 3.3 shows the administration building in *Newbie at CUHKSZ*. Through an open world-based role-playing game (RPG) story, the life of a freshman here is presented to the players. The player will explore the campus, communicate with the staff, and finish missions as a "newbie" of CUHKSZ. During gameplay, the player will get familiar with life and study in CUHKSZ. The missions contain greetings from presidents and deans, interesting stories circulated among students, photography quests testing players' eyesight, and so on. We hope these games can play a role like virtual campus orientation, which can help first-year students get used to the campus quickly and promote the university to the public.

Figure 3.3 Voxel-style administration building in *Newbie at CUHKSZ*. Source: Meta.

3.2.1 Gameplay in *Newbie at CUHKSZ*

Essentially, *Newbie at CUHKSZ* is a single-player game rather than a Metaverse, so some specific targets need to be set for the players to follow. Therefore, we designed a series of stories in the form of puzzle games to guide the students to become familiar with the university campus, explore their career interests, cultivate humanistic values, etc. In Sections 3.2.1.1–3.2.1.3, we will illustrate some representative cases to show the gameplay of *Newbie at CUHKSZ*.

3.2.1.1 University Campus Exploration

In *Newbie at CUHKSZ*, we design many tasks to guide the students in exploring the university campus. For example, the task will ask the players to find some targets (e.g. a key, a nonplayer character (NPC)) in some specific places. With the designed tasks, the frequent crossing from one place to another can quickly familiarize the players with the university campus.

On the other hand, we also designed a parkour game that allows the players to look around the university campus from a bird's-eye view. Some screenshots of the parkour game are shown in Figure 3.4. During the parkour game, the level will generate virtual stairs in the air to elevate the students to the rooftop, and the students need to reach a specific destination following the indicators with a jump and run. We also set traps, obstacles, and mazes on the stairs to enhance the entertainment experience of this task. When reaching the destination, the students will be asked to take photos of the virtual campus to complete the

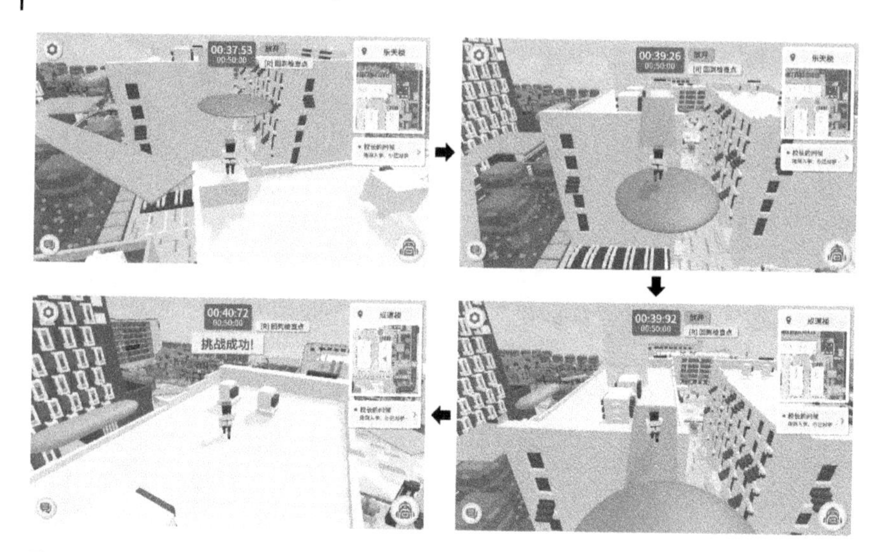

Figure 3.4 Parkour game in *Newbie at CUHKSZ*. Source: Meta.

photo-shooting challenges. From the bird's-eye view, the students can obtain a different cognition and experience of the whole campus, which even cannot be achieved in the real world.

3.2.1.2 School and Major Selection

In CUHKSZ, the newly enrolled students would not be assigned a major and school, and they must select their major after the first school year. Therefore, an important target for *Newbie at CUHKSZ* is to help the students discover their career interests and find their best-matching major and school.

In our predesigned tasks, the students are asked to search for the deans (NPCs) of different schools in CUHKSZ. Then, the players can chat with each dean to obtain information about the affiliated schools. An example is illustrated in Figure 3.5. In this example, the dean of the School of Science and Engineering (SSE) will introduce the majors of SSE to the players, such as Electrical and Computer Engineering, New Energy Science and Engineering, and so on. After that, the students can preliminarily select the school they are interested in and answer the school-related qualification questions. For instance, when the students choose the SSE, they will be asked questions about computer science, physics, mathematics, etc. In contrast, the students will be asked biology-related questions when they select the School of Life and Health Sciences (LHS). If the students correctly answer all questions, they will be recommended to join the target school because we consider the students would study their interesting knowledge by themselves.

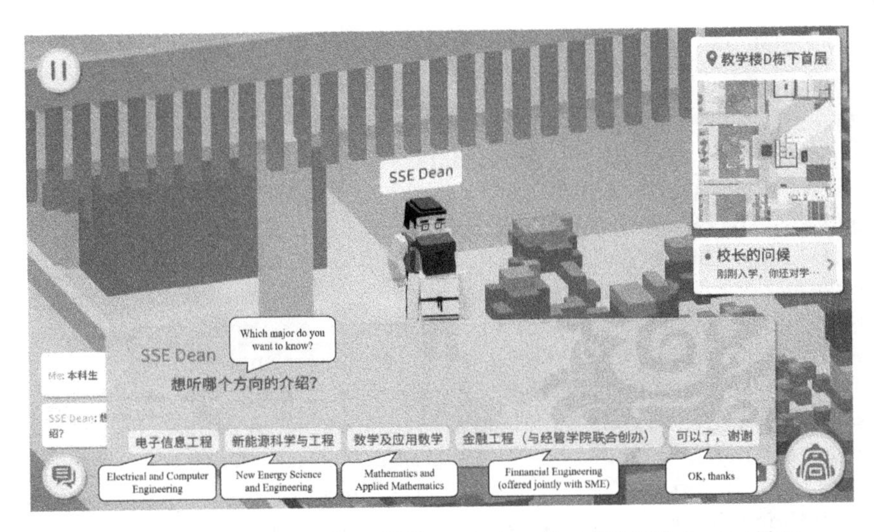

Figure 3.5 Major introduction from SSE Dean in *Newbie at CUHKSZ*. Source: Meta.

3.2.1.3 The Sense of Social Good

In *Newbie at CUHKSZ*, we also designed puzzle games based on real campus stories and tales to cultivate the students' sense of social good. For example, we collaborated with the university animal welfare organization to compose a love story between two stray cats. In this story, the students are asked to help a cat take care of his wife. This puzzle game intends to cultivate the awareness of wild animal protection engagingly because many wild animals might appear on campus, e.g. cats, birds, fishes.

Also, we collaborated with the security department on a story regarding computer theft to arouse safety awareness and social responsibility. As shown in the left image of Figure 3.6, a student lost his laptop in the library, so the task requires the players to seek out his stolen computer by completing corresponding puzzles. After completing this task, the security staff will emphasize safety awareness to the players, as shown in the right image of Figure 3.6.

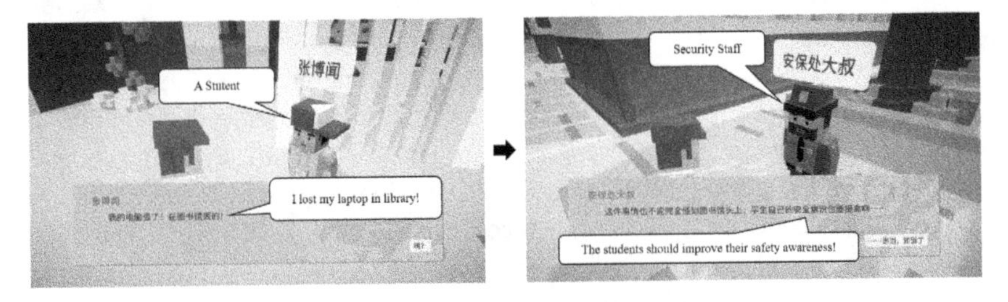

Figure 3.6 Task about safety awareness in *Newbie at CUHKSZ*. Source: Meta.

3.2.2 Limitations and Iteration

According to the feedback from the incoming students enrolled in 2020, they mentioned that the implemented *Newbie at CUHKSZ* effectively helped them have a comprehensive understanding of their university, even though they never went on site due to the COVID-19. However, we also noticed and realized some limitations or misunderstandings of *Newbie at CUHKSZ*.

(1) Single-Player Game: The most intuitive limitation of a single-player game is the lack of society [19] because there only exists social relationships with virtual NPCs in a single-player game (more discussions of social effect in single-player games can refer to [13, 17]). However, the university is a complete society in which all participants have their corresponding roles, rights, obligations, lifestyle, etc., so a single-player virtual campus cannot satisfy the social requirements of the students. Therefore, an iteration direction of the virtual campus is to provide a complete social experience for students and faculties.

(2) Game Target: In our consideration, the game target is a significant difference between a game and a Metaverse [4]. In fact, the predesigned puzzle games in *Newbie at CUHKSZ* indeed guided the players to explore the virtual campus. However, after finishing these game targets, the players seem to have nothing to do except hang out in the virtual space. Moreover, some predesigned puzzle games in *Newbie at CUHKSZ* seem redundant as a virtual campus orientation, but the players must pass them to reach the next level.

(3) Participatory Involvement: As we mentioned above, gameplay of *Newbie at CUHKSZ* is almost composed by the provided tasks, which show limited extension space for the players to participate. Therefore, another iteration direction of *Newbie at CUHKSZ* is to design some mechanisms to enhance the players' experience of participatory involvement. As we mentioned in the first point, an intuitive approach is to create a more complete social experience. On the other hand, user-generated content (UGC) has been playing a highly necessary role in video games. It could give the players a sense of participatory involvement and increase their enjoyment [6].

(4) Voxel-Style Virtual World: The implemented *Newbie at CUHKSZ* is a voxelized version of CUHKSZ campus built in *Unity* with a 3D modeling software *MagicaVoxel*.[8] However, we noticed many comments on social media that said, "*Newbie at CUHKSZ* is nothing remarkable because it is just some buildings constructed in *Minecraft*." The similar comments indeed disappointed our student builders, which drove us to consider other artistic styles in the next version. However, it does not mean that we have a bias against voxel-style. On the contrary, these comments reflect the success of *Minecraft*, which builds a deep impression of voxel-style in ordinary people's minds. We also think highly of

8 https://ephtracy.github.io/

the voxel-style in future Metaverse applications due to its many advantages. For example, voxel-style objects are simple and intuitive for UGC creation and modification with low learning cost, and most voxel-style avatars are abstract, which could effectively avoid the uncanny valley problem, etc. For a more detailed discussion about UGC in video games, the readers can kindly refer to Duan et al. [6]. Overall, the adjustment of artistic style is also an iteration target for the next version.

(5) Language: Since the original target users are only incoming students in China, we only designed the Chinese user interface and interaction in *Newbie at CUHKSZ*. However, the target of the university campus Metaverse has a broader scope, including many international students and faculties, so the next version should be an English version to satisfy the international requirement.

3.3 CUHKSZ Metaverse

According to the limitations of *Newbie at CUHKSZ* mentioned in Section 3.2.2, the essential target of the new prototype is to build a university campus Metaverse. Therefore, we implemented a blockchain-driven exemplary system, *CUHKSZ Metaverse*, for demonstration and future social experiments.

The proposed *CUHKSZ Metaverse* is also developed on *Unity*, and the application can be deployed on smartphones and PCs, and the browser-based cloud streaming will also be considered in the future. Further, the 3D models are constructed using *Blender*.[9] Unlike *Newbie at CUHKSZ*, *CUHKSZ Metaverse* applied low poly artistic style, a polygon mesh in 3D computer graphics with a relatively small number of polygons [3], to construct the virtual world. Figure 3.7 illustrates the low poly style administration building in *CUHKSZ Metaverse*. Moreover, *CUHKSZ Metaverse* can provide the on-campus students with an interactive Metaverse, a mixed environment where students' actions in the real world could correspondingly affect the virtual world and vice versa.

In this section, we will discuss the design of *CUHKSZ Metaverse* and introduce some key components of the university campus Metaverse.

3.3.1 Three-Layer System Design

The design of *CUHKSZ Metaverse* follows a three-layer Metaverse architecture, as shown in Figure 3.8, including (from bottom to top): infrastructure layer, interaction layer, and ecosystem layer. This architecture is relatively concise compared with other Metaverse architectures, e.g. a seven-layer Metaverse

9 www.blender.org/

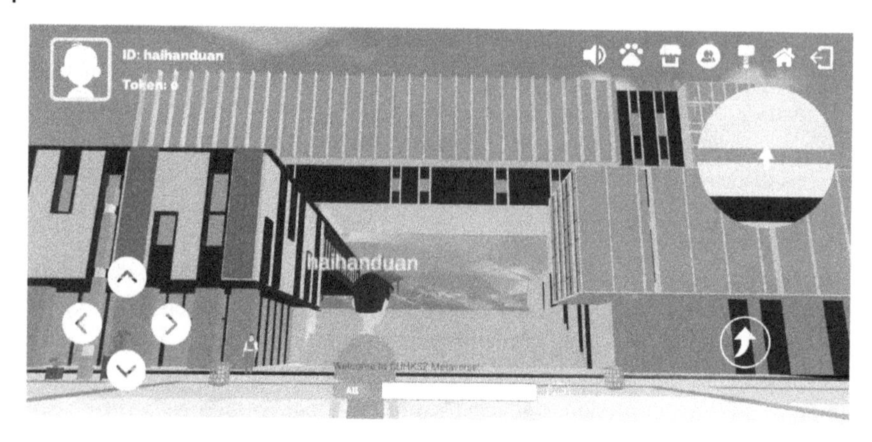

Figure 3.7 Low poly administration building in *Newbie at CUHKSZ*. Source: Meta.

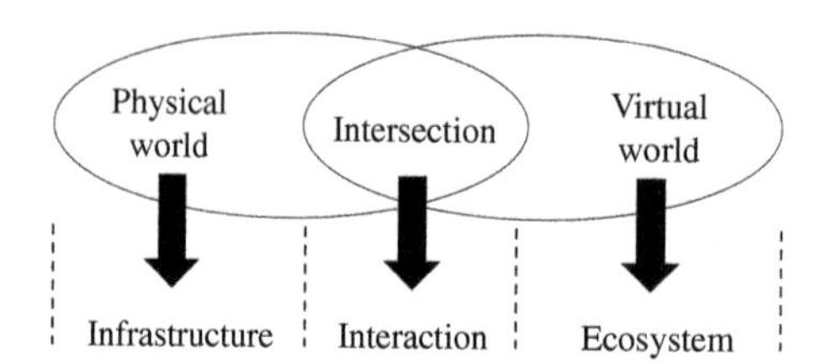

Figure 3.8 Three-layer Metaverse architecture of *CUHKSZ Metaverse*.

architecture proposed by Radoff [14], where the layers from bottom to top are: infrastructure, human interface, decentralization, spatial computing, creator economy, discovery, and experience. However, the essential requirements of a university campus prototype are basically included in our three-layer Metaverse architecture in Figure 3.8.

In Figure 3.8, the left two circles denote the virtual world and physical world with an intersection between the two worlds, and the structure of these two circles corresponds to three layers in the right part. The infrastructure layer contains the fundamental requirements for supporting the operation of a virtual world, including computation, communication, blockchain, and storage. Moreover, to bridge the physical and virtual worlds, we consider that the immersive user experience, digital twins, and content creation are essential components in the interaction layer. At the top layer, the ecosystem can provide a breathing and parallel living world that continuously serves all of the world's inhabitants, e.g. the social experiences, economics.

In fact, this three-layer architecture is our preliminary consideration of the Metaverse because the development of the Metaverse is still in the early stages, so its architecture does not have a consistent definition in either academia or industry. More information about the three-layer architecture can refer to

Duan et al. [4]. In Section 3.3.2, we will introduce the implementation details of each layer combined with some key components for the readers to understand *CUHKSZ Metaverse* better.

3.3.2 Campus Metaverse Prototype

In this section, we will introduce the implementation of each layer in *CUHKSZ Metaverse*. Figure 3.9 shows the key components of *CUHKSZ Metaverse*, where the readers can overview the proposed prototype. The details of the components will be discussed layer by layer in the remaining of Section 3.3.2.

3.3.2.1 Infrastructure Layer

The infrastructure layer mainly contains the components that are fundamentals to maintain the routine operation of a Metaverse. In fact, the Metaverse is a large-scale networked multimedia system, which needs to support the connection of multiple players, so the computational power and communication network are basically required. However, the above two components are general technologies for any Internet application, so they are not the key features of *CUHKSZ Metaverse*. More importantly, the blockchain is a crucial component of *CUHKSZ Metaverse*.

In *CUHKSZ Metaverse*, we applied FISCO-BCOS[10] as the blockchain infrastructure. The blockchain is introduced to support the decentralized ecosystem by deploying the smart contract and connecting with the Metaverse system (the detailed applications and utilization of the blockchain system will be discussed in Section 3.3.2.3). Specifically, FISCO-BCOS is an open source high-performance financial-grade consortium blockchain platform developed by WeBank.[11] FISCO-BCOS can provide rich features, including group architecture, pluggable consensus mechanisms, privacy protection algorithms, etc. FISCO-BCOS also provides

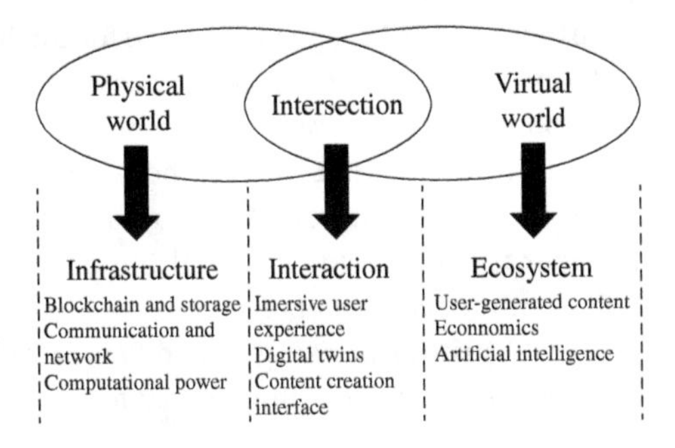

Figure 3.9 Key components of *CUHKSZ Metaverse*.

10 http://fisco-bcos.org/
11 www.webank.com/'#/home

an information port that can directly demonstrate the information of the consortium blockchain [7]. Moreover, Solidity[12] is utilized as the smart contract programming language to support the blockchain-based ecosystem. The introduction of consortium blockchain in *CUHKSZ Metaverse* benefits the system operation because the high cost (e.g. Ethereum Gas Fee[13]) of transactions in the public blockchain is hard to afford by students.

3.3.2.2 Interaction Layer

The components of the interaction layer would be directly perceived by the users, which highly decides the user experience. In fact, human–computer interaction (HCI) is an independent discipline about the design, evaluation, and implementation of interactive computing systems for human use and the study of phenomena surrounding them [9], which is also a necessary part of Metaverse applications. Specifically, we present some representative components of *CUHKSZ Metaverse* to reflect our consideration of HCI in the Metaverse.

(1) Metaverse Viewer: The Metaverse viewer is the most commonly used interface for users to interact with *CUHKSZ Metaverse*. To implement the Metaverse viewer, we envision a cross-platform future that various devices would connect to *CUHKSZ Metaverse*, including smartphones, PCs, browser-based cloud streaming, etc. In this version, we first adopt the smartphone as the inaugural platform due to the following consideration: (i) it is convenient to promote the Metaverse concept to our target users, since most students and faculties are used to taking part in social activities through a smartphone; (ii) the smartphones can support the Metaverse connection from a mobile platform, which provides continuous user experiences without environmental restriction; (iii) almost current smartphones equipped with lots of sensors, such as camera, global positioning system (GPS), gyroscope, LiDAR, which retains considerable potential for further study and innovation on interactive approaches; (iv) many factories of smartphones have provided VR devices that used their smartphones as the headset display, so the smartphones can be a good interface for the future version with VR experience.

A screenshot of our Metaverse viewer is shown in Figure 3.10. In this figure, we annotated the essential factors, e.g. the user-controlled avatar, buttons. Like most RPG mobile games, the operation logic is very intuitive, and button icons are also meaningful and understandable, making *CUHKSZ Metaverse* amateur-friendly. Some essential buttons and their corresponding applications will be introduced in detail later.

(2) Ubiquitous Sensing-Based Service: Currently, traditional keyboards and mouses are the most common way to interact with users in existing Metaverse

---

12 https://github.com/ethereum/solidity
13 https://etherscan.io/gastracker

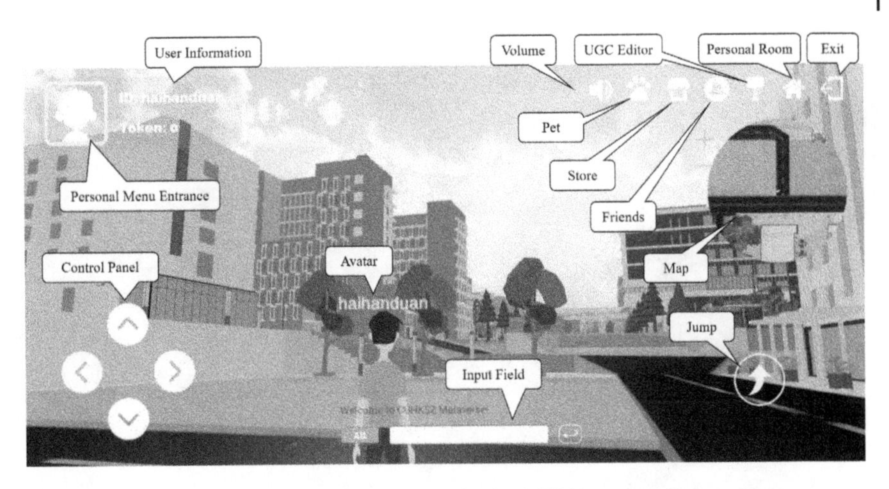

Figure 3.10 Screenshot of Metaverse viewer in *CUHKSZ Metaverse*. Source: Meta.

applications, e.g. *CryptoVoxels*,[14] which lacks an immersive user experience. *CUHKSZ Metaverse* intended to bridge the physical and virtual worlds. Thus, we design some interesting applications based on ubiquitous sensing technologies so that the users' activities in the physical world can correspondingly impact the virtual world.

For example, in *CUHKSZ Metaverse*, we utilize localization information provided by GPS as a source of sensing input. Specifically, we featured a location-based incentive mechanism to maximize the social welfare of the on-campus students as follows: the students may start the Metaverse viewer in Power-Saving mode and voluntarily report their GPS location to earn tokens of *CUHKSZ Metaverse*, as shown in the left part of Figure 3.11. Moreover, the token production rates are different for different positions. For instance, in the university library or classroom, the students can earn more tokens compared with the dormitories. Apparently, this approach was intended to encourage students to leave their dormitories and study at the library. A detailed token map is shown in Figure 3.12, which illustrates the different token production rates in different places. In fact, the incentive mechanism design of the token production rates is also a meaningful topic that is worthy of further study. Moreover, according to their position information, the students will automatically join the corresponding chat room to chat with nearby students, as shown in the right part of Figure 3.11.

Furthermore, we also designed an interesting mechanism in *CUHKSZ Metaverse*, which could change the day and night according to the physical time in CUHKSZ. Figure 3.13 illustrates the difference between day and night, and the system would automatically turn all lights on. In the future iteration, more

14 www.voxels.com/

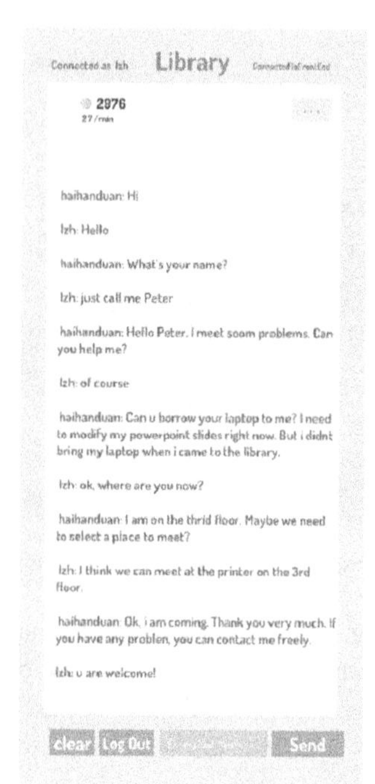

Figure 3.11 Power-saving mode and regional chat room in *CUHKSZ Metaverse*. Source: Meta.

Figure 3.12 Map of token production rates in *CUHKSZ Metaverse*. Source: Meta.

Figure 3.13 Day and night in *CUHKSZ Metaverse*. Source: Meta.

exciting applications based on ubiquitous sensing and Internet of Things (IoT) can be added to *CUHKSZ Metaverse*. For example, the weather can be changed according to reality, the school buses can travel the campus based on their physical position information, and the lectures can be simultaneously given as digital twins of reality. Also, it would be exciting if the operation in *CUHKSZ Metaverse* could map to the physical world. For instance, a question from the students in the virtual classroom can be sent to the physical classroom, and the responses from the teachers can also be sent back to the Metaverse, which could present a classroom digital twins prototype [1].

(3) User-Generated Content Creation Interface: Generally, the Metaverse is an evolving virtual world with unlimited scalability and interoperability for expansion, in which operators need to construct the basic elements, and users could arbitrarily create UGC [4]. Therefore, efficient content creation is a necessary component for the interaction between users and the Metaverse.

In *CUHKSZ Metaverse*, we also intended to adopt UGC as a core user gameplay and mechanism. However, it is still a challenge for an average user to create customized items in a 3D environment, as 3D object quality relies heavily on professional 3D modeling knowledge and practical experience. To address this issue, we designed and implemented an easy-to-use voxel-style UGC editor, as shown in Figure 3.14. Due to the advantages of the voxel-style editor mentioned in Section 3.2.2, an untrained user can learn to easily make simple 3D items with voxels in minutes, as in creating UGC in *Minecraft*. Based on the created simple 3D items, the editor can apply artificial intelligence (AI) technologies to generate high-resolution fine-grained models and transform the models to low poly style (e.g. using generative adversarial networks (GANs) [8] like DECOR-GAN [2]), which better fits the artistic style of the *CUHKSZ Metaverse*. Afterward, the ownership of created models will be confirmed as nonfungible

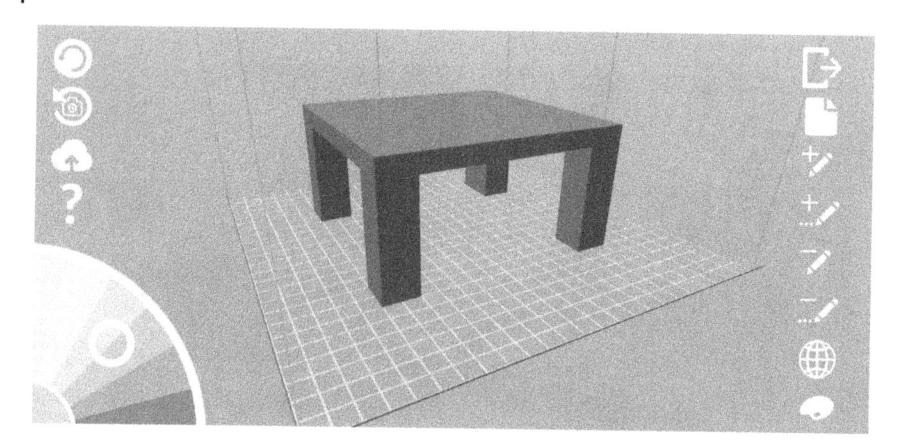

Figure 3.14 User-generated content editor in *CUHKSZ Metaverse*. Source: Meta.

tokens (NFTs) [11] on the blockchain for more usage (e.g. trading, exhibition, and collection).

Moreover, we are also conducting a study on an interesting topic about the uniqueness and digital scarcity of generated UGC by AI technologies from coarse-grained models to fine-grained models. In fact, digital assets can represent the users' personalities, lifestyles, and experiences. For example, the profile picture (PFP) is a typical case of digital assets that highly emphasizes uniqueness. Therefore, the users may hope to independently and solely possess the ownership of their created UGC. However, the simplicity of UGC creation would lead to weak distinctions between the generated fine-grained UGC by GANs because the drafts created by users may be very similar. Therefore, we proposed Crypto-dropout, a specially designed dropout used in the generative neural networks, which could cause pseudorandom disturbance based on the hash value of user information to guarantee the uniqueness of generated results. More details can refer to Duan et al. [5]. In the following update, the Crypto-dropout will be integrated into our UGC editor to guarantee the uniqueness of UGC in *CUHKSZ Metaverse*.

On the other hand, *CUHKSZ Metaverse* also provides a pet system to enrich the users' experience, as shown in Figure 3.15. In fact, the pet system is also a UGC editor with limited freedom compared with the voxel-style UGC editor. In the pet system, the users are only allowed to customize the color map of a predefined pet model (e.g. a cat in Figure 3.15), which is the simplest way to create UGC. In the future iteration of *CUHKSZ Metaverse*, the pet system's degree of freedom will be expanded. For example, we intend to allow the users to modify the breed, appearance, model, and action of their pets. However, it is hard to find a good trade-off between the granularity of editing and the difficulty of the editor, which is a general and promising interdisciplinary research topic that covers computer vision, computer graphics, computer-aided design (CAD), and HCI.

Figure 3.15 Pet editor in *CUHKSZ Metaverse*. Source: Meta.

3.3.2.3 Ecosystem Layer

The ecosystem can provide a breathing and parallel living world that continuously serves all the world's inhabitants. Specifically, as we mentioned in Section 3.2.2, *Newbie at CUHKSZ* lacks the components in the ecosystem layer, which cannot provide social experiences to the players. In *CUHKSZ Metaverse*, we built a relatively complete ecosystem based on blockchain and AI technologies. Then we will discuss some representative components to illustrate our design.

(1) Token-Driven Ecosystem: For a modern Metaverse, the token-driven ecosystem is a key element that makes the Metaverse different from a single-player game. *CUHKSZ Metaverse* employs blockchain-based tokens to feature a fair and transparent ecosystem, which is regarded as monetary representations of the community.

According to the predefined rules, the token's production rate may vary for distinct users subject to the residents' actions and performance in both the virtual and physical worlds, e.g. the location-based incentive mechanism, which was discussed in Section 3.3.2.2. The tokens can be utilized in various activities, such as trading in the official store, trading UGC with other players, voting, and so on. Figure 3.16 illustrates the store in *CUHKSZ Metaverse*. In this store, we uploaded some models as the "System goods" that users can buy for collection or decorating their personal rooms. Also, all users can trade their NFTs created using the UGC editor on this market, which could provide the users with liquidity for their tokens and UGC.

On the other hand, we also hope the token-driven ecosystem in *CUHKSZ Metaverse* can lead to real value exchange with the physical world. Therefore, some exciting activities can be conducted with the physical world. For example, cooperating with the campus canteen, the tokens in *CUHKSZ Metaverse* may be

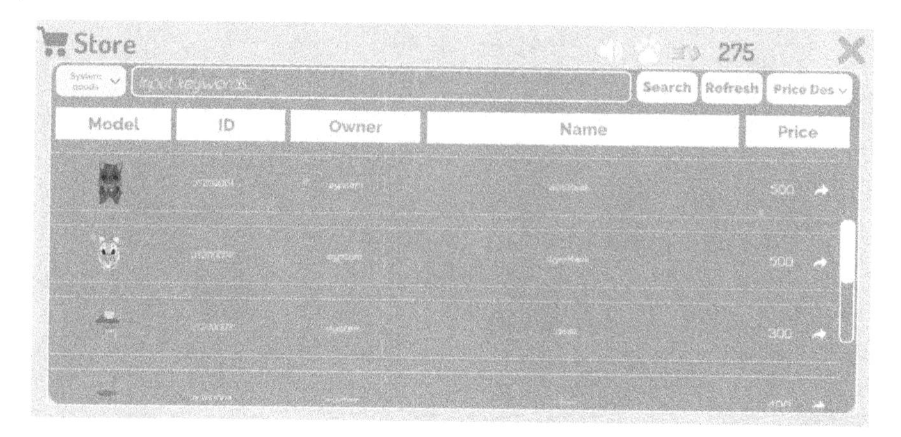

Figure 3.16 Store in *CUHKSZ Metaverse*. Source: Meta.

used to buy natural foods or drinks. Here, we also have some potential research topics like how to find a suitable price. Moreover, the owner of an NFT with specific attributes (e.g. highest trading price in a weak, most favorite item by other users) can obtain a physical object which is generated by 3D printing technology based on the NFT.

(2) Autonomous Governance: In the design of *CUHKSZ Metaverse*, we highly emphasized the necessity of social experience, so we intended to provide an autonomous governance mechanism to students and faculties based on the blockchain system. Specifically, *CUHKSZ Metaverse* enforces autonomous governance by introducing a Delegated Proof of Stake (DPoS) [16] voting protocol for any motions or proposals to revise predefined rules. Since our prototype is a university campus Metaverse, *CUHKSZ Metaverse* encourages students to establish the student union and elect a committee, which could efficiently make decisions for some matters representing other users. Moreover, there is an online forum maintained by CUHKSZ students named *LGULife*,[15] which is a very convenient community for discussing the proposals about *CUHKSZ Metaverse*. This series of work effectively assure autonomous governance and maintain the operation and iteration of *CUHKSZ Metaverse*.

In fact, *CUHKSZ Metaverse* is a preliminary study of the autonomous governance on university campus Metaverse, which could only provide limited reference to the related works because the autonomous governance in the decentralized Metaverse is a significantly large research area that covers lots of disciplines, e.g. sociology, psychology, economy [10, 20]. Therefore, we believe efficient and complete solutions need the joint effort of various communities.

15 www.lgulife.com/

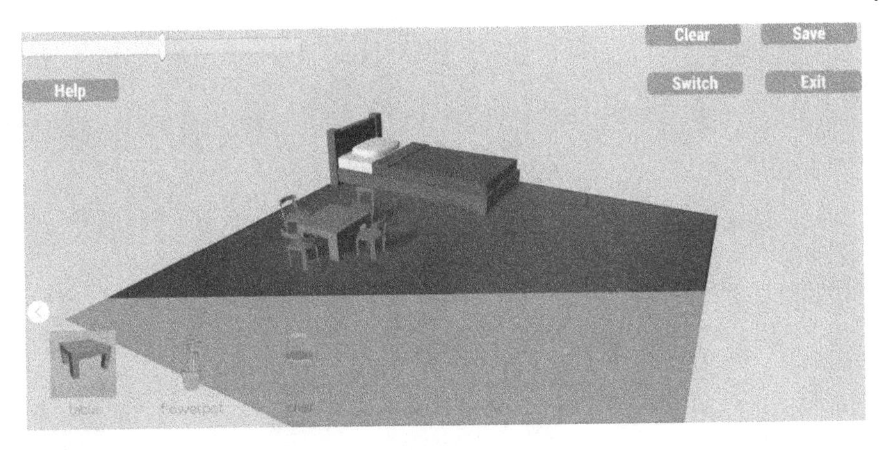

Figure 3.17 Personal room decoration in *CUHKSZ Metaverse*. Source: Meta.

(3) Personal Room and UGC Display: As we mentioned in Section 3.3.2.2, *CUHKSZ Metaverse* provides a UGC creator that allows users to create their customized items, and the Metaverse residents can trade the UGC with others. However, only possessing the UGC in their virtual backpack obviously weakens the social attributes of the users' creativity and imagination. Therefore, *CUHKSZ Metaverse* allows the users to have a personal room, in which the users can freely decorate the room using the possessed NFTs. Figure 3.17 shows the interaction process of personal room decoration, where the users can drag their NFTs from the backpack and place them in the room. In this figure, we used the table created by the UGC editor and four chairs bought from the store to decorate the room. Moreover, Figure 3.18 shows the personal room of the Metaverse viewer. In fact, the mechanism of the personal room is similar to the virtual land system of business Metaverse projects (e.g. *CryptoVoxels*, *The Sandbox*,[16] etc.), while the "land" in *CUHKSZ Metaverse* is distributed by the system and is not allowed to trade for better equality. On the other hand, the decoration layout of the personal room can also be regarded as a kind of UGC, which is worthy of further development and study.

In addition, a question is raised with the personal room: how can we enter other users' rooms? In *CUHKSZ Metaverse*, we provided the billboard system, a channel for the users to give full play to their personalities in public. As illustrated in Figure 3.19, a user may rent a specific area in the virtual campus and set their customized billboard, which could display their information or links to their personal rooms. In this way, we believe that the billboard system could build a naturally social experience for the users of *CUHKSZ Metaverse*.

---

16 https://www.sandbox.game/en/

Figure 3.18 Personal room in *CUHKSZ Metaverse*. Source: Meta.

Figure 3.19 Billboard system in *CUHKSZ Metaverse*. Source: Meta.

(4) AI-Driven Metaverse Observer: Unlike the Metaverse viewer discussed in Section 3.3.2.2, we developed an AI-driven Metaverse observer to track ongoing intriguing events that happened in *CUHKSZ Metaverse*. The Metaverse observer has a wider and higher vision that could cover an area of *CUHKSZ Metaverse*. Moreover, the Metaverse observer can automatically recommend ongoing intriguing events to users by tracking and analyzing real-time operation data of *CUHKSZ Metaverse*, and the events with a high flow of users are more likely to be recommended. Figure 3.20 shows an example around the basketball court, where a basketball match may be playing in the physical world. Then, many players may drive their avatars to the basketball court in the virtual world, so they can meet each other and make friends in both the physical and virtual worlds. Therefore,

Figure 3.20 An example of the basketball court in *CUHKSZ Metaverse*. Source: Meta.

the Metaverse observer can provide global information for users and audiences to better capture the happening events in *CUHKSZ Metaverse*. In current version, we are evaluating the performance of the proposed observer system and the performance will be published on academy in the future.

3.4 Conclusions and Future Research Directions

This chapter introduces the iteration procedure from *Newbie at CUHKSZ* to *CUHKSZ Metaverse*, which provides the readers with an intuitive understanding of the difference between a single-player game and a Metaverse. Moreover, this chapter discusses the key components of *CUHKSZ Metaverse* according to a three-layer Metaverse architecture, including the infrastructure layer, interaction layer, and ecosystem layer. The motivation of this chapter is to give the readers a relatively complete view of a practical Metaverse implementation, which could be helpful for the readers to understand other Metaverse projects, applications, and implementations.

In fact, the development of our prototype *CUHKSZ Metaverse* is still in a very early stage and so do the concept of Metaverse. Thus, there are many potential research directions that can be worthy of further studies during the iteration of the prototype. In this section, we list the following future research directions for the readers:

(1) Infrastructure Layer: Currently, the *CUHKSZ Metaverse* prototype has some problems in the infrastructure layer that should be fixed, since the system was developed by our students rather than professional engineers. For example, the server is hard to hold large-scale concurrent connections due to the limitation of infrastructure. Moreover, the efficiency of the utilized blockchain platform can also be improved, including the implementation of smart contracts, operation nodes, etc. Therefore, the fundamental support of *CUHKSZ Metaverse* has much potential for improvement.

(2) Interaction Layer: For enhancing the user experience of *CUHKSZ Metaverse*, there are also many effective and novel methods that can be introduced as new components of the prototype. For example, more IoT connections can be embedded in *CUHKSZ Metaverse*, e.g. the lights in classrooms can be controlled by people in both the physical world and the virtual world. On the other hand, how to facilitate the UGC creation procedure of normal users is a promising research direction, since the Metaverse has unlimited extension space so the construction of the Metaverse will highly depend on the users' creativity and innovation. For example, multimodel input can be utilized in the UGC creation procedure of the Metaverse, e.g. the text-to-image method (such as DALL-E2 [15]) can highly improve the efficiency of UGC creation and also enhance the interests of participants. More importantly, how to better apply VR/AR interaction approaches can be a good research direction in the future, which will determine the social and gameplay experience of the users.

(3) Ecosystem Layer: The ecosystem layer of *CUHKSZ Metaverse* also needs further iteration regarding a mature society model. For instance, we consider the token-driven ecosystem is an essential component of a sustainable Metaverse system. However, the current *CUHKSZ Metaverse* lacks a corresponding circulation of consumption and token creation, which is still our primary iteration direction. Behind the problem mentioned above, we believe the economic system and mechanism design of the Metaverse will be a promising research direction in the future because the Metaverse is a brand new form that is different from the existing world, which will raise many exciting studies to improve the ecosystem of the Metaverse.

Acknowledgement

Thanks for all developers of *Newbie at CUHKSZ* and *CUHKSZ Metaverse* at the Chinese University of Hong Kong, Shenzhen:

- *Newbie at CUHKSZ*: Changfeng Chen, Hanzhe Zhang, Jiazhen Liu, Liya Ge, Oushuo Huang, Pinyi Wang, Taifeng Fu, Tengfei Wang, Wei Wu, Yian Chen, Yichang Liu, Yilin Wang, Yining Tang, Ziheng Mo.
- *CUHKSZ Metaverse*: Honghao Chen, Nanjun Yao, Qiuhong Chen, Tong Chen, Xiangyu Xu, Yifan Zhao, Yiyan Hu, Yuyang Liang, Zexin Lin, Zhen Ren, Zhonghao Lin.

- Valuable suggestions and advice: Hao Cui, Hao He, Hao Wu, Jiaye Li, Jinhan Sun, Lehao Lin, Sizheng Fan, Yu Chen.

Special thanks for the kind help and support:

- Xiao Wu, White Matrix Inc.
- Zikai Alex Wen, Hong Kong University of Science and Technology (Guangzh-ou).

Bibliography

1 Karan Ahuja, Deval Shah, Sujeath Pareddy, Franceska Xhakaj, Amy Ogan, Yuvraj Agarwal, and Chris Harrison. Classroom digital twins with instrumentation-free gaze tracking. In *Proceedings of the 2021 CHI Conference on Human Factors in Computing Systems*, pages 1–9, 2021.

2 Zhiqin Chen, Vladimir G. Kim, Matthew Fisher, Noam Aigerman, Hao Zhang, and Siddhartha Chaudhuri. DECOR-GAN: 3D shape detailization by conditional refinement. In *Proceedings of the IEEE/CVF Conference on Computer Vision and Pattern Recognition*, pages 15740–15749, 2021.

3 Dariush Derakhshani and Randi L. Derakhshani. *Introducing 3ds MAX 2008*. John Wiley & Sons, 2008.

4 Haihan Duan, Jiaye Li, Sizheng Fan, Zhonghao Lin, Xiao Wu, and Wei Cai. Metaverse for social good: A university campus prototype. In *Proceedings of the 29th ACM International Conference on Multimedia*, pages 153–161, 2021.

5 Haihan Duan, Xiao Wu, and Wei Cai. Crypto-dropout: To create unique user-generated content using crypto information in metaverse. In *2022 IEEE 24th International Workshop on Multimedia Signal Processing (MMSP)*. IEEE, 2022.

6 Haihan Duan, Huang Yiwei, Zhao Yifan, Huang Zhen, and Wei Cai. User-generated content and editors in video games: Survey and vision. In *2022 IEEE Conference on Games (CoG)*. IEEE, 2022.

7 Sizheng Fan, Hongbo Zhang, Yuchen Zeng, and Wei Cai. Hybrid blockchain-based resource trading system for federated learning in edge computing. *IEEE Internet of Things Journal*, 8(4):2252–2264, 2020.

8 Ian Goodfellow, Jean Pouget-Abadie, Mehdi Mirza, Bing Xu, David Warde-Farley, Sherjil Ozair, Aaron Courville, and Yoshua Bengio. Generative adversarial nets. *Advances in Neural Information Processing Systems*, 27, 2014.

9 Thomas T. Hewett, Ronald Baecker, Stuart Card, Tom Carey, Jean Gasen, Marilyn Mantei, Gary Perlman, Gary Strong, and William Verplank. *ACM SIGCHI Curricula for Human-Computer Interaction*. ACM, 1992.

10 Christoph Jentzsch. Decentralized autonomous organization to automate governance. *White paper, November*, 2016.

11 Logan Kugler. Non-fungible tokens and the future of art. *Communications of the ACM*, 64(9):19–20, 2021.

12 Lei Ling, Gui Juntao, and Ding Xi. The design and implementation of the 3D virtual campus models. In *2017 4th International Conference on Systems and Informatics (ICSAI)*, pages 1747–1751. IEEE, 2017.

13 Erica L. Neely. The ethics of choice in single-player video games. In *On the Cognitive, Ethical, and Scientific Dimensions of Artificial Intelligence*, pages 341–355. Springer, 2019.

14 Jon Radoff. The metaverse value-chain. [EB/OL]. https://medium.com/building-the-metaverse/the-metaverse-value-chain-afcf9e09e3a7. Accessed: 2021-04-07.

15 Aditya Ramesh, Prafulla Dhariwal, Alex Nichol, Casey Chu, and Mark Chen. Hierarchical text-conditional image generation with clip latents. *arXiv preprint arXiv:2204.06125*, 2022.

16 Sheikh Munir Skh Saad and Raja Zahilah Raja Mohd Radzi. Comparative review of the blockchain consensus algorithm between proof of stake (POS) and delegated proof of stake (DPoS). *International Journal of Innovative Computing*, 10(2), 2020. doi: 10.11113/ijic.v10n2.272.

17 Jaroslav Švelch. The good, the bad, and the player: The challenges to moral engagement in single-player avatar-based video games. In *Ethics and game design: Teaching values through play*, pages 52–68. IGI Global, 2010.

18 Tomoko Tateyama, Asuka Kigami, Shun Nishikawa, Tetsuro Katada, and Shimpei Matsumoto. Construction of virtual campus guide system using mobile phone. In *2018 7th International Congress on Advanced Applied Informatics (IIAI-AAI)*, pages 765–768. IEEE, 2018.

19 April Tyack and Peta Wyeth. Exploring relatedness in single-player video game play. In *Proceedings of the 29th Australian Conference on Computer-Human Interaction*, pages 422–427, 2017.

20 Shuai Wang, Wenwen Ding, Juanjuan Li, Yong Yuan, Liwei Ouyang, and Fei-Yue Wang. Decentralized autonomous organizations: Concept, model, and applications. *IEEE Transactions on Computational Social Systems*, 6(5):870–878, 2019.

4

Wireless Technologies for the Metaverse

Hongliang Zhang[1], Shiwen Mao[2], and Zhu Han[3]

[1] *School of Electronics, Peking University, Beijing, China*
[2] *Department of Electrical and Computer Engineering, Auburn University, Auburn, AL, USA*
[3] *Department of Electrical and Computer Engineering, University of Houston, Houston, TX, USA*

After reading this chapter, you should be able to:

- Learn the important role that wireless technologies play in the Metaverse.
- Have a clear picture on the standardization of wireless technologies for the Metaverse, especially for extended reality (XR).
- Know how to solve wireless problems for the Metaverse and potential future research directions.

4.1 Introduction

In the Metaverse, users can have immersive experiences in the virtual world provided by extended reality (XR) services, driven by the social interaction shifts recently [21]. First, online meeting is becoming frequent due to the pandemic, and it will be important to provide immersive interactions in the digital world. Second, with an augmented reality (AR) glass, users could link all the related information with the physical world, making them a "super hero" to react promptly [18]. For example, when you meet a new guy in a conference, the AR glass can show you his Google Scholar profile so that you can know his research interest and could have a deep understanding during the conversation with him. Motivated by the aforementioned use cases, it is reasonable to envision that the Metaverse will play an important role in our daily lives in the near future. To put such a vision into practice, it poses high data rate requirements on supporting wireless

Metaverse Communication and Computing Networks: Applications, Technologies, and Approaches, First Edition.
Edited by Dinh Thai Hoang, Diep N. Nguyen, Cong T. Nguyen, Ekram Hossain, and Dusit Niyato.
© 2024 The Institute of Electrical and Electronics Engineers, Inc. Published 2024 by John Wiley & Sons, Inc.

networks for the transmission of high-resolution XR videos and satisfying the user's Quality-of-Experience (QoE).

On the other hand, the recent development of wireless communications and multiaccess edge computing (MEC) techniques enable users to access the virtual environment anytime and anywhere, which is an important driving force of Metaverse [13]. The standardization of XR support in New Radio (NR) has kicked off in the Third Generation Partnership Project (3GPP) [5], aiming to provide performance enhancements for XR services over the NR. Moreover, due to the faster response of the MEC compared to cloud computing, seamless immersive experiences can be provided in the wireless edge-empowered Metaverse [25].

With the deployment of edge servers, XR contents can be processed before transmission based on the available computation and communication resources, aiming to maximize the QoEs for these users. However, taking the compression as an example, compressing more data to save communication resources requires more computation resources. The coupling of communication and computation makes the resource management for the Metaverse not trivial. Without addressing these implementation issues from the communication and computation perspectives, it will be extremely difficult to achieve the vision of the Metaverse given before.

In this chapter, we will study how the wireless techniques influence the Metaverse. The organization of this chapter is as follows: in Section 4.2, we first review the standardization in 3GPP to show the NR support of XR, especially the traffic models and key performance indicators (KPIs). In Section 4.3, we provide a case study to show how to address the problems in location-dependent AR services in the wireless edge-enabled Metaverse. Finally, conclusions and future directions are provided in Section 4.4.

4.2 XR over NR: Standardization in 3GPP

The standardization of XR support via NR in 3GPP can be dated back to 2016, with Service and System Aspect (SA) working group (WG) on 5G service requirements for high-rate and low-latency XR applications [2]. The work was continued in 2018, by defining the relevant traffic characteristics in [4] and providing various XR applications in [1]. In parallel, SA2 ("System Architecture and Services") standardized new 5G quality of service identifiers (5QI) to support interactive services including XR [3]. In what follows, we shall first introduce the XR use cases.

4.2.1 Selected XR Applications

In [1], over 20 XR use cases are identified, which makes the performance evaluation challenging. As a result, these use cases are then categorized into three types in [5]: virtual reality (VR), AR, and cloud gaming (CG), as defined:

- **VR:** VR generates a virtual world where the user is fully immersed. In general, VR services are realized by viewport-dependent streaming. Transmission of new VR contents can be triggered by user movements and the demand for the next portion of the 3D video.
- **AR:** AR merges virtual objects with the live 3D view of the real world, creating a realistic personalized environment that the user interacts with. AR services rely on not only expensive motion detection sensors, but also cameras mounted on AR glasses to upload videos. Therefore, AR could require a video stream in both the uplink (UL) and the downlink (DL).
- **CG:** CG refers to an interactive gaming application executed at the cloud server. Therefore, heavy computations are offloaded from the CG device, thus relaxing the requirements at the user side.

4.2.2 Traffic Models

As traffic model is one of the principal elements needed when simulating applications, we also introduce three selected 3GPP-adopted traffic models [5]. The three above applications, i.e. VR, AR, and CG, have their own features in the data streams, as summarized in Table 4.1.

- **Video:** Video stream is the most important flow for all the considered applications. To keep a reasonable complexity, a single data packet in the model represents multiple IP packets corresponding to the same video frame. The average interarrival time is an inverse of the frame rate in frames per second (fps, i.e. 60 fps leads to 16.6 ms).
- **Motion:** Another important stream is motion updates sent by the XR device in the UL. This stream includes (i) user pose information update obtained from sensors; and (ii) auxiliary information, e.g. user's location, and/or commands from the user to the server.
- **Aggregated Audio and Video:** In addition to video, audio data can also be a separate stream. As the audio data are relatively small compared to the video data, modeling this stream is not mandatory. On the other hand, the frame rate of audio is higher than that of video, which may be important for the power consumption consideration.

Table 4.1 3GPP traffic model [5].

Traffic stream	DL/UL	Use case	Frame rate	Average data rate	PDB
Video	DL	AR VR	**60 fps** 30, 90, 120 fps	**30, 45 Mbit/s,** 60 Mbit/s@60 fps	10 ms
		CG		**8, 30 Mbit/s,** 45 Mbit/s@60 fps	15 ms
	UL	AR	60 fps	**10**, 20 Mbit/s @60 fps	30 ms
Motion	UL	AR VR CG	250 fps	0.2 Mbit/s	10 ms
Audio+ video	DL	AR/VR/CG	100 fps	0.756 Mbit/s,	30 ms
	UL	AR		1.12 Mbit/s	

If multiple values are given, the bold ones are typical values.

4.2.3 Major XR-Specific 3GPP KPIs

In 3GPP, the approved metrics for two major XR KPIs are capacity and power consumption [5].

4.2.3.1 Capacity

The study of capacity could be useful for understanding the limitation of current NR systems in supporting XR applications and the potential directions for future necessary enhancements to better support XR. For capacity, there are two metrics as listed:

User Satisfaction: A user is declared to be satisfied if all the considered streams meet their own packet error rate (PER) and packet delay budget (PDB) requirements, i.e. more than a certain percentage of packets are successfully transmitted within a given air interface PDB.

System Capacity: With the user satisfaction defined above, system capacity is defined as the maximum number of users per cell with at least $Y\%$ of users being satisfied. Here, $Y = 90$ for the baseline, and other values of Y can be evaluated optionally.

4.2.3.2 Power Consumption

The study of power consumption could be useful for understanding (i) the limitation of existing NR systems in supporting XR applications; and (ii) the potential directions for future necessary enhancements to improve power efficiency.

The metric for power evaluation is the user's power consumption for XR data transmission, which is indicated by the power saving gain (PSG). The PSG is determined from A: the power consumption of a power saving scheme where a specific power saving scheme defined in 3GPP Release15/16/17 is used, for example, the user can skip monitoring Physical Downlink Control Channel (PDCCH) for a certain duration as indicated in the network's PDCCH skipping command; and B: the power consumption of a baseline (AlwaysOn) case where the user is always on to keep monitoring PDCCH, even when it is not receiving/transmitting XR data. Mathematically, it is defined as

$$PSG = \frac{B - A}{B} \times 100\%. \tag{4.1}$$

Since the user's PSG typically comes with the loss in capacity (i.e. more precisely, the loss in the satisfied user equipment [UE] ratio), it also needs to be considered jointly with power consumption/PSG.

Limited to the communication requirements as discussed above, wireless edge computing technique is also adopted to process the XR video contents in order to save some communication resources. In Section 4.3, we will give a case study to show how to manage the communication and computation resources.

4.3 Case Study: Location-Dependent AR Services in the Wireless Edge-Enabled Metaverse

In this case study, we take location-dependent AR services in wireless edge-enabled Metaverse systems as an example. In such a system, AR users do not walk very fast as they will enjoy their AR contents. As a result, the environment can be considered as unchanged during a time slot. Within this time slot, each user needs to perform simultaneous localization and communication first to estimate its location and request AR contents from edge server. This information will be regarded as constant within this time slot. Due to limited computation and communication resources, the edge server needs to adaptively change the resolution of AR contents to provide good experience to Metaverse users.

4.3.1 System Model

As shown in Figure 4.1, we consider a wireless edge-enabled Metaverse system for location-dependent AR services. In this system, there exists a base station (BS) equipped with M antennas, denoted by $\mathcal{M} = \{1, \ldots, M\}$, and an edge server is deployed at the BS to provide AR services for Metaverse users. K Metaverse users, denoted by $\mathcal{K} = \{1, \ldots, K\}$, run AR applications, and the BS pushes video streams

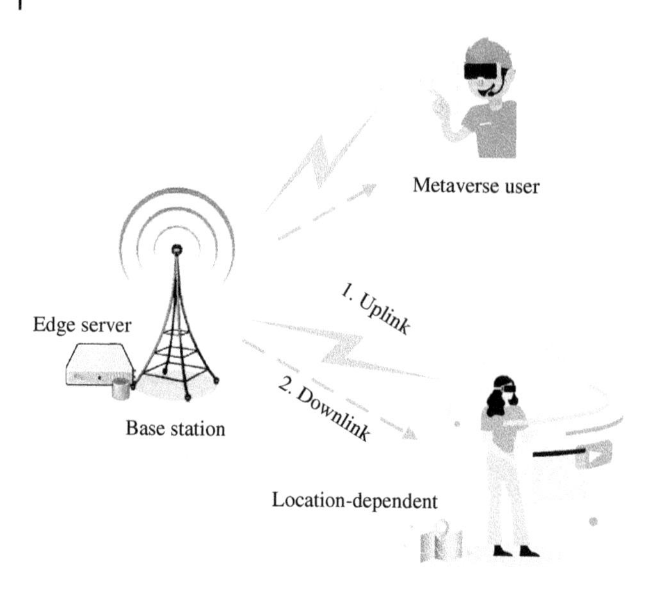

Figure 4.1 System model considered in this case study.

to these Metaverse users periodically. We assume that the AR device held by each user is equipped with N antennas. The bandwidth owned by the system is divided into K channels, each with bandwidth W, and each channel is used to serve one user. As different users occupy orthogonal channels, there is no interference among users.

The system is operated in a synchronized manner. To be specific, the timeline is divided into several time slots, and we have the following two phases in each time slot:

- **Uplink Phase:** In this phase, the user will perform simultaneous localization and transmission. Specifically, the AR device will transmit wireless signals to update its own position and orientation. At the same time, it will also inform the BS of starting video streaming according to the estimated results.
- **Downlink Phase:** After the uplink phase, the edge server and the BS will process a fixed number of frames for the video and transmit the processed video frames back to the user. They need to take the computation and communication resources into consideration when processing these video frames.

It is worthwhile to point out that the virtual objects will remain unchanged within the considered time window. In other words, the total data volume to be transmitted within the time window is fixed. However, for some certain AR applications, the total amount of data will be large or the served number of users could be large [24]. Therefore, we consider an adaptive data transmission scheme for AR data. In other words, we consider the nontrivial case where the data cannot be fully transmitted within a time slot and only a portion of the data can be transmitted. Moreover, the data transmitted in previous time slots will be cached in the headset, and there is no need to retransmit it again in the following time slots. Details of these two phases will be presented in Sections 4.3.1.1 and 4.3.1.2. Note that these two phases are discussed for time slot t, and the time index is omitted unless mentioned.

4.3.1.1 Uplink Phase

In the uplink phase, each Metaverse user will send wireless signals for simultaneous communications and localization, as elaborated below:

Communication: To guarantee that the BS can receive the data request from each user, the received signal-to-noise ratio (SNR) should be larger than a predefined threshold. To be specific, denote the channel between the BS and user k as $\boldsymbol{H}_k \in \mathbb{C}^{N\times M}$ and the transmitted signals from user k as $\boldsymbol{x}_k \in \mathbb{C}^{(A+1)\times 1}$, where A is the number of single-antenna anchor nodes for localization. We also define the beamformer for user k as $\boldsymbol{W}_k \in \mathbb{C}^{N\times(A+1)}$. Therefore, the received signal at the BS from user k can be expressed as

$$\boldsymbol{y}_k = \boldsymbol{H}_k^T \boldsymbol{W}_k \boldsymbol{x}_k + \boldsymbol{n}_k, \tag{4.2}$$

where $\boldsymbol{n}_k$ is the addictive Gaussian noise with zero mean and variance σ^2.

Let $|\boldsymbol{x}_k|^2 = 1$, and the SNR constraint can be expressed as

$$\gamma_k = \frac{(\boldsymbol{H}_k^T \boldsymbol{W}_k)(\boldsymbol{H}_k^T \boldsymbol{W}_k)^H}{\sigma^2} \geq \eta^k, \tag{4.3}$$

where η_k is the SNR threshold for AR user k, which is related to the running AR service.

Moreover, subject to the power budget for each user, the beamformer vector should satisfy

$$\mathrm{Tr}(\boldsymbol{W}_k \boldsymbol{W}_k^H) \leq P_{max}, \tag{4.4}$$

where P_{max} is the power budget for each user.

Localization: At the same time, each user also generates beams for location estimation [32]. Let the channel between user k and anchor node i be $\boldsymbol{h}_{ik} \in \mathbb{C}^{1\times N}$, and thus, the received signal can be expressed as

$$y_{ik} = \boldsymbol{h}_{ik} \boldsymbol{W}_k \boldsymbol{x}_k + n_{ik}, \tag{4.5}$$

where n_{ik} is also the addictive Gaussian noise with zero mean and power σ^2. The SNR for each anchor node can be expressed as

$$\gamma_{ik} = \frac{(\boldsymbol{h}_{ik} \boldsymbol{W}_k)(\boldsymbol{h}_{ik} \boldsymbol{W}_k)^H}{\sigma^2}. \tag{4.6}$$

Each user can gather the signals received at these anchor nodes and estimate its location from these signals. However, the estimation will not be perfect and will definitely cause some estimation errors. To be specific, the location estimation error can be expressed as

$$e_k = \mathbb{E}[|\boldsymbol{l}_k - \tilde{\boldsymbol{l}}_k|^2], \tag{4.7}$$

where $\tilde{\boldsymbol{l}}_k$ is the estimated location and $\boldsymbol{l}_k$ is the ground-truth. For any unbiased estimator of position, the estimation error is bounded by the Cramér-Rao bound [17]. To be specific, the estimation error can be approximated as

$$e_k \approx \mathrm{tr}(\boldsymbol{J}_k^{-1}), \tag{4.8}$$

where $\boldsymbol{J}_k$ is the Fisher information matrix (FIM). According to the results in [16], the FIM can be defined as

$$\boldsymbol{J}_k = \frac{8\pi^2 W^2}{c^2} \sum_{i=1}^{A} \gamma_{ik} \boldsymbol{q}_{ik} \boldsymbol{q}_{ik}^T, \tag{4.9}$$

where c is the propagation speed, and $\boldsymbol{q}_{ik} = [\cos\phi_{ik}, \sin\phi_{ik}]^T$ with ϕ_{ik} being the angle between user k and anchor i, which is a constant obtained by estimating the angle of arrival (AoA).

For orientation, we assume that the headset is equipped with an accelerometer, gyroscope, and magnetometer, and the orientation can be obtained through these sensors, which are assumed to be perfectly detected.

4.3.1.2 Downlink Phase

In the downlink phase, the BS needs to transmit the data of virtual objects to these Metaverse users. However, limited by communication resources, the BS might not be able to transmit all the data to users within one time slot. Instead, the BS only sends the data which might be within the field-of-view (FoV) of the user relating to its position, and thus the transmitted data will change according to the movement (including position and orientation) of the user, as shown in Figure 4.2. On the other hand, even though the BS only transmits the data that might be within the FoV, the bandwidth might still not be sufficient to transmit the frames of the video with the highest resolution that the BS could have. Instead, the BS will transmit a portion of the video frames according to the available communication resources by adjusting the resolution of the video frames. As different parts of a video frame contribute differently to the QoE, their corresponding resolutions might vary. In the following, we first explain how the system changes the resolution of the video.

As shown in Figure 4.3, the video frame for each user is partitioned into I 3D tiles and only part of these tiles will fall inside the user's FoV. Here, we define $\mathcal{I}^k = \{1, \dots, I\}$ as the set of tiles for user k, and I is the number of tiles. However, at a certain position and orientation, as different tiles contribute differently to the

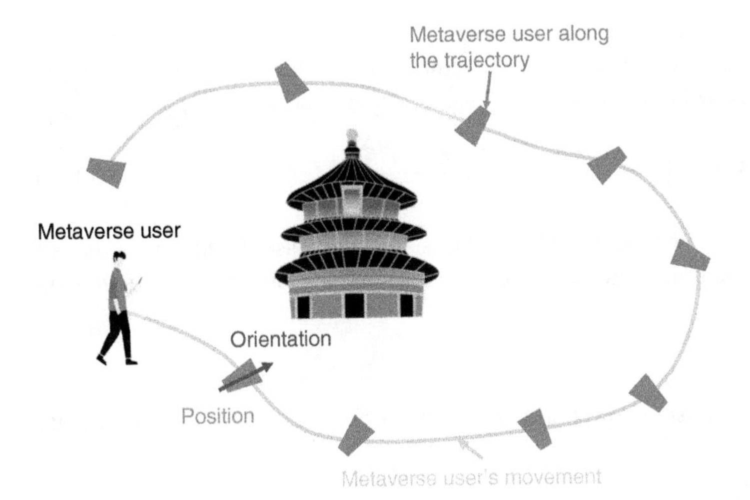

Figure 4.2 Adaptive AR communications and Metaverse user movement.

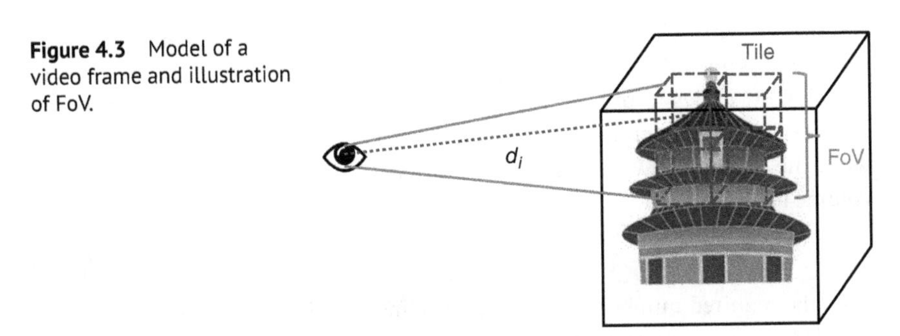

Figure 4.3 Model of a video frame and illustration of FoV.

user's QoE, the resolutions for different tiles should vary. For user k, define the resolution for the i-th tile as $\beta_i^k \in (0, B]$, where B is the highest resolution that a tile could be [20]. Therefore, the resolution for each tile should be adjusted according to the system computation and communication resources. It is worth pointing out that the maximum resolution level for each tile will change in different time slots. To be specific, the transmitted video frames will be cached in the headset, and thus those data do not need to be retransmitted. Based on these facts, we elaborate on the computation and communication models in the following.

Communication Model: The volume of transmitted data for each tile is proportional to the resolution of this tile [29]. Define κ as the coefficient, and thus, the data traffic to be transmitted to user k can be expressed as

$$D_k = \sum_{i \in \mathcal{I}^k} \kappa \beta_i^k. \tag{4.10}$$

Let the transmit power of antenna $m \in \mathcal{M}$ for user k be p_m^k, and $\boldsymbol{p}^k = [p_1^k, \dots, p_M^k]^T$ be the transmit power vector at the BS for user k. Therefore, the transmission rate between the BS and user k can be expressed as

$$R_k = W \log_2 \left(1 + \frac{\boldsymbol{H}^k (\boldsymbol{p}^k)^{1/2} (\boldsymbol{H}^k (\boldsymbol{p}^k)^{1/2})^H}{\sigma^2} \right), \tag{4.11}$$

where σ^2 is the power of the additive Gaussian white noise. To guarantee that all the data can be transmitted to the user, the data rate should meet the following constraint:

$$D_k \leq R_k. \tag{4.12}$$

Moreover, the transmit power of each antenna over all channels cannot exceed its energy budget, i.e.

$$\sum_{m \in \mathcal{M}} \sum_{k \in \mathcal{K}} p_m^k \leq P. \tag{4.13}$$

Computation Model: To adjust the resolution of the video frame, the edge server needs to utilize its central processing units (CPUs). Let the number of overall CPU cycles to process the videos be V within a time slot. According to the video model shown in Figure 4.3, for user k, the original data volume is κIB while the transmitted volume is D_k as shown in (4.10). Therefore, the compressed volume is

$$C_k = \kappa IB - D_k. \tag{4.14}$$

As the required number of CPU cycles is linear with the compressed data volume [11], let ν be the data bits that can be processed in a CPU cycle. The required number of CPU cycles for user k can be expressed as

$$\mu_k = \frac{C_k}{\nu}. \tag{4.15}$$

Therefore, to ensure that all the computation tasks can be completed, we have

$$\sum_{k \in \mathcal{K}} \mu_k \leq V. \tag{4.16}$$

4.3.2 Problem Formulation

In this part, we elaborate on how the location estimation error influences the Metaverse user's QoE. We first introduce the QoE model to quantify the performance of the Metaverse system and then formulate the minimum QoE maximization problem.

4.3.2.1 Impact of Estimation Error

As we have introduced before, the location estimation cannot be perfect and will cause some errors here. As a result, the estimation error will change the probability that each tile falls within the FoV. In the following, we will show the details about how the estimation error influences the probability.

In this chapter, we assume that the FoV consists of $b_w \times b_h$ tiles. However, as shown in Figure 4.4, the estimation error of position will cause the shift of the FoV. Therefore, we use the probability for each tile that falls in the FoV to quantify the impact of the estimation error. Intuitively speaking, if the estimation error is smaller, the tiles within the FoV associated with the estimated position will have a higher probability to be shown in the FoV.

We assume that the ground-truth position is within the error range centered at the estimated one [26]. To be specific, we divide the range of location error into several discrete levels so that the corresponding FoV is shifted left (right) or up (down), as illustrated by positions 1 and 2 in Figure 4.4. Moreover, the orientation including azumith and evaluation angles will also influence the shift of the FoV as discussed below.

Without loss of generality, we assume that the position error range is divided into L levels with a fixed gap ΔL, denoted by $\mathcal{L}$. We define the azumith and evaluation angles for user k as θ_k and φ_k, respectively. We also assume that the position follows a certain known distribution within the error range, denoted by $\boldsymbol{f}^p$. For tile i, we define x_l^i to indicate whether this tile is in the possible FoV l under azumith angle θ_k and evaluation angle φ_k,[1] where

$$x_l^i = \begin{cases} 1, & \text{if tile } i \text{ is in the possible FoV } l; \\ 0, & \text{otherwise.} \end{cases} \tag{4.17}$$

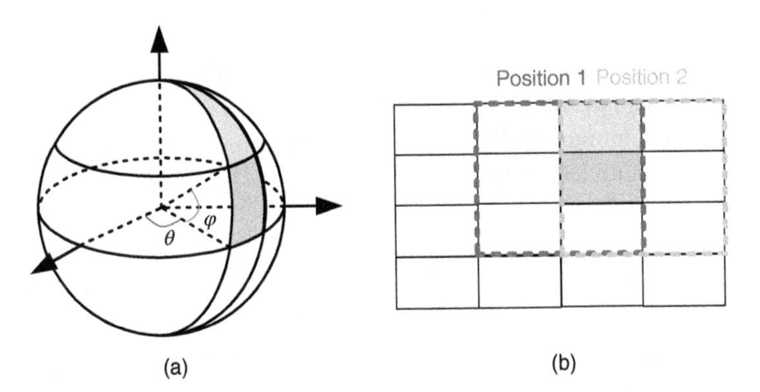

Figure 4.4 Spherical mapping of tiles: (a) Sphere and (b) plane.

1 This indicates that the FoV is associated with the l-th position in $\mathcal{L}$.

Therefore, the probability that tile i falls in the FoV can be expressed as

$$f_i = \sum_{l \in \mathcal{L}} x_l^i f_l^p, \tag{4.18}$$

where f_l^p is the probability that the l-th location is the ground-truth.

4.3.2.2 QoE Model

According to the earlier discussions, different tiles contribute to the total QoE differently since its probability to be presented in the FoV varies. Before presenting the optimization problem, we need to first model the QoE contribution of each tile.

According to the results in [15], the QoE contributed by tile i for user k can be expressed as

$$Q_i^k = \omega_i^k(\beta_i^k + \bar{\beta}_i^k)f_i^k. \tag{4.19}$$

Here, ω_i^k is the quality weight for the tile, which could be obtained from historical data. In other words, this parameter shows the importance of tile i in each video frame for user k. $\bar{\beta}_i^k$ is the data of this tile that has been stored in the headset, and f_i^k is the probability that tile i falls within the FoV of user k, which is calculated by (4.18).

Therefore, the overall QoE for user k can be expressed as

$$Q^k = \tilde{Q} + s\log_2\left(\sum_{i \in \mathcal{I}^k} Q_i^k\right) - \log_2\left(\sum_{i \in \mathcal{I}^k} \omega_i^k(B + \bar{\beta}_i^k)f_i^k\right), \tag{4.20}$$

where $\tilde{Q}$ is a constant representing the maximum QoE that the system can achieve, and s is a scaling factor.

4.3.2.3 Problem Statement

As communication and computation resources are limited, it is not possible to guarantee the QoE for all the Metaverse users. For fairness among these users, we aim to maximize the minimum QoE among all the users, which tries to align the QoEs of different users under limited resource constraints [12]. Mathematically, the problem can be written as

$$\begin{aligned} &\max_{\{\beta_i^k\},\{\mathbf{p}^k\},\{\mathbf{W}_k\}} \min_{k \in \mathcal{K}} Q^k, \\ &\text{s.}t. \ (4.3),\ (4.4),\ (4.7),\ (4.12),\ (4.13),\ \text{and}\ (4.16). \end{aligned} \tag{4.21}$$

Problem (4.21) is a max-min problem, which is not easy to deal with. To address this issue, we first introduce an auxiliary variable η, and replace the objective as the maximization of ϵ. To guarantee that ϵ is the minimum of Q^k among $\mathcal{K}$, we need to add the following constraint:

$$Q^k \geq \epsilon, \forall k \in \mathcal{K}. \tag{4.22}$$

Moreover, the uplink and downlink phases are coupled, which makes the problem hard to tackle. To address this issue, we decouple this problem into two subproblems:

(1) Uplink Subproblem: In the uplink phase, the subproblem can be transformed into the minimization of the estimation error range, which is positively related to the QoE. Mathematically, as each Metaverse user independently optimizes its waveform, the optimization problem can be written as

$$\begin{aligned} &\min_{\boldsymbol{W}_k} e_k, \\ &\text{s.}t. \ (4.3) \text{ and } (4.4). \end{aligned} \tag{4.23}$$

(2) Downlink Subproblem: In this subproblem, we can obtain the localization error range according to the optimized waveform in the previous phase $\{\boldsymbol{W}_k\}$. After that, the BS needs to determine the transmit power at each antenna and the resolution for each tile. Mathematically, the optimization problem can be written as

$$\begin{aligned} &\min_{\{\beta_i^k\},\{\boldsymbol{p}^k\},\epsilon} \quad -\epsilon, \\ &\text{s.}t. \ (4.12), \ (4.13), \ (4.16), \ \text{and } (4.22). \end{aligned} \tag{4.24}$$

4.3.3 Algorithm Design

In this subsection, we will elaborate on how to solve the aforementioned two subproblems.

4.3.3.1 Waveform Design Algorithm for the Uplink Subproblem

In this subproblem, we aim to minimize the estimation error under the SNR constraint and power budget. It is easy to check that the constraints are convex. Therefore, we elaborate on how to tackle the objective e_k.

According to the results in [19], such a problem can be converted into a semidefinite program (SDP). To be specific, we first introduce an auxiliary matrix $\boldsymbol{Z}_k$ and add another constraint as

$$\boldsymbol{Z}_k \succeq \boldsymbol{J}_k^{-1}. \tag{4.25}$$

Since $\boldsymbol{J}_k$ is a positive semidefinite matrix, due to the property of Schur complement, the inequality is equivalent to

$$\begin{bmatrix} \boldsymbol{Z}_k & \boldsymbol{I} \\ \boldsymbol{I} & \boldsymbol{J}_k \end{bmatrix} \succeq 0. \tag{4.26}$$

Therefore, the problem for user k can be rewritten as

$$\begin{aligned} &\min_{\boldsymbol{W}_k,\boldsymbol{Z}_k} \operatorname{Tr}(\boldsymbol{Z}_k), \\ &\text{s.}t. \ (4.3), \ (4.4), \ \text{and } (4.26). \end{aligned} \tag{4.27}$$

Such a problem is an SDP, which can be solved by existing convex optimization techniques [8].

4.3.3.2 Joint Resolution and Transmit Power Optimization Algorithm for the Downlink Subproblem

In this subproblem, we use the waveform obtained in the uplink subproblem to calculate the estimation error range. In other words, the probability that each tile falls within FoV f_i^k is constant in this problem as defined in Section 4.3.2.1.

As the constraints define a convex set, the problem can be solved by existing convex optimization techniques. In the following, we introduce the algorithm with more details.

Let λ_1^k, λ_2^m, λ_3, and λ_4^k be the Lagrangian multiplier corresponding to (4.12), (4.13), (4.16), and (4.22). Therefore, the Lagrangian can be expressed as

$$\begin{aligned} U = &-\epsilon + \sum_{k\in\mathcal{K}} \lambda_1^k(D_k - R_k) + \sum_{m\in\mathcal{M}} \lambda_2^m \left(\sum_{k\in\mathcal{K}} p_m^k - P \right) \\ &+ \lambda_3 \left(\kappa IBK - \sum_{k\in\mathcal{K}} D_k - \nu V \right) + \sum_{k\in\mathcal{K}} \lambda_4^k(\epsilon - Q_k), \end{aligned} \tag{4.28}$$

and the Lagrangian dual problem can be written as

$$\max_{\{\lambda_1^k\},\{\lambda_2^m\},\lambda_3,\{\lambda_4^k\}} \min_{\{\beta_i^k\},\{\mathbf{p}^k\},\epsilon} U. \tag{4.29}$$

We can solve the dual problem iteratively by decomposing it into master and slave subproblems as follows:

(a) Slave Subproblem: According to the Karush–Kuhn–Tucker (KKT) conditions, we can obtain the optimal solution by equaling the first derivative of the Lagrangian function U over the variables to 0. Specifically, we have $\frac{\partial U}{\partial p_m^k} = 0$, $\frac{\partial U}{\partial \beta_i^k} = 0$, and $\frac{\partial U}{\partial \epsilon} = 0$. After some transformation, we have

$$\mathrm{Tr}((\boldsymbol{H}^k)^H \boldsymbol{P}(\boldsymbol{H}^k)) = \frac{\lambda_1^k W(\boldsymbol{h}_m^k)^H \boldsymbol{h}_m^k}{\lambda_2^m} - \sigma^2, \tag{4.30}$$

$$\sum_{i\in\mathcal{I}^k} \omega_i^k(\beta_i^k + \bar{\beta}_i^k) f_i^k = \frac{(\lambda_1^k - \lambda_3)\kappa}{\lambda_4^k \omega_i^k f_i^k \ln 2}, \tag{4.31}$$

$$\sum_{k\in\mathcal{K}} \lambda_4^k = 1. \tag{4.32}$$

We can see that p_m^k and β_i^k can be obtained through solving linear programs defined in (4.30) and (4.31), respectively. In (4.32), it gives a condition on the Lagrangian multipliers. For ϵ, it can be set as the minimum Q_k among $k \in \mathcal{K}$.

(b) Master Subproblem: Once the results of the slave subproblem are obtained, the solution of the dual problem can be obtained by a subgradient method [7]. The update rule is given as

$$\lambda_1^{k,(t+1)} = \left[\lambda_1^{k,(t)} - \delta_1(D_k - R_k)\right]^+, \forall k \in \mathcal{K}, \tag{4.33}$$

$$\lambda_2^{m,(t+1)} = \left[\lambda_2^{m,(t)} - \delta_2\left(\sum_{k\in\mathcal{K}} p_m^k - P\right)\right]^+, \forall m \in \mathcal{M}, \tag{4.34}$$

$$\lambda_3^{(t+1)} = \left[\lambda_3^{(t)} - \delta_3\left(\kappa IBK - \nu V - \sum_{k\in\mathcal{K}} D_k\right)\right]^+, \tag{4.35}$$

$$\lambda_4^{k,(t+1)} = w_k^{t+1}\left[\lambda_4^{k,(t)} - \delta_4(\epsilon - Q_k)\right]^+, \forall k \in \mathcal{K}, \tag{4.36}$$

where t is the iteration indicator, $[x]^+ = \max\{0, x\}$, w_k^{t+1} is weighting factor to satisfy (4.32), and δ_1, δ_2, δ_3, and δ_4 are step sizes to guarantee convergence.

Remark 4.1 Subproblem (4.24) has a non-empty feasible set only when

$$\sum_{k\in\mathcal{K}} R_k^* \geq \kappa IBK - \nu V, \tag{4.37}$$

where $\sum_{k\in\mathcal{K}} R_k^*$ is the maximum sum-rate under the power budget constraint (4.13).

Proof: According to (4.14)–(4.16), we have

$$\sum_{k\in\mathcal{K}} D_k \geq \kappa IBK - \nu V. \tag{4.38}$$

On the other hand, we can derive from (4.12) that

$$\sum_{k\in\mathcal{K}} R_k \geq \sum_{k\in\mathcal{K}} D_k. \tag{4.39}$$

This indicates that we can find a feasible solution $\{D_k\}$ as long as the maximum value of $\sum_{k\in\mathcal{K}} R_k$ is guaranteed to be larger than $\kappa IBK - \nu V$. This completes the proof.

4.3.4 Simulation Results

In this subsection, we evaluate the performance of the proposed scheme.

4.3.4.1 Settings

The simulation settings are presented in Figure 4.5, a virtual object is placed in the center of the room, and we have $K = 5$ Metaverse users who walk around the virtual object in a circular trajectory as suggested in [24] whose radius is 10 m. $A = 3$ single-antenna anchor points for sensing are randomly located in a circle whose radius is 50 m. The distance between the virtual object and the BS is set to 150 m, and the height of the BS is set to 10 m. According to [22], we assume that the possible positions are uniformly distributed within the error range.

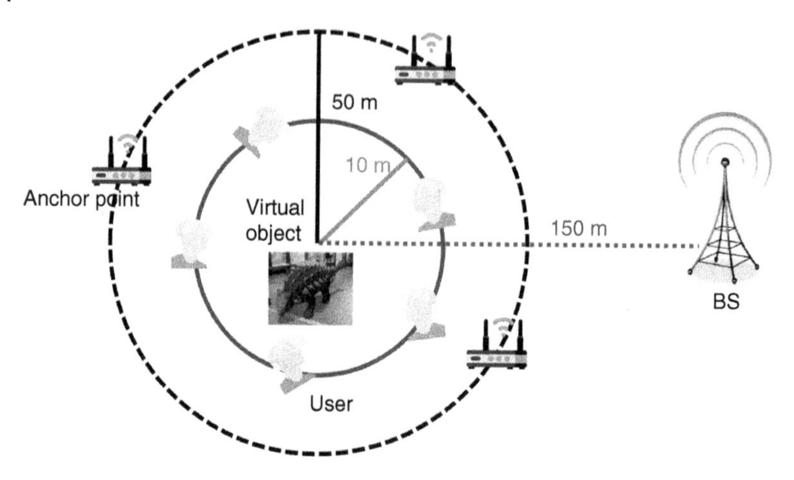

Figure 4.5 Evaluation settings. Source: Meta.

Table 4.2 Weight for tiles.

Row \ Column	1	2	3	4	5	6
1	0.001	0.002	0.006	0.008	0.005	0.001
2	0.011	0.026	0.115	0.137	0.049	0.015
3	0.032	0.052	0.158	0.180	0.061	0.045
4	0.006	0.011	0.032	0.033	0.006	0.008

For video transmission, the video frames are segmented into chunks of one second length, i.e. the duration for each time slot is one second [31], and we consider a total duration of $T = 10$ s. The video is encoded by FFMPEG with X.264 and its best quality is 4K,[2] corresponding to a bitrate of 35 Mbps. In other words, $B = 3.686 * 10^5$ and $\kappa = 2.086$. The FoV is divided into four rows and six columns, and the weights for the tiles are given in Table 4.2 [10]. The maximum QoE constant is $\tilde{Q} = 30$ and scaling factor is $s = 10$.

For the uplink phase, we assume that the AR device is equipped with 4 antennas, and the BS is equipped with 16 antennas. The channel is Rayleigh faded where the decay factor is 3.76 and power gains factor is −17.7 dB. For the downlink phase, the CPU capacity for the edge server is $V = 6 \times 10^{10}$ cycles/second and the required CPU cycles for the compression task is $\mu = 2000$ cycles/bit [27]. Other simulation settings are listed in Table 4.3 [30].

2 4K video means each frame has 4096 × 2160 pixels.

Table 4.3 Parameters for simulation.

Parameter	Value
Power budget for each user P_{max}	20 dBm
Transmission bandwidth	1 MHz
Carrier frequency	1.9 GHz
Noise figure	5 dB
SNR threshold	0 dB
Error tolerance level π	0.1
Power budget for the BS P	23 dBm

4.3.4.2 Evaluation Results

To evaluate the performance of the proposed scheme, we also present the results of the following two benchmark schemes:

- Fixed Location: the edge server only transmits the VR content according to the estimated location without considering estimation error. The resolution adjustment, communication, and computation resource management schemes are the same as that in the proposed scheme.
- Uniform Resolution: the edge server does not recognize the weight of each tile and makes resolutions of all the possible tiles within the FoV the same. The resolution needs to be optimized according to the available computation and communication resources.

Figure 4.6 shows how QoE changes for different number of users K. In general, the system QoE, i.e. the minimum QoE among all the users, decreases with the number of users. The reasons are twofold: (i) The interference among users will make the estimation less accurate. As a result, more data need to be transmitted; and (ii) the communication and computation resources for each user diminish with more users. On the other hand, comparing with the benchmark schemes, we can observe that the proposed scheme achieves the best performance, indicating the effectiveness of the proposed scheme. In comparison with the fixed location scheme, our scheme is similar to the idea of robust optimization [9], which allocates communication and computation resources to those FoVs with a lower probability to reduce the QoE drop cased by the estimation error. Moreover, compared with the uniform resolution scheme, we can learn that proper resolution matching, i.e. transmitting tiles with higher weight with higher resolution[3] will result in a higher total QoE. Furthermore, we can also observe that the QoE obtained by

3 The idea is similar to the maximum ratio combining in wireless communications [6].

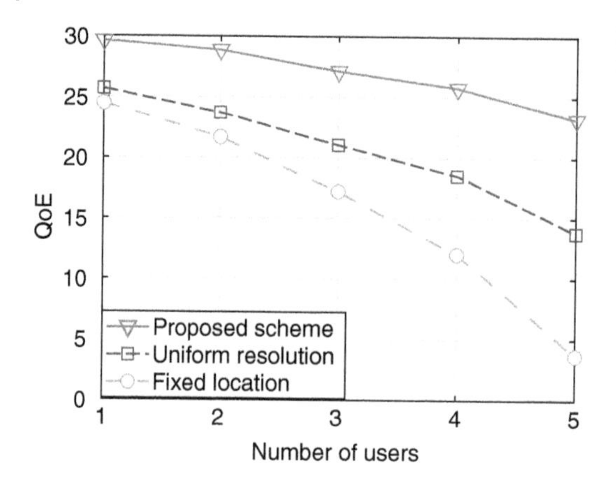

Figure 4.6 QoE vs. number of users K.

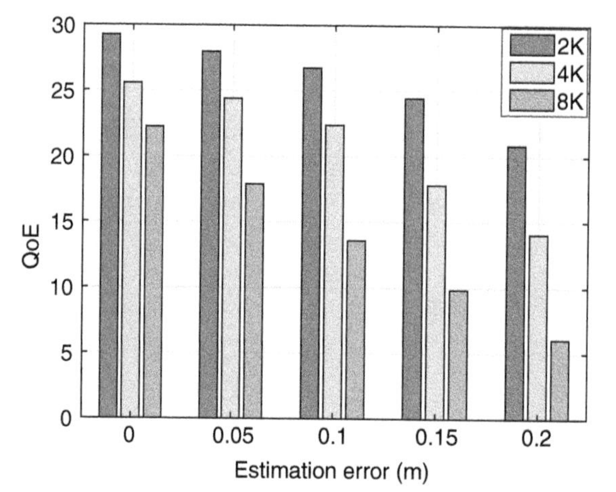

Figure 4.7 QoE vs. estimation error with different video quality levels.

the uniform resolution scheme is higher than that obtained by the fixed location scheme, indicating that the estimation error has more significant impact on the QoE than resolution matching.

In Figure 4.7, we plot the QoE vs. estimation error range with different video quality levels, i.e. 2K, 4K, and 8K. It is observed that the QoE decreases with a larger estimation error range. We can also learn that the drop rate is increasing with a larger estimation error. This is because a larger range results in a larger potential FoV range, and thus more data are required to be transmitted. Moreover, we can observe that the QoE will decrease for higher-quality videos with limited communication resources. The reason under this observation is that more data need to be transmitted to maintain the same QoE level while the communication resource might not be sufficient to support more data transmissions.

4.4 Conclusions and Future Research Directions

In this chapter, we have introduced the state-of-the-art standardization progress in 3GPP on the NR support of XR. Moreover, we have provided a case study to show how to transmit AR videos in wireless edge-enabled Metaverse. This case study shows that the communication resources are still the bottleneck for the Metaverse. In the following, we will outline several research directions on wireless technologies for Metaverse.

- **Distributed Caching:** To immerse users in a shared virtual world, Metaverse needs to provide plenty AR/VR contents. To reduce the transmission overhead, caching of these contents on edge servers or devices could be a promising solution [23]. However, the Metaverse poses several challenges on caching. First, the contents might be very large so that it would be impossible to cache them in a single server. Therefore, caching schemes should be carefully investigated to ensure the uniformity of the Metaverse content across geographically distributed edge servers. Second, unlike most existing caching schemes that mainly focus on user's interest and popularity, the contents in the Metaverse should also be related to the location, i.e. should be semantic, which requires the development of novel caching schemes.
- **Haptic Communications:** In addition to visual and auditory content, the Metaverse also needs haptic and kinematic interaction to provide a fully immersive experience [28]. This information is very sensitive to latency, which poses a very strict end-to-end latency requirement to the Metaverse. Moreover, haptic communications are vulnerable to network jitter, and thus some content prediction schemes are necessary to guarantee the network reliability.
- **Synchronization:** Metaverse is a mixture of physical and virtual worlds. Therefore, it is an important issue to maintain synchronization between physical and virtual worlds. Digital twin is a possible solution to address this issue [14]. Through a digital mapping from a physical world, users can easily control and calibrate the physical world through the Metaverse to achieve synchronization.
- **Synchronization:** Metaverse is a mixture of physical and virtual worlds. Therefore, it is an important issue to maintain synchronization between physical and virtual worlds. Digital twin is a possible solution to address this issue [14]. Through a digital mapping from a physical world, users can easily control and calibrate the physical world through the Metaverse to achieve synchronization.
- **Wireless Technologies Integration:** In the Metaverse, diverse requirements are necessary to provide immersive interactions in the digital world, e.g. data rate, latency, and reliability [18]. Therefore, it will be important to fully utilize

the communication and computation resources in the system and integrate advance wireless technologies, e.g. reconfigurable intelligent surfaces, millimeter wave/Terahertz communications, and semantic communications, to meet these requirements.

Acknowledgment

Z. Han's work is supported in part by the NSF under grant CNS-2107216 and CNS-2128368. S. Mao's work is supported in part by the NSF under grant CNS-2148382.

Bibliography

1 3GPP. Extended Reality (XR) in 5G. Technical Report (TR) 26.928, December 2020. Version 16.1.0.
2 3GPP. Service requirements for the 5G system. Technical Specification (TS) 22.261, December 2021. Version 18.5.0.
3 3GPP. System architecture for the 5G System (5GS). Technical Specification (TS) 23.501, December 2021. Version 17.3.0.
4 3GPP. Typical traffic characteristics of media services on 3GPP networks. Technical Report (TR) 26.925, September 2021. Version 17.0.0.
5 3GPP. Study on XR (Extended Reality) evaluations for NR. Technical Report (TR) 38.838, January 2022. Version 17.0.0.
6 Kyung Seung Ahn and Robert W. Heath. Performance analysis of maximum ratio combining with imperfect channel estimation in the presence of cochannel interferences. *IEEE Transactions on Wireless Communications*, 8(3):1080–1085, 2009.
7 Stephen Boyd, Lin Xiao, and Almir Mutapcic. Subgradient methods. *Lecture Notes of EE392o, Stanford University, Autumn Quarter*, 2004:2004–2005, 2003.
8 Stephen Boyd, Stephen P. Boyd, and Lieven Vandenberghe. *Convex Optimization*. Cambridge university press, Cambridge, UK, 2004.
9 Yongsheng Cao, Demin Li, Yihong Zhang, Qinghua Tang, Amin Khodaei, Hongliang Zhang, and Zhu Han. Optimal energy management for multi-microgrid under a transactive energy framework with distributionally robust optimization. *IEEE Transactions on Smart Grid*, 13(1):599–612, 2021.
10 Jacob Chakareski, Ridvan Aksu, Xavier Corbillon, Gwendal Simon, and Viswanathan Swaminathan. Viewport-driven rate-distortion optimized 360° video streaming. In *Proceedings of IEEE International Conference on Communications (ICC)*, pages 1–6, 2018.

11 Dawei Chen, Choong Seon Hong, Li Wang, Yiyong Zha, Yunfei Zhang, Xin Liu, and Zhu Han. Matching-theory-based low-latency scheme for multitask federated learning in MEC networks. *IEEE Internet of Things Journal*, 8(14):11415–11426, 2021.

12 John M. Danskin. The theory of max-min, with applications. *SIAM Journal on Applied Mathematics*, 14(4):641–664, 1966.

13 Sahraoui Dhelim, Tahar Kechadi, Liming Chen, Nyothiri Aung, Huansheng Ning, and Luigi Atzori. Edge-enabled metaverse: The convergence of metaverse and mobile edge computing. *arXiv preprint arXiv:2205.02764*, 2022.

14 Yue Han, Dusit Niyato, Cyril Leung, Dong In Kim, Kun Zhu, Shaohan Feng, Sherman Xuemin Shen, and Chunyan Miao. A dynamic hierarchical framework for IoT-assisted digital twin synchronization in the metaverse. *IEEE Internet of Things Journal*, 10(1):268–284, 2022.

15 Wei Huang, Lianghui Ding, Guangtao Zhai, Xiongkuo Min, Jenq-Neng Hwang, Yiling Xu, and Wenjun Zhang. Utility-oriented resource allocation for 360-degree video transmission over heterogeneous networks. *Digital Signal Processing*, 84:1–14, 2019.

16 Seongah Jeong, Osvaldo Simeone, Alexander Haimovich, and Joonhyuk Kang. Beamforming design for joint localization and data transmission in distributed antenna system. *IEEE Transactions on Vehicular Technology*, 64(1):62–76, 2014.

17 Steven M. Kay. *Fundamentals of Statistical Signal Processing: Estimation Theory*. Prentice-Hall, Inc., Englewood Cliffs, NJ, 1993.

18 Latif U. Khan, Zhu Han, Dusit Niyato, Ekram Hossain, and Choong Seon Hong. Metaverse for wireless systems: Vision, enablers, architecture, and future directions. *arXiv preprint arXiv:2207.00413*, 2022.

19 William Wei-Liang Li, Yuan Shen, Ying Jun Zhang, and Moe Z. Win. Robust power allocation for energy-efficient location-aware networks. *IEEE/ACM Transactions on Networking*, 21(6):1918–1930, 2013.

20 Jie Li, Cong Zhang, Zhi Liu, Wei Sun, and Qiyue Li. Joint communication and computational resource allocation for QoE-driven point cloud video streaming. In *ICC 2020-2020 IEEE International Conference on Communications (ICC)*, pages 1–6. IEEE, 2020.

21 Wei Yang Bryan Lim, Zehui Xiong, Dusit Niyato, Xianbin Cao, Chunyan Miao, Sumei Sun, and Qiang Yang. Realizing the metaverse with edge intelligence: A match made in heaven. *arXiv preprint arXiv:2201.01634*, 2022.

22 Horacio A. B. F. Oliveira, Eduardo F. Nakamura, Antonio A. F. Loureiro, and Azzedine Boukerche. Error analysis of localization systems for sensor networks. In *Proceedings of the 13th Annual ACM International Workshop on Geographic Information Systems*, pages 71–78, 2005.

23 Georgios S. Paschos, George Iosifidis, Meixia Tao, Don Towsley, and Giuseppe Caire. The role of caching in future communication systems and networks. *IEEE Journal on Selected Areas in Communications*, 36(6):1111–1125, 2018.

24 Xukan Ran, Carter Slocum, Yi-Zhen Tsai, Kittipat Apicharttrisorn, Maria Gorlatova, and Jiasi Chen. Multi-user augmented reality with communication efficient and spatially consistent virtual objects. In *Proceedings of International Conference on emerging Networking EXperiments and Technologies*, pages 386–398, Barcelona, Spain, December 2020.

25 Pengfei Wang, Zijie Zheng, Boya Di, and Lingyang Song. HetMEC: Latency-optimal task assignment and resource allocation for heterogeneous multi-layer mobile edge computing. *IEEE Transactions on Wireless Communications*, 18(10):4942–4956, 2019.

26 Wolfram Wiesemann, Daniel Kuhn, and Melvyn Sim. Distributionally robust convex optimization. *Operations Research*, 62(6):1358–1376, 2014.

27 Ding Xu, Qun Li, and Hongbo Zhu. Energy-saving computation offloading by joint data compression and resource allocation for mobile-edge computing. *IEEE Communications Letters*, 23(4):704–707, 2019.

28 Minrui Xu, Wei Chong Ng, Wei Yang Bryan Lim, Jiawen Kang, Zehui Xiong, Dusit Niyato, Qiang Yang, Xuemin Sherman Shen, and Chunyan Miao. A full dive into realizing the edge-enabled metaverse: Visions, enabling technologies, and challenges. *arXiv preprint arXiv:2203.05471*, 2022.

29 Wanting Yang, Xuefen Chi, Linlin Zhao, and Zehui Xiong. QoE-based MEC-assisted predictive adaptive video streaming for on-road driving scenarios. *IEEE Wireless Communications Letters*, 10(11):2552–2556, 2021.

30 Hongliang Zhang, Yun Liao, and Lingyang Song. D2D-U: Device-to-device communications in unlicensed bands for 5G system. *IEEE Transactions on Wireless Communications*, 16(6):3507–3519, 2017.

31 Yuanxing Zhang, Pengyu Zhao, Kaigui Bian, Yunxin Liu, Lingyang Song, and Xiaoming Li. DRL360: 360-degree video streaming with deep reinforcement learning. In *IEEE INFOCOM 2019-IEEE Conference on Computer Communications*, pages 1252–1260. IEEE, 2019.

32 Hongliang Zhang, Boya Di, Kaigui Bian, Zhu Han, H. Vincent Poor, and Lingyang Song. Toward ubiquitous sensing and localization with reconfigurable intelligent surfaces. *Proceedings of the IEEE*, 110(9):1401–1422, 2022.

5

AI and Computer Vision Technologies for Metaverse

Thien-Huynh The[1], Quoc-Viet Pham[2], Xuan-Qui Pham[3], Tan Do-Duy[1], and Thippa Reddy Gadekallu[4]

[1] *Department of Computer and Communication Engineering, Ho Chi Minh City University of Technology and Education, Ho Chi Minh City, Vietnam*
[2] *School of Computer Science and Statistics, Trinity College Dublin, Dublin, Ireland*
[3] *ICT Convergence Research Center, Kumoh National Institute of Technology, Gumi, Korea*
[4] *School of Information Technology and Engineering, Vellore Institute of Technology, Tamil Nadu, India*

After reading this chapter, you should be able to:

- Explain the roles of artificial intelligence (AI) in the Metaverse and how AI can improve users' immersive experience.
- Describe how AI intensively contributes many technical aspects when developing virtual worlds in the Metaverse.
- Investigate the potential AI-based applications developed in the Metaverse.
- Describe several computer vision tasks which can help to shine the appearance of virtual worlds.
- Discuss the promising Metaverse application domains empowered by computer vision.
- Identify current challenges and open research directions of AI and computer vision for the Metaverse development.

5.1 Introduction

In the last decades, we have witnessed the birth of numerous Metaverse projects, mostly serving 3D gaming, which is fueled by the evolution of hardware (such as big data storage infrastructure, centralized and decentralized computing platforms, wireless communication networks, built-in precise sensor technology,

Metaverse Communication and Computing Networks: Applications, Technologies, and Approaches, First Edition.
Edited by Dinh Thai Hoang, Diep N. Nguyen, Cong T. Nguyen, Ekram Hossain, and Dusit Niyato.
© 2024 The Institute of Electrical and Electronics Engineers, Inc. Published 2024 by John Wiley & Sons, Inc.

lab-on-a-chip devices, and graphic processing unit) and the advancement of software (such as resource allocation and offloading in communications, language processing, computer vision, and human–computer interaction) to build and progressively enrich the virtual worlds in quantitative and qualitative aspects. Being superior to the traditional Metaverse modality with limited immersive technologies and experiences caused by insufficient data and inadequate computational capacity, the new one has a great capability to design a free-and-smart environment for enterprises (where users are provided various AI-based built-in computer vision tools to produce creative contents, modify objects, and assign functionalities). Indeed, besides satisfying a very high visualization with advanced computer vision technologies, a new standard Metaverse platform should meet some key features, such as security, privacy, persistence, all-time connection and synchronization, financial allowance, decentralization, and interoperability. Nowadays, AI technology, with machine learning (ML) algorithms and deep learning (DL) architectures, has been seen in many diversified aspects of the Metaverse [34], along with the computer vision technologies to shine the appearance of the virtual worlds. It can be said that while computer vision is considered the face of the Metaverse (where users can enjoy the beauty of virtual worlds), AI is the soul of the Metaverse by uplifting its intelligence to the next level.

5.1.1 Main Contributions

The roles of AI and computer vision in building a smart virtual world have been seen in some previous studies; however, a comprehensive discussion on the roles and benefits of AI and computer vision for the foundation and the development of the Metaverse that relies on their existing technologies are currently missing. To fill this gap, we deliver this chapter to convey AI and computer vision for the Metaverse in the technical and application aspects as follows:

- **AI for the Metaverse**: Besides giving the preliminary of AI, including conventional and recently advanced techniques, we, in turn, depict the tentative role of AI in the Metaverse with natural language processing, blockchain, networking, digital twins (DTs), and neural interface. After that, the application aspects of AI for the Metaverse are reviewed with healthcare, manufacturing, smart cities, gaming, and other auxiliary ones.
- **Computer vision for the Metaverse**: This part first provides state-of-the-art methods to deal with several principal computer vision tasks (e.g. image classification, object detection, image segmentation, facial recognition, human pose estimation, and scene reconstruction and video motion analysis), which are widely applied in the Metaverse to build any virtual worlds. Second, some potential applications based on computer vision that can be deployed in the Metaverse will be carried out.

- **Open research directions**: We disclose some open research directions for AI and computer vision when being applied to build a virtual world in the Metaverse.

5.1.2 Chapter Organization

The organization of this chapter is as follows. This chapter first conveys the preliminaries of AI techniques, from traditional algorithms to modern learning architectures, and several key processing tasks of computer vision, including image-based and video-based processing. In parallel, we discuss various existing applications of AI and computer vision, which can be developed in the virtual worlds to enrich the quantity and quality of diverse available services of the Metaverse. Finally, we argue about some research directions to enhance the trustworthiness of AI models and the performance of complex computer vision tasks.

5.2 AI for the Metaverse

In this section, a wide spectrum of AI, including conventional ML algorithms to recently modern DL architectures that embrace different learning mechanisms for various learning tasks, is briefly conveyed. After that, the role of AI in the Metaverse is exploited by investigating the state-of-the-art AI-powered methods for several technical and application aspects. The content of this section is outlined in Figure 5.1.

5.2.1 Preliminary of AI

This part reviews some widely used AI/ML algorithms that are potentially exploited for the Metaverse. Basically, AI algorithms can be grouped into conventional and advanced techniques, which are leveraged to deal with three principal problems in data science: clustering, classification, and regression.

5.2.1.1 Conventional Techniques

Numerous conventional AI/ML algorithms are usually distinguished based on data available for utilization for learning, i.e. supervised learning, unsupervised learning, semisupervised learning, and reinforcement learning.

Supervised learning: The algorithms in this learning group disclose and learn the input–output relations over a mapping function, in which the label of input data for learning models is provided; that means each input

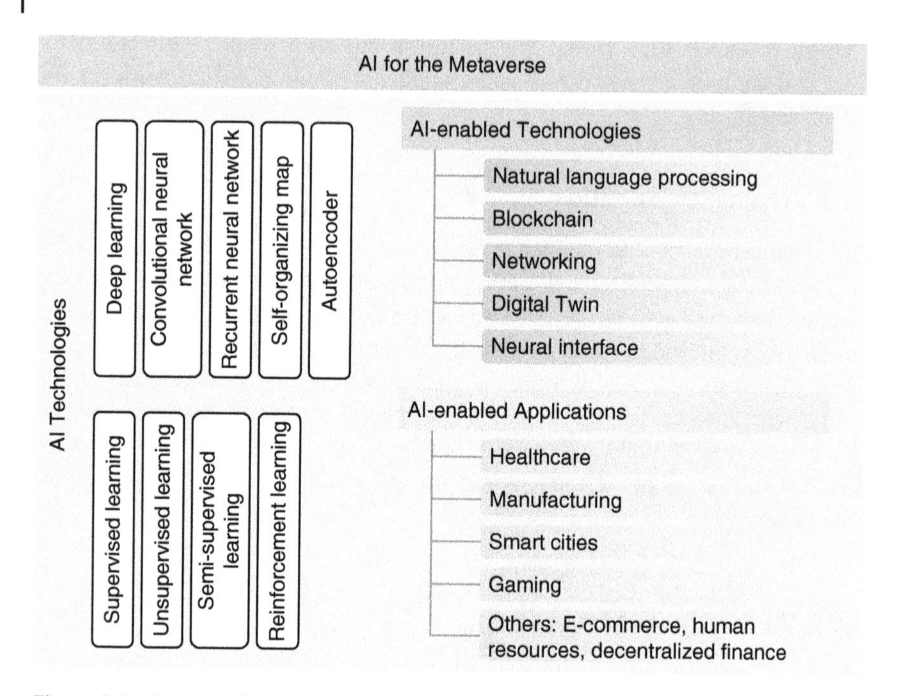

Figure 5.1 Outline of AI for the Metaverse, including the brief survey of AI technologies and the discussion of AI-enabled cutting-edge technologies and AI-powered applications potentially deployed in the Metaverse.

sample in the training set is marked with an answer (so-called label) to allow classifying or predicting the outcome of an unforeseen input sample based on the trained/learned model [38]. Several classification (by annotating/assigning/labeling a sample in the test set into a discrete class/group) and regression (by expressing the relation between dependent and independent variables on continuous data, usually in analog systems) tasks using supervised learning can be found in the real world, such as handwritten digit classification and scene recognition. Many common algorithms are found for supervised learning, such as decision trees, random forest, Naïve Bayes, k-nearest neighbor, and support vector machine.

Unsupervised learning: The algorithms for unsupervised learning are designed for unlabeled data analysis; this means the statistical model of pattern recognition can be established with unlabeled data in the training set. Typically, these algorithms cannot be applied directly to solve classification and regression problems, but they have the capability of discovering and modeling hidden patterns to find out data groups/clusters without the need for human intervention [2]. Some unsupervised learning algorithms, such as hierarchical clustering, k-means clustering, association rule, and principal component analysis, are usually applied to data mining tasks like dimensionality reduction and sample clustering.

Semisupervised learning: This learning fashion is introduced to encounter the disadvantages of supervised learning (i.e. cost reduction of labeling activities) and unsupervised learning (i.e. a limited range of applications) [71]. In semisupervised learning, a model is trained with labeled and unlabeled datasets simultaneously over two primary steps: blindly clustering feature-sharing data using an unsupervised learning algorithm and applying available labeled data to annotate unlabeled data [76]. Some typical algorithms of semisupervised learning are the generative model, graph-based model, and self-training scheme.

Reinforcement learning (RL): To deal with uncertain and complex environments, RL algorithms are designed to make a sequence of decisions based on the association learning of single agent or multiple agents, in which a suboptimal solution can be reached with many trials and errors [41]. From the beginning with random trials to ending with sophisticated tactics and superhuman skills, the goal of RL is to maximize the reward and minimize the penalty to agents. Relying on searching mechanisms with numerous trials, RL refers to as one of the most effective manners to imply machine creativity.

In the Metaverse's perspective, supervised learning algorithms can contribute a wide range of functionalities, such as video-based action recognition to control an avatar freely in virtual worlds, automatic user authentication, and verification using bioinformatics to access the Metaverse, speed recognition, and semantic analysis to facilitate interactive communications between users and intelligent agents. On the other hand, unsupervised ML techniques without the need of labeled data are responsible for discovering user interests and behaviors in personalized recommendation systems, visualizing unlabeled data over diagrams, images, graphs, and charts to monitor and maintain activity logs, and anomaly detection to protect the user from threats and diversified cyberattacks.

5.2.1.2 Advanced Techniques

In the last decade, DL – a subset of ML featured by multilayered artificial neural networks to achieve the state-of-the-art performance of many classification and regression tasks in a wide spectrum of applications [30, 33, 59], has been fueled by semiconductor technology in aggressively increasing the computing power of graphic processing unit (GPU). Here, some well-known deep architectures are investigated, including recurrent neural network (RNN), convolutional neural network (CNN), self-organizing map (SOM), and autoencoder.

Recurrent neural network: As one of the foundational neural networks, RNN is featured by feedback connections linked with preceding layers in the architecture besides having feed-forward connections as in regular neural networks. The feedback computing flow allows RNNs to preserve the memory of past inputs and process models timely. Some advanced architectures are introduced based on RNN by improving its computing units, such as long short-term memory (LSTM) and gated recurrent unit (GRU).

Convolutional neural network: As one of the most successful deep networks in dealing with several challenging image processing and computer vision problems, CNN can learn complex patterns from high-dimensional unstructured data (e.g. audio signals, images, videos, and wireless signals) with the principles of linear algebra (e.g. matrix multiplication). In the architecture, CNN is generally specialized by multiple convolutional layers to compute meaningful features, from coarse to fine, and some fully connected layers to compute output class probabilities. Several advanced CNN architectures (e.g. VGG, GoogleNet, ResNet, DenseNet, Inception, and EfficientNet) are introduced with higher accuracy and lower complexity for image classification and other computer vision tasks.

Self-organizing map: SOM is a deep network designed for unsupervised learning to find the cluster of data points. In a regular architecture, the nodes are initialized with random values of weights and then trained to seek the node with the least Euclidean-based input–output distance for annotating as the best matching unit. The deep SOM algorithm and its counterparts have been applied to solve engineering problems and have lately extended to data analysis.

Autoencoder: As a special type of neural networks, an autoencoder network learns an input–output pattern with compression and decompression functions. Concretely, the input layer of this network encodes data into hidden layers by an encoding function for compression, thus reserving less but more relevant information in hidden layers. At the end of this network, the output layer reconstructs the original data with a decoding function. Accordingly, the difference between the original input data and the reconstructed output data is estimated in the training phase over an error function. With the backward propagation scheme, autoencoder learning is suitable for self-supervised learning tasks.

5.2.2 Roles of AI in the Metaverse

For the development of secure, smart, scalable, and reliable virtual worlds in the Metaverse, AI has to play an essential role in several cutting-edge technologies, such as data processing, data management, resource management, modeling, and neural interface, to bring the best immersive experiences to users while ensuring that everyone can enjoy all the services in the Metaverse with an acceptable price. In this part, state-of-the-art AI-powered approaches are investigated for being potentially applied to the Metaverse, and an in-depth discussion on how to enhance user experience with AI should be given accordingly.

5.2.2.1 Data Processing

One of the most prominent roles of AI for the Metaverse is data processing, which is intensively applied to deal with the large-scale heterogeneous data generated by users. In the virtual worlds, users communicate with others and intelligent agents by typing with keyboard and speaking with microphone, where the collected data are represented as text and speech for processing hereafter. For example, natural language processing (NLP) is mostly responsible for multiple communication types within the Metaverse by processing a large amount data, which is enabled by AI. Typically, NLP encompasses a variety of computational models and learning procedures to automatically analyze and understand human languages over the speech and text modality. In fact, NLP also considers various topics, such as speech-to-text and text-to-speech transformation, conversation design, voice branding, text prediction, voice-based multilanguage, and multicultural analysis. In the Metaverse, NLP plays as the core technology for designing intelligent virtual assistants (a.k.a. chatbots) and renovating the natural dialogue capability of nonplayer characters (NPCs). AI-based NLP enables chatbots to understand complicated-and-long human conversations in various realistic scenarios, especially overcoming the challenging problems of varying undertones and dialects. Moreover, chatbots can immediately answer nuanced questions and selectively learn personalized patterns from the interaction with users. However, it exposes difficulties of dealing with different languages of the users from different countries with diversified cultures.

As one of the most fundamental tasks of NLP, language modeling aims to predict multiple words and linguistic units by extracting syntactic and semantic relations of preceding words and phrases, which is the core functionality in machine translation and text recommendation. In [14], several deep networks with key-value attention structures were designed for language modeling. Relying on the performance benchmark with the Wikipedia corpus database, RNNs and LSTM networks released higher accuracy and lower memory in use if compared with other deep models. The work [8] modernized a regular LSTM network with residual connections to effectively deal with text perplexity. Some recent works improved standard CNNs with dense connection and residual connection to overcome the long-term dependencies in long-and-complicated sentences and short paragraphs [58] and also upgraded the regular structure of RNNs with bidirectional structures and gated connection to effectively handle word-aware and character-aware language modeling issues [44]. To be familiar with users from different countries around the world, some deep models are designed for multiple languages (e.g. English, German, French, and Chinese) besides solving specific tasks, such as prefixes and suffixes identification, hyphenated words recognition, and misspelled word detection [37].

Several conventional ML and recent DL models have been adopted for various sentence-based tasks (e.g. question type classification and sentiment analysis), which require sophisticated feature engineering techniques (e.g. feature extraction and feature selection) to allow virtual assistant modules in the Metaverse to be more efficient, reliable, and adaptable with a variety of dialogues in practice. Natural language generation is an optional function, usually embedded in advanced chatbot systems, which gives the virtual assistants a reasoning capability to generate task-specific conversation-oriented answers. Some mixture RNN-CNN models were exploited to automatically give short descriptions for image captioning and long descriptions for virtual question answers. In the Metaverse, the combination of multiple NLP techniques, such as text parsing, semantic analysis, context retrieval, and dialogue interpretation and creation, should be adopted to fully support text-based and speech-based human–chatbot interactive experiences. Notably, the typing-aided communication may reduce the immersive quality in some long and complicated conversations.

5.2.2.2 Data Management

Building a Metaverse experience that is safe to service providers and users means getting ahead of threats and risks of data management issues. Managing and utilizing data generated in the Metaverse can be complicated and may require advanced protocols to prevent informative data from unauthorized access or manipulation without room for errors and other corrupted problems. Blockchain emerges as a potential solution to address multiple challenging issues in data management, such as data security, data sharing, data traceability, and immutability. For example, blockchain can simplify the data management of trusted information for government agencies to access and manipulate critical public-sector data while preserving the security and privacy of informative data. As known, blockchain, relying on cryptography techniques, allows one to record transactions and track digital assets interconnected in a business network with immutability and impenetrability features. For the Metaverse, blockchain can meet the requirements of security (i.e. being much relevant in the decentralization of storage and processing nodes), trust (i.e. a token-based mechanism to secure storage devices capable of transmitting virtual contents, personal data, and encrypted keys), decentralization (i.e. allowing a massive number of independent nodes to synchronize), smart contract (i.e. performing economic regularization and many other relations between different users in a Metaverse ecosystem), and interoperability (i.e. enabling the joint interoperability and functioning of different systems without any barrier). In this context, AI with the capability of detecting, identifying, and classifying various cyber-attacks on the Metaverse can combine with blockchain [10] to improve the performance of data management in the Metaverse.

In the last decade, many works have taken advantage of blockchain and AI to improve the security and privacy of data acquisition, storage, and transfer activities in big data infrastructures (i.e. between edge devices and the data center), which can be realized as a promising solution for a secure Metaverse [46]. For example, the work [73] investigated the pros and cons of conventional ML algorithms and DL networks for cyber-attack detection in blockchain-based networks. Besides, some other concerns were taken into consideration, including AI-aided smart contract examination, reward mechanism, and cost-efficient on-chain pattern learning. Several privacy-preserving and secure IoT frameworks were developed by combining blockchain with the proof-of-work mechanism and ML with data engineering (e.g. data transformation and normalization) to strongly combat cyber-attacks in smart city networks [39], in which some supervised learning algorithms are applied to accurately detect frauds and predict threats. Recently, DL has been integrated with blockchain to deal with many challenging security and privacy problems in big data. For instance, the work [80] introduced DeepChain, an innovative DL-aided blockchain framework for data privacy and integrity preservation. Multiagent deep RL was adopted to secure the mobile offloading in edge computing-based blockchain frameworks [53] and the vehicular crowdsensing in blockchain-based Internet of Vehicular systems [79].

Federated learning (FL), a collaborative ML deployed in decentralized systems for data privacy preservation by training statistical AI models with local data collected at the edge and then collaboratively aggregating parameters at the central server, was leveraged for trustworthy data sharing among different parties in a blockchain network [49]. To guarantee the privacy of massive IoT collected by heterogeneous devices, FL should be combined with blockchain in many data curation tasks, including acquisition, storage, management, and transmission. In the Metaverse, malicious attacks can occur frequently, and users' sensitive and private information (e.g. identities and digital assets) can be leaked outside for unfriendly purposes. In blockchain networks, interoperability is another important concern where there exists collaboration between different data infrastructures. In this context, integrating blockchain with FL in decentralized systems was recently implemented in computing resource monitoring, allocation, and management applications [54], in which some problems of centralized systems are addressed, such as untrustworthy servers, external cyber-attacks, and server malfunctions.

5.2.2.3 Resource Management

In the Metaverse, a massive number of users should be served with pervasive networking access over wireless networks, which allows users around the world to access the virtual worlds anytime with always-on, stable connectivity, thus revealing several issues in resource management. The last decade has witnessed the birth of many innovative technologies, such as open radio access network (O-RAN) and massive ultrareliable and low latency communications (mURLLC), to improve the overall performance of resource utilization and management of wireless communications and networking systems, thus ensuring a reliable connection with extremely high throughput as well as ultralow latency for best-experiencing comfort [70]. In this context, many ML and DL models have shown great potential to effectively resolve existing problems, such as intelligent radio resource allocation and mobile data offloading, in modern networks like fifth-generation (5G) and beyond. For example, the work [3] adopted RL to address the resource sliding problem in enhanced mobile broadband (eMBB) and URLLC, in which the complex patterns of heterogeneous resource allocation and scheduling are learned collaboratively with network states and channel conditions to facilitate the real-world environment. In [23], RL demonstrated the superiority of other learning mechanisms in terms of joint subcarrier-power management and multiagent allocation, reducing latency and enhancing reliability in IoT networks accordingly. A novel radio resource management approach was proposed in [6] with a distributed risk-aware ML-based mechanism for effectively supervising the transmission of scheduled and nonscheduled data traffic in URLLC communications.

Recently, DL has been adopted to improve the performance of different tasks in URLLC, such as intelligent spectrum management, channel estimation, and dynamic traffic prediction, which help to improve the immersive experience on mobile devices. For instance, the authors in [31] and [75] designed two advanced deep networks with CNN architectures, namely MCNet and SCGNet, for automatic modulation classification (i.e. to identify the modulation type of incoming signal at the receiver accurately) in the physical layer, enhancing the spectrum utilization efficiency accordingly. For the use of mobile devices to access Metaverse and enjoy built-in services, a large amount of data generated by densely connected users should be transferred over dynamic wireless channels, which highly demand efficient spectrum utilization approaches to satisfy a high-speed and reliable connection, thus guaranteeing high-quality video streaming and mitigating lagged phenomena in experiencing 3D worlds. For the improvement of regular ML-based channel state information (CSI) estimation, while keeping an acceptable complexity, the work [50] introduced a supervised DL framework with a two-stage training mechanism via a combination of CNN and LSTM networks, which in turn enables robust, stable connectivity in varying wireless conditions. In [24], a 3D CNN architecture was specially designed for intelligent cellular traffic forecasting in modern communications, in which the deep model takes into consideration both the short-term and long-term traffic data patterns to effectively perform in varying real-world traffic flow scenarios. In conclusion, AI can be referred to as a powerful tool to address many challenging issues in future wireless networks, thus allowing users to experience high-class services and applications with stable connectivity and nearly no latency on their mobile devices.

5.2.2.4 Modeling

The Metaverse can bring us closer to completely stimulated virtual reality (VR) through a disruptive transformation; however, it may require a digitalized copy of the real world as a gateway to provide users with fully connected, deeply immersive, and fascinating 3D experiences. Many businesses and enterprises are currently working on Metaverse-based concepts and policies to introduce new possibilities and experiences for digitally driven consumers. They attempt to model dimensionally precise real-life spaces into the Metaverse virtual mirror world by deploying DTs with as many original features as possible. DT is generally defined as a ground-breaking digital replication of real-world entities, which can synchronize operational assets, processes, and systems with the real world at different levels and be featured by many regular functions, such as monitoring, visualization, analysis, prediction, and management. By owning several distinctive properties, DT is realized as one of the fundamental implementation sectors of the Metaverse and therefore plays the role of access gateway for users to enter the virtual worlds by creating exact replications of reality, including facility and functionality perspectives. For instance, systems technicians can operate 3D replications of complex modules, processes, and even systems at different levels (e.g. descriptive, informative, interpretative, predictive, and autonomous) and purposes (e.g. technical training, education, and commercial customization), in which AI empowers the performance of various physical analysis tasks over remotely virtual replications (see Figure 5.2).

In [15], a robust DT framework was developed with a multilayer perception neural network as the core technology to verify sensory data, estimate fault conditions, and detect fault sensors. Relying on the combination of AI and immersive technologies, the work [78] introduced a cutting-edge DT framework for monitoring and analyzing welder behaviors in the automated automotive manufacturing process, in which traditional ML algorithms (e.g. decision tree and support vector machine) were adopted to learn the complex welder behaviors patterns based on the bidirectional data flow between immersive devices and automated robots. For smart urban agriculture services and applications, a hybrid cyber-physical framework [21] was built with many virtual replications of farming facilities, products, and procedures, in which some advanced ML algorithms are responsible for every production stage, from observation to diagnosis, decision, and action. For industrial IoT, DTs were built to capture physical characteristics and simulate the operational functionalities of industrial components, modules, and a whole system in real time. The bias between a real entity and its digital replication can be solved with a trustworthy aggregation mechanism of FL and a multiagent

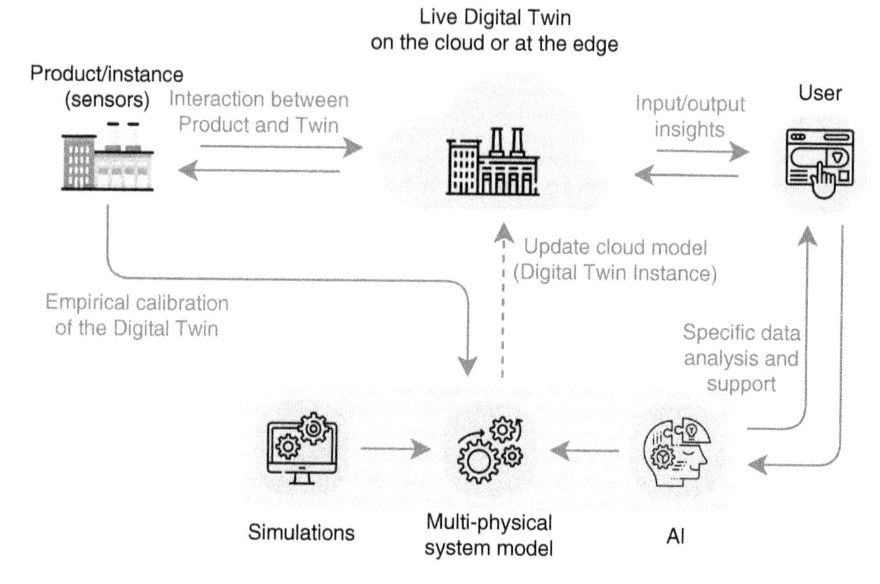

Figure 5.2 A physical product/prototype can be integrated with DT via IoT and wireless technologies, thus allowing to collect sensory data from operating physical products and then transfer it to the cloud/edge, which then learns the multiphysical system model with AI as the core technology.

learning policy of deep RL. With AI-aided DTs, a virtual world in the Metaverse can be built as a set of many digital replications projected from real entities with full operation and functionality.

5.2.2.5 Neural Interface

One of the most popular manners that allow users to interact with digital objects and immerse virtual worlds in the Metaverse is over VR devices, such as headsets and glasses. However, many technology companies have been paying attention to the neural interface, a.k.a. brain–machine interface or brain–computer interface, which creates a ground-breaking interaction between human and wearable devices with nearly no boundary. In principle, a brain–computer interface (BCI) system can detect and record neural signals using external electrodes or optical sensors wearing on the skull and other body components. The neural signals are then processed for pattern analysis and recognition with the help of AI before making decisions to respond to neural stimulation [9]. Many potential systems can revolutionize how users interact, communicate, and control their avatars in the virtual worlds over neural signals; however, low accuracy of such AI-aided systems for brain signal-based complex single tasks and also simple sequential tasks is still the main obstacle.

In the last decades, many works have exploited ML algorithms and DL architectures to analyze and understand neural signals in BCI systems. In [26], a semisupervised learning approach was studied for brain signal classification, in which online supervised learning improves the performance of some use cases (e.g. mental analysis and motor imagery) and offline unsupervised learning reduces the signal annotating cost of unlabeled samples. A regular ML framework, including electroencephalography (EEG) features extraction and conventional classification algorithms (e.g. logistic regression, support vector machine, and Naive Bayes), was adopted in [1] to interpret brain activities and then compile them into communication and control commands. With a variety of architectures along with a great capability of multiresolution signal analysis, DL has been manipulated to improve the overall performance of BCI systems, especially being superior in dealing with large-messy-confusing data. In [51], the work adopted a capsule network (CapsNet) to boost the accuracy of event-related potential detection in BCI systems while ensuring the practicality and compatibility with different common spellers in the cognitive neuroscience domain. In the future, neural interface technologies promisingly revolutionize human–machine interactions and modernize the immersive modality, thus enabling users to interact with the virtual worlds in a touchless manner, but commercializing mind-control devices and delivering them to regular users with a comfortable price are crucial concerns in next few years.

5.2.3 AI for Metaverse Applications

This part discusses AI-assisted works from the perspectives of four applications potentially deployed in the Metaverse, including healthcare, manufacturing, smart cities, and gaming, besides some auxiliary ones, such as E-commerce, human resource, and decentralized finance.

5.2.3.1 Healthcare

Some revolutionary techniques like big data, edge computing, 5G, and AI have been recently adopted collaboratively to address many challenging issues in the healthcare and medical domains [60], accordingly improving the overall quality and performance of services and applications and, so far, reducing the cost of health services to patients and the operation cost to healthcare institutions while expanding high-quality medical care to low-income people. In this context, the Metaverse can provide patients with the convenience of healthcare monitoring and diagnosis over built-in remote services in the 3D virtual worlds while allowing health technicians and medical experts to effectively analyze big data and accurately make decisions with the help of AI. Moreover, the combination of AI and modern immersive technologies creates more opportunities for healthcare training and medical education, such as virtual reality surgery training, rehabilitation with virtual reality, and communication with patients.

In many wearable-based healthcare and wellness applications, AI is usually applied for automatic classification and prediction by learning complex patterns from sensory data. For example, for daily living assistance and early health risk awareness, an innovative ML-based physical activity recognition method [32] was developed with a fused feature extraction mechanism (i.e. combining hand-crafted features and deep features) to boost the accuracy of indoor and outdoor activities recognition. In [63], a fall detection method using wearable devices (e.g. smartphones and smartwatches with built-in sensors like accelerometer and gyroscope) was studied with a hierarchical CNN architecture. Some other deep models with RNN and LSTM were also considered for sensor-based early health attention and assessment, such as heart failure detection and unhealthy behavior awareness. DL with CNN architectures has been extensively adopted to address various challenging tasks in medical image analysis with less requirement of specialized knowledge and technical experience. For example, the work [27] designed an advanced CNN architecture, as the combination of a backbone network and a twofold feature augmentation mechanism, to increase the accuracy of diabetic retinopathy severity recognition using fundus images and with wide-field swept-source optical coherence tomography angiography.

5.2.3.2 Manufacturing

With an ongoing wave of the industrial revolution, digital transformation has been coming to pass in many manufacturing sectors, where digitalization allows factory operators and production engineers to analyze and understand physical entities. Unlike digital transformation in the physical world, the Metaverse can project all physical entities in the real world to a virtual manufacturing world by the foundation of real interaction and persistence. By collaboratively applying cutting-edge technologies, such as wireless communication, AI, and DT, the manufacturing Metaverse is able to modernize digital operation and production, in which AI is the core technology to develop a smart factory with a variety of services and applications, such as early failure estimation, production line optimization, and self-maintenance.

In manufacturing, the shorting production lifecycle and increasing diversity of products have caused the high expense of frequent system configuration and production line setup. The work [16] developed a symbiotic hybrid human–machine framework for production ramp-up, in which the ramp-up experiences of operators and the sensory data collected by edge devices were collaboratively learned by Q-learning models to find an optimal ramp-up strategy. Quality inspection is one of the most important scopes in many high-quality manufacturing systems, but it suffers from some difficulties, including varying operation conditions and manufacturing environments. For automatic fault detection and production diagnosis, many deep networks with different architectures, from RNN to LSTM and CNN, have been adopted to deliver reliable manufacturing solutions [40]. Designing an optimal serial production line is always one of the top priorities to boost the productivity of the whole manufacturing system and uplift the business. In [28], predicting nonlinear production progress was formulated as an optimization model that is handled by a deep network for discovering the short-term and long-term patterns of production states. In summary, through the virtual representations of physical entities in the Metaverse, industrial manufacturing efficiency is improved in different aspects, such as speeding up production progress, predictive maintenance of production lines, managing operational risk, and presenting high transparency to manufacturers and customers.

5.2.3.3 Smart Cities

In smart cities, useful information about citizens' opinions and requirements is collected and gathered through various sources, such as handwriting reports and social media, spreading in various scopes and domains. Relying on the citizens' feedback, city governments should make decisions about which services should be eliminated or improved with some changes in policy and operation, which in turn enhances the quality of services and eases the processing procedure of documents. By integrating several revolutionary digital technologies into the Metaverse platform, smart cities can provide more smart interactive public services to citizens. For example, several smart services, such as monthly utility payments, transportation schedule checking, automatic parking registration, smart home control, smart community portal, healthcare declaration, and other cultural–educational–civic services, can be done via a service gateway deployed in the virtual worlds. The social impacts and benefits of technologies to the smart cities are currently limited due to platforms' dispersion and fragmentation; hence, the Metaverse as an all-in-one service gateway can become a promising solution to broaden the presence and popularity of smart services in citizens' daily life.

For smart cities, AI has shown great significance in obtaining intelligence and automation in smart services. In [57], AI played the core technology (among others, such as neural interface, VR, and IoT) in a smart home control system using a steady-state visual evoked potential-based BCI approach, in which the brain signals were analyzed by ML algorithms for control pattern recognition and command–stimulation response. The work [83] introduced an AI-based hybrid transportation management system to monitor and control urban traffic, in which the traffic condition in the real world can be analyzed and predicted by adopting a multiagent reinforcement learning approach on virtual replications, being projected from physical entities with full functionality. The massive growth of industrialization and urbanization has caused many environmental issues, and air pollution is currently attracting much more attention because of its negative effects on human health. In [17], a high-accuracy forecasting method was proposed with a hybrid deep network (i.e. combine CNN and LSTM architectures to comprehensively extract underlying features and interdependence of multivariate sequential sensory data) for early air pollution warning. Despite facing more challenges, such as asynchronous development of infrastructures and heterogeneous skills of citizens, building a Metaverse ecosystem for smart cities with all administrative services (e.g. education, culture, environment, healthcare, transportation, and other civic services) will significantly save the budget expenditures.

5.2.3.4 Gaming

Gaming which has been receiving huge funds from big techs and venture capitals around the world, and attracting a massive number of young players across different gaming platforms, from console to mobile and personal computer (PC), is always the prime application sector of the Metaverse and the best way to access the enjoyable virtual worlds. In the Metaverse for gaming, players can experience various game styles (e.g. real-time strategy, role-playing, and multiplayer online battle arena) and can earn money from their collectible items over tradeable activities on in-game markets. In the gaming industry, AI is one of the most valuable techniques that have a significant impact on the way of developing and modernizing video games.

Besides challenging missions and attractive stories, video games should have more beautiful in-game worlds that incite studios and developers to adopt AI to enhance the intelligence of NPCs, especially in the scope of reacting to players' actions reasonably. The role of AI in games has been investigated across many topics [81], such as NPC behavior analysis and learning, tactical planning, player reaction modeling, procedural content creation, player–NPC interaction design, AI-assisted story design, and AI in open-world games. For in-game scenarios modeling and decision-making, the work [20] presented a comprehensive survey on AI adoption in intelligent video games for PC. This survey found that several common ML algorithms (e.g. decision tree, fuzzy logic, rule-based system, ensemble learning, and Markov model) were frequently adopted to handle some key tasks in game development, such as game flow modeling, immersive gaming experience evaluation, gameplay customization and adaptation, and NPC behavior modeling and reaction planning. Some game studios have applied AI for testing in gameplay design and development, for example, a metamorphic testing scheme was designed with a decision tree algorithm to effectively handle a massive number of possible moving strategies in artificial chess games [43]. In a nutshell, gaming is always an indispensable application sector in the Metaverse to attract more users and players who want to find joy in games or challenge themselves, in which AI is the most important material to make games more logical as well as more intelligent to players.

5.2.3.5 Others

In addition to healthcare, manufacturing, smart cities, and gaming, there are some potential AI-aided Metaverse applications, such as E-commerce, human resources, and decentralized finance.

- **E-commerce**: Nowadays, many worldwide consumer brands have been integrated into the Metaverse by collaborating with big techs to build a digital shopping ecosystem where mainstream consumers can enjoy seamless shopping experiences (i.e. filling a gap between online shopping and offline shopping) with immersive devices. Retailers are currently hinted at the personalizing shopping experience of customers (e.g. behavior and demand) to grow revenue, in which AI can help analyze customers' shopping behavior and recommend the most proper products.
- **Human resources**: For several years, some big tech companies like LG and Samsung have built a specific Metaverse for human resource-related activities, such as recruitment and training. For example, job applicants will log in to the Metaverse and control their avatars to join such recruitment events as immersive interviews with committees and interactions with chatbots enabled by AI-aided NLP technology. Furthermore, a virtual working environment built in the Metaverse can revolutionize the future working style and workplace, in which AI can contribute in many aspects, such as automatic employee training and intelligent project progress analysis, to enhance productivity and liberate employees from physical barriers.
- **Decentralized finance (DeFi)**: Referring to as cryptocurrency-based finance, DeFi can provide several decentralized authority-based services (e.g. leading, yield farming, staking, and insurance) over blockchain-based smart contracts to guarantee high security and privacy. DeFi services are usually supplied by decentralized applications (so-called Dapps, which are completely built on open-sourced distributed platforms) and controlled by AI-assisted automated market makers to keep an all-time DeFi ecosystem operation via liquidity pools. In other words, users can provide liquidity to available pools operated by an AI-assisted supervision mechanism of decentralized exchanges to earn rewards or incentives. Some common market marker algorithms are myopically optimizing, reinforcement learning, utility-maximizing with risk attributes, and logarithmic market scoring rules.

5.3 Computer Vision for the Metaverse

Computer vision is a subfield of AI that enables computers and systems to extract meaningful information from digital images, videos, and other visual inputs and consequently take actions or generate recommendations based on that information. If AI enables computers to think, computer vision enables them to see, observe, analyze, and understand complex scenarios in the real world. Computer vision works much the same as human vision, except humans have a head start. Human sight has the advantage of lifetimes of context to train how to tell objects apart, how far away they are, whether they are moving, and whether there is something wrong with an image. Computer vision trains machines to perform these functions, but it has to do it in much less time with cameras, data, and algorithms rather than retinas, optic nerves, and the visual cortex. Because a system trained with AI to inspect products or watch a production asset can analyze thousands of products or processes a minute, noticing imperceptible defects or issues, it can quickly surpass human capabilities. Computer vision is popularly used in industries ranging from energy and utilities to manufacturing and automotive, especially the foundation and importance of computer vision are getting more and more significant. Currently, the whole market of computer vision, including software and hardware, is continuing to grow and expand to many other domains with numerous applications. In the Metaverse, computer vision plays an essential role in building an attractive virtual world in a 3D perspective view, thus allowing users to have the best immersive experience. Furthermore, computer vision allows immersive devices such as virtual reality headsets to automatically recognize and understand visual information about surrounding objects and events, hence enabling to build more accurate and reliable virtual and augmented environments. The content of this section is outlined in Figure 5.3, in which we distinguish different fundamental computer vision tasks in the field and then discover potential applications of computer vision in the Metaverse.

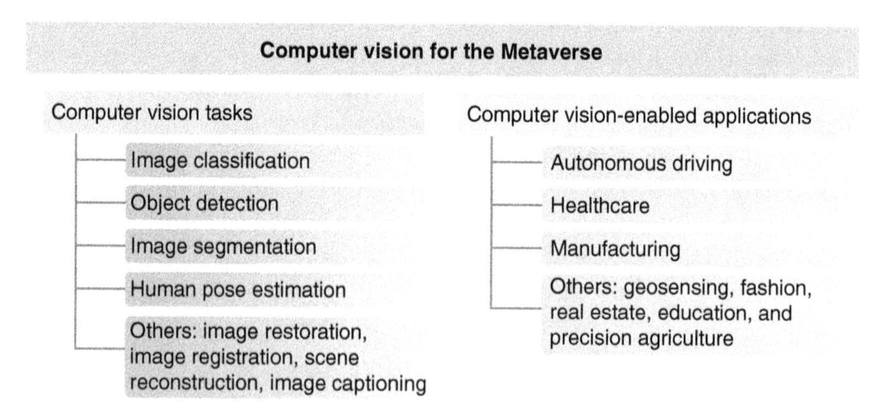

Figure 5.3 Outline of computer vision for the Metaverse, including the investigations of computer vision tasks and its applications potentially deployed in the Metaverse.

5.3.1 Fundamental Computer Vision Tasks

Typical computer vision tasks are about making computers analyze and understand digital images and video as well as other visual data collected from the real world. This can involve a series of steps, such as extracting, processing, and analyzing meaningful information from such inputs to make decisions. The evolution of machine vision witnessed a large-scale formalization of complicated problems into multiple solvable problem statements, thus helping scientists and researchers identify problems distinctly and work on them with AI efficiently. The most popular computer vision tasks associated with AI are image classification, object detection, semantic segmentation, facial recognition, human pose estimation, and scene reconstruction.

5.3.1.1 Image Classification

As one of the most studied topics ever since the ImageNet data was released in 2010, image classification is the task of classifying images into a set of predefined classes/groups/categories using supervised learning (i.e. training a classification model on a set of sample images with given classes). As opposed to complex topics, like object detection and image segmentation, which have to localize and compute local features of detected interest regions, image classification deals with processing the entire image and assigning the most appropriate label. Expected to have only one class for each image, a trained model of image classification returns a prediction probability about the image classes. Many conventional image classification methods have adopted the regular ML-based framework with global feature extraction [36], i.e. for visual content description with texture, color, and shape information of the entire image by a single vector.

The performance of image classification is affected by many factors; therefore, designing a universal framework to deal with practical problems is nearly impossible. In [48], a literature review recommended an appropriate image-processing framework with the prerequisite of a successful image classification into a thematic map, in which the effective use of multiple features extracted from images and the selection of the most suitable ML-based classification model are significant for improving the accuracy of classification. Furthermore, this review argued that nonparametric classification methods, such as decision trees, neural networks, and knowledge-based classification, have presented more superiorities over others in multisource image classification. In the review work [65], the adoption of DL with CNN architectures to image classification was conveyed, covering the practical design and implementation method, from predecessors to recent state-of-the-art DL systems. In the last decade, many well-known deep backbone architectures of convolutional networks (ConvNet) have been introduced to deal with the image category classification task [5], including AlexNet, VGG,

Inception, GoogleNet, ResNet, DenseNet, and EfficientNet. When applying these deep networks for object recognition and scene understanding in the Metaverse, it is necessary to retrain them with a new dataset of computer-generated 3D images because the 3D virtual world in the Metaverse is mostly created from software instead of capturing from the physical world. Despite presenting a considerable accuracy improvement compared to ML algorithms, DL-based approaches have remained challenging, including insufficient explainability and interpretability, geometric invariance, mobile development with limited computing resources, multilabeled image classification, and theoretical gap between biological neural networks and regular CNNs. Although the role of image classification in the Metaverse is not clearly expressed, it still shows some benefits in automatic scene recognition (i.e. returning the label of a visual frame captured in the virtual worlds) and nonfungible token art classification. With different use cases considered in the Metaverse, image classification can be deployed as an auxiliary function to provide general information as the label of a whole scene.

5.3.1.2 Object Detection

Object detection refers to the detection and localization of multiple objects with bounding boxes in images or videos. Different from image classification with the inference of only one label or class for one image, object detection basically seeks class-specific details in an image and identifies objects along with their detailed, meaningful information whenever they appear simultaneously. It means that multiple objects with their relative locations are delivered accordingly. These classes can be animals, humans, vehicles, or anything on which the detection model has been trained by supervised learning. Traditional object detection methods use local feature detection and extraction techniques (e.g. Harris, SIFT, HOG, and FAST) to detect local features in an image and classify them based on ML algorithms; however, they are time-consuming and inaccurate when dealing with a large-messy-confusing image dataset. In the last decade, DL models, e.g. You Only Look Once (YOLO) and Region with CNN feature (R-CNN), have achieved ground-breaking performance in terms of correct detection rate while meeting the real-time processing criteria. Notably, different from the ML framework that requires a feature extraction step, images decomposed from video streaming in the Metaverse are passed directly into deep models for detecting 3D objects in a scene, from which the detailed information (such as label, shape, size, color, texture, and other descriptions) can be extracted and viewed on the screen of head mounted displays.

Several region-based methods have been introduced based on the generic object detection framework for localizing and then classifying detected objects (annotated by rectangular bounding boxes). For instance, R-CNN [22] adopted a selective research mechanism to generate some 2,000 region proposals from the input image (e.g. anchor boxes are also considered as region proposals), thus enhancing the quality of candidate bounding boxes and improving the accuracy of object detection accordingly with high-level object features. In line with the region-based framework, some works have tried to refine preidentified region proposals only to preserve a limited number of the most informative ones [61]. Although some improvements in the accuracy of object detection were obtained by R-CNN and its variants, they still consume very high computational cost and memory capacity, which becomes the bottleneck in real-time applications. In this context, one-step frameworks based on global regression/classification approaches can reduce the prediction time in the inference stage by straightly mapping pixels to bounding boxes and calculating class probabilities. Two well-known methods that follow the one-step framework are YOLO [66], and single short multibox detector (SSD) [45]. Remarkably, YOLO can greatly reduce the computation of both object detection and recognition by suppressing all low-probability-scores bounding boxes using nonmaximal suppression. In summary, besides a significant imbalance between annotated objects as foreground and nonobject presentation area as background, most of the traditional CNN-based object detection frameworks are limited by some practical issues, such as scale variation and occlusion/truncation in two-dimensional (2D) images. As known, the virtual environment in the Metaverse is built with a variety of virtual units (e.g. single object and multiple object modules), in which users represented as avatars can pick up and manipulate virtual objects in many physical ways, consequently changing their positions and appearances; therefore, object detection should be designed to effectively deal with the increasing number of object categories, the variation of objects' scale and viewpoint, the varying illumination of the surrounding environment, and the limited view angle of the camera.

5.3.1.3 Image Segmentation

Similar to image classification and object detection, image segmentation has been a fundamental computer vision problem from the early days of the field. Realized as an essential component of many visual analysis and understanding systems, image segmentation aims to partition an image (or a video frame) into multiple objects and segments and plays a central role in a variety of applications, ranging from video surveillance to autonomous driving and medical image analysis. Basically, image segmentation can be formulated as a pixelwise classification problem with semantic labels (semantic segmentation), individual objects partition (instance segmentation), and combination (panoptic segmentation). In particular, semantic segmentation performs pixelwise labeling/annotating with a set of object categories for all image pixels; hence, it requires more specific demands than whole-image classification. Instance segmentation broadens the scope of semantic segmentation by detecting and delineating each object of interest in a single image.

Numerous image segmentation methods have been proposed in the literature, from rudimentary methods, e.g. thresholding, histogram-based bunding, region-growing, k-means clustering, watershed methods, to more sophisticated methods, e.g. active contours, graph cuts, Markov random fields, and sparsity-based approaches. In the last decades, DL models based on the CNN architecture have established a new era of image segmentation with remarkable performance improvements, i.e. obtaining the highest segmentation accuracy on popular benchmarks. Regarding the deep architectures for image segmentation, some prominent networks include fully convolution networks (FCNs), encoder–decoder-based models, convolutional networks with graphical models, multiscale and pyramid networks-based models, R-CNN-based models, dilated convolutional models and DeepLab family, RNN-based models, attention-based models, and generative models and adversarial training mechanism, and CNN with active contour models [52]. As a pioneering work in DL-based semantic segmentation, FCNs [47] were designed with only convolutional layers, which produced the output as a segmentation map (instead of classification scores in image classification) having the same size as the input image. To manipulate more useful scene-level semantic context globally, a combination of CNNs and fully connected conditional random fields (CRFs) was presented in [12] for boundaries localization with higher segmentation accuracy, hence improving the overall performance on the challenging PASCAL VOC 2012 dataset with additional contextual information. Furthermore, one of the most popular deep networks for semantic image segmentation is encoder–decoder architecture with the core technique of deconvolution (a.k.a. transposed convolution). In the original architecture [55], there are two parts: an encoder adopting VGG to generate a map of pixelwise multiclass probabilities and a decoder identifying the pixelwise class labels using deconvolution and unpooling layers to gain a final segmentation mask. SegNet [7], a fully convolutional encoder–decoder architecture as shown in Figure 5.4, was proposed for semantic segmentation with a novel decoder that upsamples its low-resolution feature maps based on the pooling indices extracted in the encoder. In [67], R-CNN was originally renovated for image segmentation instead of for object detection, with an additional region proposal network to infer bounding box information extracted within a region of interest (ROI). Notably, some advanced extensions of R-CNN have been studied to effectively handle the instance segmentation problems, i.e. simultaneously performing object detection and semantic segmentation. On the other hand, in the Metaverse, more robust and real-time semantic segmentation methods are needed to understand the

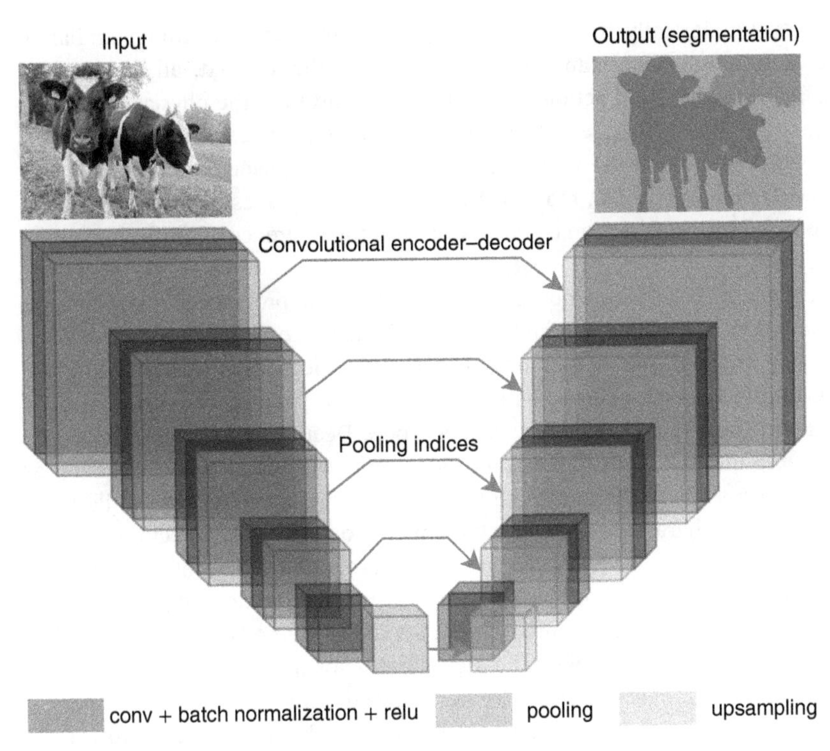

Figure 5.4 The SegNet architecture in Waldo93 / 207 images / Pixabay.

pixel-level information in 3D virtual worlds over immersive devices. In addition, some adaptive image segmentation approaches are highly recommended for deployment to deal with the diversity and complexity of virtual objects, scenes, and avatars. In some scenarios of convolved virtual–physical worlds, semantic segmentation helps distinguish the pixels of virtual objects from real objects or even the real background.

5.3.1.4 Human Pose Estimation

The main objective of human pose estimation is to determine the relative 2D/3D coordinates of human joints from either raw natural images, depth images, videos, or skeleton data as acquired by built-in motion capture devices like Microsoft Kinect cameras. This is a very challenging task because of the wide range of human silhouettes and appearances of human posture in realistic conditions, besides other issues such as occlusion, varying illumination, and cluttered background. With the assistance of VR systems, the human posture of users can be determined by VR cameras and 360 VR video cameras that are able to automatically tracking motions with a built-in algorithms, similar to Microsoft Azure Kinect and Intel Realsense. In this context, all the motions (e.g. hand-waving) and actions (e.g. jumping) of users in the physical world are projected to the Metaverse and presented by avatars; hence, some human–object interactive actions in the virtual world can be accomplished without using the VR controller. Before DL, most of the conventional pose estimation approaches were based on the detection of body parts with the inference of artificial pictorial structures.

In the era of DL, depending on the manner of processing input images, human pose estimation methods can be divided into two categories: holistic methods and part-based methods. The holistic methods typically model the pose estimation task globally without learning the patterns of individual body parts and their spatial relations. In this group, DeepPose [74] was emphasized as a representative holistic method that formulates human pose estimation as a joint regression problem to address with deep neural networks, while there is no need to characterize any graphical model or body part detector explicitly. However, holistic-based methods have suffered from inaccuracy of high-sensitive (i.e. being confused with others in various viewpoints), and complicated poses in the real world. On the contrary, the part-based methods detect and localize all body parts individually with the incorporation of spatial information and the reference of a graphical model (that was trained with a set of artificial poses). The work [13] adopted local part patches and background patches to train a deep network using CNN architecture to learn the conditional probabilities of part presence in an image or a video frame and their spatial relations. Moreover, in [72], an efficient high-resolution representation learning approach for human pose estimation was introduced with a sophisticated-designed CNN architecture, namely high-resolution net (HRNet). Compared with existing deep networks for pose estimation, HRNet presented a more accurate keypoint heatmap estimation with the preservation of high-resolution feature representations owing to a multiscale fusion mechanism. More recently, with the advancement of 3D vision sensors and the popularity of 3D cameras, 3D human pose estimation has attracted much more attention with a wide range of civil and industrial applications. A comprehensive review of 3D pose estimation methods can be found in [77]. When developing a high-accuracy, low-cost approach to estimate and track human poses, some issues should be taken into consideration, including self-occlusion of body parts, multiuser interactive scenarios, and a vast number of complex poses in realistic conditions. In the 3D environments, 3D human hand tracking and 3D body pose estimation models have shown the potential for VR services and applications in the Metaverse, such as action classification in user-vs.-NPC interaction and activity recognition in freely immersive gaming.

5.3.1.5 Others

Besides the aforementioned topics, some other auxiliary tasks are involved in the computer vision domain.

- **Image restoration**: Image restoration refers to the process of recovering an image from its degraded or corrupted version. Depending on corruption types, such as noise and lens blurring in the image capturing process, JPEG compression in postprocessing, and haze and motion blurring in nonideal conditions photography, there will have different appropriate methods for them, mostly relying on image processing techniques and AI algorithms (e.g. DL with generative adversarial networks). In the Metaverse, image restoration helps recover and improve the quality of degraded images, which are potential to apply in NFT art pictures and photos. Moreover, image restoration is widely applied in VR and augmented reality for handling color attenuation, texture degradation, and blurring issues when displaying images and other digital content.
- **Image registration**: Image registration (a.k.a., fusion or synthesis) aims to align multiple scenes into a single integrated image and helps effectively address some common geometric issues such as image rotation, scale, and skew when overlaying images. Image registration is usually adopted in medical and satellite imagery from multiple camera resources to geometrically align, connect, and combine many adjacent images into a single larger image. As focusing on the combination and alignment of multiple images based on geometric processing algorithms, image registration has been widely exploited to generate 3D images from multiple 2D images, thus enriching the digital contents for viewing and visualization with VR devices or other immersive screen-based devices. Besides, this technique can be used to generate a single spherical panoramic image from two or more nearly images captured by different camera sources, thus providing a seamless viewing experience (e.g. immersive 360° viewing experience) to users in 3D virtual worlds.
- **Scene reconstruction**: As one of the most complex and challenging problems in computer vision, scene reconstruction refers to the 3D rendering and reconstruction of multiple objects from a photograph. Most of the existing scene reconstruction methods roughly operate by creating a point cloud at the surface of objects (e.g. using 3D scanner or laser scanner) and reconstructing a linked mesh from this cloud. Many studies have focused on generating 3D models directly from 2D images with voxel-based object reconstruction using autoencoders and their counterparts, while other works can exploit deep networks for depth estimation for 3D scene reconstruction from images acquired from different view angles. In the Metaverse, scene reconstruction allows virtual world designers to create 3D scenes from 2D images or videos, thus saving time and cost when dealing with a huge design workload to build

an open world. Furthermore, 3D construction can be applied to convert 2D objects to 3D objects in NFT products.

- **Image captioning**: Relying on the intersection of computer vision and NLP, image captioning is the task of describing the context of an image in words. Most of the existing image captioning applications have followed an encoder–decoder framework, in which an input image should be encoded into an intermediate informative representation and decoded into a descriptive text sequence. Image captioning can exploit scene understanding and object detection to gain useful information for better image description. Image captioning can automatically generate concise and meaningful captions and detailed descriptions for digital content, such as objects and scenes in 3D virtual worlds and NFT art pictures. In some Metaverse platforms integrating social networks and media, where millions of images are generated and posted daily, image captioning helps accelerate the automatic subtitle creation procedure to reduce the workload of content executives.

5.3.2 Computer Vision for Metaverse Applications

5.3.2.1 Autonomous Driving

The emergence of autonomous driving will have a significant impact on our travel life, but the development and implementation of autonomous driving in practice cannot be obtained overnight because of some challenges, such as expensive hardware, imperfect software, and scarcity of road test conditions. Especially self-driving systems require massive amounts of road data for learning complicated road patterns in realistic conditions to make the right decision regarding every driving scenario. Although several cities have welcome autonomous driving tests, the scenes and road environments are simple, consequently making autopilot systems silly and nearly impossible to evaluate on realistic road conditions. To this end, the Metaverse can create an artificial 3D virtual road perspective, closely similar to a real one, based on the key enabler of computer vision for autonomous driving where various complicated scenarios can be realized, and a massive amount of road data can be acquired effortlessly to commercialize autopilot systems soon.

Nowadays, computer vision has contributed as an important enabling technology to autonomous driving, in which several challenging issues on the road can be addressed based on fundamental computer vision tasks, e.g. object detection for moving vehicle detection, image classification for traffic signal recognition, and semantic segmentation for scene understanding, as illustrated in Figure 5.5. For instance, the work [4] developed a vision-based driving assistance system for scene recognition and awareness using video frames acquired by a dashboard camera, in which the saliency maps containing moving objects, isolated from

Figure 5.5 Computer vision for the Metaverse of autonomous driving. Source: Meta.

the scene background with bounding boxes, are used to classify road surface by adopting a maximum likelihood classification algorithm. For on-road applications, localizing surrounding objects accurately in an image plane is important to improve the accuracy of distance estimation, safe trajectory prediction, and driving scheduling. In [64], an efficient object localization method was proposed by exploiting a deep ConvNet and an iterative ROI pooling mechanism to improve the accuracy of object detection and localization while presenting a comparative processing speed if compared with other existing DL-based models. To overcome the environmental limitation of object detection at nighttime, the work [68] proposed an online image-to-image translation technique with transfer learning on an additional nighttime image dataset. To analyze and understand the outdoor road scenes for autonomous driving, the work [11] introduced a range-aware real-time semantic segmentation network by learning 3D LiDAR point clouds with an autoencoder–decoder architecture. Replying on the benchmarks on several real datasets, the proposed method improved the instance-level segmentation accuracy for small-far objects while keeping an acceptable speed of around 19 frames/s. A comprehensive survey about the state-of-the-art datasets, methods, and challenges of object detection and semantic image segmentation for autonomous driving was summarized in [19], which provides a whole picture of self-driving technology assisted by computer vision to be deployed in the Metaverse.

5.3.2.2 Healthcare

Smart healthcare is always the first priority in our modern life, which enables us to monitor health status in real-time, early detect health risks, and timely diagnose diseases to provide effective treatment plans. Although many telehealth and telemedicine systems have been developed to overcome crucial problems of traditional physical systems (e.g. increasing cost for in-person office visits) by partly transforming to remote processes, they cannot provide a completed immersive healthcare experience in semivirtual environments besides lacking a service hub like hospitals and healthcare centers. To this end, the health and medical Metaverse can be considered a promising approach to enhance patient care in ways we are yet to imagine, especially as we realize the more and more importance of computer vision in medical image processing scope.

The last decade has witnessed the modernization of medical image processing thanks to the advancement of computer vision when DL with various advanced architectures creates a significant performance improvement besides the capability of dealing with many specialized medical image formats. For instance, recently, an accurate and low-cost method [56] was proposed to localize anatomical landmarks in medical images, e.g. coronary computed tomography angiography scan, magnetic resonance scan, and cephalometric X-rays where the relative coordinates of multiple detected landmarks were obtained by performing regression and classification simultaneously for every image patch. In [35], a semantic medical image segmentation method, powered by deep ConvNets, was developed to accurately track the tumor volume changes in lung computed tomography images, thus conducting an efficient lung cancer diagnosis. In addition, ACDC@LungHP (Automatic Cancer Detection and Classification in Whole-slide Lung Histopathology) challenge [42] was introduced as the benchmark of different computer-aided methods of image-based lung cancer diagnoses.

The 3D virtual world in the Metaverse can become a seamless space for several 3D medical image analysis services, which are expected to provide more accurate diagnoses than 2D-based methods. In [62], a 3D motion estimation method was developed for left ventricular dynamics, containing the meaningful information of the myocardial borders, in which a deformable image registration and an unscented Kalman smoother were adopted for both long-axis and short-axis cine cardiac magnetic resonance images before performing track-to-track fusion to achieve 3D estimates. For medical image registration in the 3D space, the work [25] studied a deep graph regularization network incorporated with discrete dense displacement mapping technique to overcome the problem of large deformations of very deep networks having straightforward architectures. The medical image segmentation topic has been extended to the 3D space in order to improve the accuracy of disease progress diagnosis, which allows to speed up therapeutic schedules. Most recently, in [82], a high-performance 3D medical image segmentation method was proposed with a deep ConvNet to learn more interslice context information and accordingly increase the accuracy of segmentation.

Nowadays, for medical skill training and education, besides the traditional training ways to acquire surgical skills through actual practice in the operation theater, some innovative simulation-based surgical platforms powered by advanced computer vision techniques have emerged as a ground-breaking medium for training and testing skills. Indeed, with surgical simulation platforms, trainees and medical students have more opportunities to exercise safely before facing realistic situations in the operation theater. In this context, computer vision techniques are used to assess the quality of the surgery, and the Metaverse can deploy such kinds of medical skill training services for sustainable education.

5.3.2.3 Manufacturing

The manufacturing industry has been adopting various intelligent automation solutions; however, the direct deployment of these solutions in realistic environments may expose some undesired issues. To this end, building a whole manufacturing process in a 3D virtual world of the Metaverse can help overcome some issues such as defect detection, product assembly, and predictive maintenance, wherein computer vision can show a great opportunity to revolutionize various sectors of the manufacturing process.

As a primary sector of industry 4.0 automation, most of the manufacturing industry has been implementing computer vision for fully automated products and component assembly along with management processes. Computer vision systems always monitor and supervise robotic arms, and guide employees in a whole assembly line. To obtain a nearly perfect accuracy in detecting defects in manufactured products, it requires systems to monitor and detect defects on a microscale. In this context, a computer vision-powered application can gather real-time image sequences from cameras to analyze using ML algorithms, consequently detecting defects by collating with the predefined quality standards. Some manufacturing processes suffer from critical temperatures and poor environmental conditions, thus conducting equipment deformation. Computer vision systems can constantly supervise the equipment based on various metrics [18]. If any deviation from metrics suggests corrosion detected by computer vision systems (e.g. object detection and segmentation), a warning message will be sent to the respective managers and operators to carry out maintenance activities proactively.

Employees working in the manufacturing industry have undergone dangerous conditions with a high risk of injury. Not adhering to safety and security requirements can cause some serious accidents, possibly leading to death. Although most manufacturing companies have installed cameras to monitor and supervise employee movements and activities, the supervision is done manually and sometimes abandoned. An AI-powered computer vision system is recommended for these situations [29] where different computer vision tasks are collaboratively performed to achieve high accuracy and reliability, such as moving human detection, semantic segmentation, human pose estimation, and physical human action/activity recognition. Another role of computer vision in manufacturing is inventory management (e.g. counting stock, maintaining inventory status in warehouses, and giving notice of under-demand material). In some massive warehouses, a computer vision system can locate stock based on automatic detection and recognition of product barcode [69]. The motion of a player in the physical world should be tracked by using some specialized-designed camera devices featured by deep imaging sensors, in which a human pose can be fully projected to a 3D skeleton of user's virtual presentation in the 3D virtual worlds of the Metaverse. Accordingly, the physical actions and activities of players will be synchronized exactly with those of their avatars for real-time processing, thus making physical interactions and experiences more realistic, especially when incorporated with VR technology.

5.3.2.4 Others

Besides healthcare, autonomous driving, and smart manufacturing, computer vision can be found as a key-enabling technology in some other Metaverse applications for practical implementation.

- **GeoSensing**: The land cover information in satellite imagery is useful for various applications, such as monitoring deforestation and urbanization areas. To identify the type of land cover (e.g. forest, water, agriculture, and urban), for each satellite image pixel, land cover classification can be regarded as semantic segmentation. Virtual lands can be deployed for natural resources monitoring and management in the Metaverse-based virtual world.
- **Fashion**: Clothing parsing is a very complex task compared with others due to the large number of clothes with different shapes. It differentiates from regular object detection and scene segmentation since fine-grained clothing classification requires high-level judgment relying on the semantics of clothing, practical human poses, and the number of classes. Clothing parsing has attracted more attention from the computer vision community thanks to its potential in real-world applications such as E-commerce and retail. Virtual fashion with 3D clothing parsing can be deployed in virtual retail stores where users can buy virtual goods and receive real items, thus boosting the virtual economy in the Metaverse.
- **Real estate**: The capability of providing clients with a realistic and immersive experience is one of the functionalities of computer vision in combination with immersive technology. Real estate sellers and marketers may take benefits of these technologies by allowing clients to see and check their properties

before making a deal. Clients can explore the property and surrounding area in real-time via VR tours and potentially interact with virtual objects in the 3D space, which are enabled and assisted by scene reconstruction algorithms. As a result, both the clients and sellers can save time and money spent on in-person showings and meetings since the whole process, from advertisement to showing and making a deal, can now be done in the real estate Metaverse.

- **Precision agriculture**: As the main advantage of computer vision in precision agriculture is a nonphysical effect, it means that insightful information about crops and fields can be obtained without touching. Computer vision for smart agriculture can be realized over four aspects: (i) crop management (e.g. disease detection and irrigation), (ii) crop health analysis, (iii) fruit quality inspection, and (iv) crop and land estimation in geographic information systems.
- **Education**: The education technology industry has been leveraging computer vision techniques in many aspects, ranging from teaching to demonstration. For instance, teachers or course instructors apply computer vision-based solutions to evaluate students' learning procedures nonobstructively. Apart from this, AI vision is being exploited for many topics, such as school logistic support, knowledge acquisition, attendance monitoring, and classroom supervision. When a completely 3D virtual computer vision-based education platform is deployed in the Metaverse, many advantages can be obtained for students and teachers through cooperation with immersive technology to carry out more realistic experiences without geographical barriers.

5.4 Conclusions and Future Research Directions

In this chapter, we have investigated the role of AI and computer vision in the development of the Metaverse and explored their potential in the virtual worlds, thus enhancing user immersive experience and expanding user-centric services. While AI is one of the key enablers to make applications more automatic and intelligent, computer vision is responsible for polishing the visualization of virtual worlds to be more attractive to end users, besides solving many specific visual-based issues. Notably, in numerous modern computer vision systems, DL revolutionizes in boosting the performance of computer vision as compared with traditional image processing and conventional ML-based approaches and effectively models many complicated visual patterns on large-messy-confusing image data.

We now delineate some AI research directions in the Metaverse. Being superior to regular virtual personal assistants that are developed for a general purpose like simple dialog management, conversational AI can serve a wide range of multilevel philosophical conversations of virtual customers/employee assistants to enhance user interactive experiences. Conversational AI is a set of technologies (e.g. automatic speech recognition, language processing, advanced dialog management, and ML) that skillfully provide human-like interactions based on recognizing speech and text, understanding the speaker's intention, deciphering various languages, and responding to human-mimicking conversations via different modalities.

In many AI-assisted services and applications, decisions are made by AI agents, which are driven by some weak-interpretability and poor-explainability ML models. Consequently, developers, virtual world designers, and users cannot completely understand AI decision-making processes (e.g. how and why an AI model delivers a prediction), and mostly trust them in a blind manner. To overcome these problems, explainable artificial intelligence (XAI) is a set of tools and methods that describe AI models, analyze risky impacts, characterize model transparency, and assess outcomes. They allow users to entirely comprehend and trust the AI models with a whole process of supervision and accountability. Furthermore, with XAI, system engineers and data scientists can understand and explain what exactly is happening inside an AI model, how the results are derived from AI algorithms, and when an AI model is likely to crash.

In the Metaverse, most of the virtual services and applications are developed in the 3D space; therefore, the existing computer vision techniques should have the capability of processing 3D objects, including 3D scene segmentation and 3D human pose estimation. In this context, the extra information exploited in the depth dimension can be useful to increase the accuracy of computer vision tasks but consume a higher computing cost and larger memory resources undesirably. Furthermore, several potential Metaverse-based applications have adopted multiple computer vision tasks separately (i.e. different tasks such as object detection, scene classification, and semantic segmentation, process only a single input image by their individual processing flows), which can waste a large amount of computing resources. To this end, a collaborative computer vision system is recommended to process a single input with different tasks simultaneously, where the learning models for these tasks can cooperate together in many aspects, such as sharing the same deep network architecture and reusing the extracted attributes of object detection for scene classification and segmentation. As known that running computer vision in the cloud is heavily limiting real-time computer vision applications, edge computing-based AI, a technology that allows the moving of ML tasks from the cloud to high-performance edge devices [60], has become one of the most suitable solutions to address the scalability problems and serve more and more use cases of computer vision applications in the Metaverse.

Bibliography

1 Berdakh Abibullaev and Amin Zollanvari. Learning discriminative spatiospectral features of ERPs for accurate brain-computer interfaces. *IEEE Journal of Biomedical and Health Informatics*, 23(5):2009–2020, 2019.

2 Mohamed Alloghani, Dhiya Al-Jumeily, Jamila Mustafina, Abir Hussain, and Ahmed J. Aljaaf. A systematic review on supervised and unsupervised machine learning algorithms for data science. In *Supervised and Unsupervised Learning for Data Science*, pages 3–21. Springer, 2020.

3 Madyan Alsenwi, Nguyen H. Tran, Mehdi Bennis, Shashi Raj Pandey, Anupam Kumar Bairagi, and Choong Seon Hong. Intelligent resource slicing for eMBB and URLLC coexistence in 5G and beyond: A deep reinforcement learning based approach. *IEEE Transactions on Wireless Communications*, 20(7):4585–4600, 2021.

4 Melih Altun and Mehmet Celenk. Road scene content analysis for driver assistance and autonomous driving. *IEEE Transactions on Intelligent Transportation Systems*, 18(12):3398–3407, 2017.

5 Laith Alzubaidi, Jinglan Zhang, Amjad J. Humaidi, Ayad Al-Dujaili, Ye Duan, Omran Al-Shamma, José Santamaría, Mohammed A. Fadhel, Muthana Al-Amidie, and Laith Farhan. Review of deep learning: Concepts, CNN architectures, challenges, applications, future directions. *Journal of Big Data*, 8(1):1–74, 2021.

6 Amin Azari, Mustafa Ozger, and Cicek Cavdar. Risk-aware resource allocation for URLLC: Challenges and strategies with machine learning. *IEEE Communications Magazine*, 57(3):42–48, 2019.

7 Vijay Badrinarayanan, Alex Kendall, and Roberto Cipolla. SegNet: A deep convolutional encoder-decoder architecture for image segmentation. *IEEE Transactions on Pattern Analysis and Machine Intelligence*, 39(12):2481–2495, 2017.

8 Karel Benes, Murali Karthick Baskar, and Lukas Burget. Residual memory networks in language modeling: Improving the reputation of feed-forward networks. In *INTERSPEECH*, pages 284–288, Stockholm, Sweden, August 2017.

9 Sergio López Bernal, Alberto Huertas Celdrán, Gregorio Martínez Pérez, Michael Taynnan Barros, and Sasitharan Balasubramaniam. Security in brain-computer interfaces: State-of-the-art, opportunities, and future challenges. *ACM Computing Surveys*, 54(1):1–35, 2022.

10 Alberto Cannavò and Fabrizio Lamberti. How blockchain, virtual reality, and augmented reality are converging, and why. *IEEE Consumer Electronics Magazine*, 10(5):6–13, 2021.

11 Tzu-Hsuan Chen and Tian Sheuan Chang. RangeSeg: Range-aware real time segmentation of 3D LiDAR point clouds. *IEEE Transactions on Intelligent Vehicles*, 7(1):93–101, 2022.

12 Liang-Chieh Chen, George Papandreou, Iasonas Kokkinos, Kevin Murphy, and Alan L. Yuille. Semantic image segmentation with deep convolutional nets and fully connected CRFs. *arXiv preprint arXiv:1412.7062*, 2014.

13 Xianjie Chen and Alan L. Yuille. Articulated pose estimation by a graphical model with image dependent pairwise relations. *Advances in Neural Information Processing Systems 27 (NIPS 2014)*, 2014.

14 Michał Daniluk, Tim Rocktäschel, Johannes Welbl, and Sebastian Riedel. Frustratingly short attention spans in neural language modeling. *arXiv preprint arXiv:1702.04521*, 2017.

15 Hossein Darvishi, Domenico Ciuonzo, Eivind Rosón Eide, and Pierluigi Salvo Rossi. Sensor-fault detection, isolation and accommodation for digital twins via modular data-driven architecture. *IEEE Sensors Journal*, 21(4):4827–4838, 2021.

16 Stefanos Doltsinis, Pedro Ferreira, and Niels Lohse. A symbiotic human-machine learning approach for production ramp-up. *IEEE Transactions on Human-Machine Systems*, 48(3):229–240, 2018.

17 Shengdong Du, Tianrui Li, Yan Yang, and Shi-Jinn Horng. Deep air quality forecasting using hybrid deep learning framework. *IEEE Transactions on Knowledge and Data Engineering*, 33(6):2412–2424, 2021.

18 Shu-Kai S. Fan, Du-Ming Tsai, Fei He, Jui-Yu Huang, and Chih-Hung Jen. Key parameter identification and defective wafer detection of semiconductor manufacturing processes using image processing techniques. *IEEE Transactions on Semiconductor Manufacturing*, 32(4):544–552, 2019.

19 Di Feng, Christian Haase-Schütz, Lars Rosenbaum, Heinz Hertlein, Claudius Gläser, Fabian Timm, Werner Wiesbeck, and Klaus Dietmayer. Deep multi-modal object detection and semantic segmentation for autonomous driving: Datasets, methods, and challenges. *IEEE Transactions on Intelligent Transportation Systems*, 22(3):1341–1360, 2021.

20 Maite Frutos-Pascual and Begoñya García Zapirain. Review of the use of AI techniques in serious games: Decision making and machine learning. *IEEE Transactions on Computational Intelligence and AI in Games*, 9(2):133–152, 2017.

21 Adam Ghandar, Ayyaz Ahmed, Shahid Zulfiqar, Zhengchang Hua, Masatoshi Hanai, and Georgios Theodoropoulos. A decision support system for urban agriculture using digital twin: A case study with aquaponics. *IEEE Access*, 9:35691–35708, 2021.

22 Ross Girshick, Jeff Donahue, Trevor Darrell, and Jitendra Malik. Rich feature hierarchies for accurate object detection and semantic segmentation.

In *Proceedings of the IEEE Conference on Computer Vision and Pattern Recognition*, pages 580–587, Columbus, OH, USA, 2014.

23 Bo Gu, Xu Zhang, Ziqi Lin, and Mamoun Alazab. Deep multiagent reinforcement-learning-based resource allocation for internet of controllable things. *IEEE Internet of Things Journal*, 8(5):3066–3074, 2021.

24 Shengnan Guo, Youfang Lin, Shijie Li, Zhaoming Chen, and Huaiyu Wan. Deep spatial-temporal 3D convolutional neural networks for traffic data forecasting. *IEEE Transactions on Intelligent Transportation Systems*, 20(10):3913–3926, 2019.

25 Lasse Hansen and Mattias P. Heinrich. GraphRegNet: Deep graph regularisation networks on sparse keypoints for dense registration of 3D lung CTs. *IEEE Transactions on Medical Imaging*, 40(9):2246–2257, 2021.

26 He He and Dongrui Wu. Transfer learning for brain-computer interfaces: A Euclidean space data alignment approach. *IEEE Transactions on Biomedical Engineering*, 67(2):399–410, 2020.

27 Cam-Hao Hua, Kiyoung Kim, Thien Huynh-The, Jong In You, Seung-Young Yu, Thuong Le-Tien, Sung-Ho Bae, and Sungyoung Lee. Convolutional network with twofold feature augmentation for diabetic retinopathy recognition from multi-modal images. *IEEE Journal of Biomedical and Health Informatics*, 25(7):2686–2697, 2021.

28 Shaohua Huang, Yu Guo, Daoyuan Liu, Shanshan Zha, and Weiguang Fang. A two-stage transfer learning-based deep learning approach for production progress prediction in IoT-enabled manufacturing. *IEEE Internet of Things Journal*, 6(6):10627–10638, 2019.

29 Thien Huynh-The, Cam-Hao Hua, and Dong-Seong Kim. Encoding pose features to images with data augmentation for 3-D action recognition. *IEEE Transactions on Industrial Informatics*, 16(5):3100–3111, 2020.

30 Thien Huynh-The, Cam-Hao Hua, Jae-Woo Kim, Seung-Hwan Kim, and Dong-Seong Kim. Exploiting a low-cost CNN with skip connection for robust automatic modulation classification. In *Proceedings of the 2020 IEEE Wireless Communications and Networking Conference (WCNC)*, pages 1–6. IEEE, 2020.

31 Thien Huynh-The, Cam-Hao Hua, Quoc-Viet Pham, and Dong-Seong Kim. MCNet: An efficient CNN architecture for robust automatic modulation classification. *IEEE Communications Letters*, 24(4):811–815, Apr. 2020.

32 Thien Huynh-The, Cam-Hao Hua, Nguyen Anh Tu, and Dong-Seong Kim. Physical activity recognition with statistical-deep fusion model using multiple sensory data for smart health. *IEEE Internet of Things Journal*, 8(3):1533–1543, 2021.

33 Thien Huynh-The, Quoc-Viet Pham, Toan-Van Nguyen, Thanh Thi Nguyen, Rukhsana Ruby, Ming Zeng, and Dong-Seong Kim. Automatic modulation classification: A deep architecture survey. *IEEE Access*, 9:142950–142971, 2021.

34 Thien Huynh-The, Quoc-Viet Pham, Xuan-Qui Pham, Thanh Thi Nguyen, Zhu Han, and Dong-Seong Kim. Artificial intelligence for the metaverse: A survey. *Engineering Applications of Artificial Intelligence*, 117(Part A): 105581, 2023.

35 Jue Jiang, Yu-Chi Hu, Chia-Ju Liu, Darragh Halpenny, Matthew D. Hellmann, Joseph O. Deasy, Gig Mageras, and Harini Veeraraghavan. Multiple resolution residually connected feature streams for automatic lung tumor segmentation from CT images. *IEEE Transactions on Medical Imaging*, 38(1):134–144, 2019.

36 Leila Kabbai, Mehrez Abdellaoui, and Ali Douik. Image classification by combining local and global features. *The Visual Computer*, 35(5):679–693, 2019.

37 Kuntharrgyal Khysru, Di Jin, Yuxiao Huang, Hui Feng, and Jianwu Dang. A Tibetan language model that considers the relationship between suffixes and functional words. *IEEE Signal Processing Letters*, 28:459–463, 2021.

38 Sotiris B. Kotsiantis, I. Zaharakis, and P. Pintelas. Supervised machine learning: A review of classification techniques. *Emerging Artificial Intelligence Applications in Computer Engineering*, 160(1):3–24, 2007.

39 Prabhat Kumar, Randhir Kumar, Gautam Srivastava, Govind P. Gupta, Rakesh Tripathi, Thippa Reddy Gadekallu, and Naixue Xiong. PPSF: A privacy-preserving and secure framework using blockchain-based machine-learning for IoT-driven smart cities. *IEEE Transactions on Network Science and Engineering*, 8(3):2326–2341, 2021.

40 Hojin Lee, Hyeyun Jeong, Gyogwon Koo, Jaepil Ban, and Sang Woo Kim. Attention recurrent neural network-based severity estimation method for interturn short-circuit fault in permanent magnet synchronous machines. *IEEE Transactions on Industrial Electronics*, 68(4):3445–3453, 2021.

41 Lei Lei, Yue Tan, Kan Zheng, Shiwen Liu, Kuan Zhang, and Xuemin Shen. Deep reinforcement learning for autonomous Internet of Things: Model, applications and challenges. *IEEE Communications Surveys & Tutorials*, 22(3):1722–1760, 2020.

42 Zhang Li, Jiehua Zhang, Tao Tan, Xichao Teng, Xiaoliang Sun, Hong Zhao, Lihong Liu, Yang Xiao, Byungjae Lee, Yilong Li, Qianni Zhang, Shujiao Sun, Yushan Zheng, Junyu Yan, Ni Li, Yiyu Hong, Junsu Ko, Hyun Jung, Yanling Liu, Yu-cheng Chen, Ching-wei Wang, Vladimir Yurovskiy, Pavel Maevskikh, Vahid Khanagha, Yi Jiang, Li Yu, Zhihong Liu, Daiqiang Li, Peter J. Schüffler, Qifeng Yu, Hui Chen, Yuling Tang, and Geert Litjens. Deep learning methods for lung cancer segmentation in whole-slide histopathology images–the ACDC@LungHP challenge 2019. *IEEE Journal of Biomedical and Health Informatics*, 25(2):429–440, 2021.

43 Aisha Liaqat, Muddassar Azam Sindhu, and Ghazanfar Farooq Siddiqui. Metamorphic testing of an artificially intelligent chess game. *IEEE Access*, 8:174179–174190, 2020.

44 Bin Liu and Guosheng Yin. Chinese document classification with Bi-directional convolutional language model. In *Proceedings of the 43rd International ACM SIGIR Conference on Research and Development in Information Retrieval*, pages 1785–1788, New York, NY, US, July 2020.

45 Wei Liu, Dragomir Anguelov, Dumitru Erhan, Christian Szegedy, Scott Reed, Cheng-Yang Fu, and Alexander C. Berg. SSD: Single shot multibox detector. In *Proceedings of the European Conference on Computer Vision*, pages 21–37. Springer, 2016.

46 Yiming Liu, F. Richard Yu, Xi Li, Hong Ji, and Victor C. M. Leung. Blockchain and machine learning for communications and networking systems. *IEEE Communications Surveys & Tutorials*, 22(2):1392–1431, 2020.

47 Jonathan Long, Evan Shelhamer, and Trevor Darrell. Fully convolutional networks for semantic segmentation. In *Proceedings of the IEEE Conference on Computer Vision and Pattern Recognition*, pages 3431–3440, Boston, MA, USA, 2015.

48 Dengsheng Lu and Qihao Weng. A survey of image classification methods and techniques for improving classification performance. *International Journal of Remote Sensing*, 28(5):823–870, 2007.

49 Yunlong Lu, Xiaohong Huang, Yueyue Dai, Sabita Maharjan, and Yan Zhang. Blockchain and federated learning for privacy-preserved data sharing in industrial IoT. *IEEE Transactions on Industrial Informatics*, 16(6): 4177–4186, 2020.

50 Changqing Luo, Jinlong Ji, Qianlong Wang, Xuhui Chen, and Pan Li. Channel state information prediction for 5G wireless communications: A deep learning approach. *IEEE Transactions on Network Science and Engineering*, 7(1):227–236, 2020.

51 Ronghua Ma, Tianyou Yu, Xiaoli Zhong, Zhu Liang Yu, Yuanqing Li, and Zhenghui Gu. Capsule network for ERP detection in brain-computer interface. *IEEE Transactions on Neural Systems and Rehabilitation Engineering*, 29:718–730, 2021.

52 Shervin Minaee, Yuri Boykov, Fatih Porikli, Antonio Plaza, Nasser Kehtarnavaz, and Demetri Terzopoulos. Image segmentation using deep learning: A survey. *IEEE Transactions on Pattern Analysis and Machine Intelligence*, 44(7):3523–3542, 2022.

53 Dinh C. Nguyen, Pubudu N. Pathirana, Ming Ding, and Aruna Seneviratne. Privacy-preserved task offloading in mobile blockchain with deep reinforcement learning. *IEEE Transactions on Network and Service Management*, 17(4):2536–2549, 2020.

54 Dinh C. Nguyen, Ming Ding, Quoc-Viet Pham, Pubudu N. Pathirana, Long Bao Le, Aruna Seneviratne, Jun Li, Dusit Niyato, and H. Vincent Poor. Federated learning meets blockchain in edge computing: Opportunities and challenges. *IEEE Internet of Things Journal*, 8(16):12806–12825, 2021.

55 Hyeonwoo Noh, Seunghoon Hong, and Bohyung Han. Learning deconvolution network for semantic segmentation. In *Proceedings of the IEEE International Conference on Computer Vision (ICCV)*, pages 1520–1528, Santiago, Chile, 2015.

56 Julia M. H. Noothout, Bob D. De Vos, Jelmer M. Wolterink, Elbrich M. Postma, Paul A. M. Smeets, Richard A. P. Takx, Tim Leiner, Max A. Viergever, and Ivana Išgum. Deep learning-based regression and classification for automatic landmark localization in medical images. *IEEE Transactions on Medical Imaging*, 39(12):4011–4022, 2020.

57 Seonghun Park, Ho-Seung Cha, and Chang-Hwan Im. Development of an online home appliance control system using augmented reality and an SSVEP-based brain-computer interface. *IEEE Access*, 7:163604–163614, 2019.

58 Ngoc-Quan Pham, German Kruszewski, and Gemma Boleda. Convolutional neural network language models. In *Proceedings of the Conference on Empirical Methods in Natural Language Processing*, pages 1153–1162, Austin, TX, USA, November 2016.

59 Quoc-Viet Pham, Nhan Thanh Nguyen, Thien Huynh-The, Long Bao Le, Kyungchun Lee, and Won-Joo Hwang. Intelligent radio signal processing: A survey. *IEEE Access*, 9:83818–83850, 2021.

60 Quoc-Viet Pham, Rukhsana Ruby, Fang Fang, Dinh C. Nguyen, Zhaohui Yang, Mai Le, Zhiguo Ding, and Won-Joo Hwang. Aerial computing: A new computing paradigm, applications, and challenges. *IEEE Internet of Things Journal*, 9(11):8339–8363, 2022.

61 Pedro O. Pinheiro, Tsung-Yi Lin, Ronan Collobert, and Piotr Dollár. Learning to refine object segments. In *Proceedings of the European Conference on Computer Vision*, pages 75–91. Springer, 2016.

62 Kumaradevan Punithakumar, Ismail Ben Ayed, Abraam S. Soliman, Aashish Goela, Ali Islam, Shuo Li, and Michelle Noga. 3D motion estimation of left ventricular dynamics using MRI. *IEEE Journal of Translational Engineering in Health and Medicine*, 8:1800209, 2020.

63 Xiaoye Qian, Huan Chen, Haotian Jiang, Justin Green, Haoyou Cheng, and Ming-Chun Huang. Wearable computing with distributed deep learning hierarchy: A study of fall detection. *IEEE Sensors Journal*, 20(16):9408–9416, 2020.

64 Rakesh Nattoji Rajaram, Eshed Ohn-Bar, and Mohan Manubhai Trivedi. RefineNet: Refining object detectors for autonomous driving. *IEEE Transactions on Intelligent Vehicles*, 1(4):358–368, 2016.

65 Waseem Rawat and Zenghui Wang. Deep convolutional neural networks for image classification: A comprehensive review. *Neural Computation*, 29(9):2352–2449, 2017.

66 Joseph Redmon, Santosh Divvala, Ross Girshick, and Ali Farhadi. You only look once: Unified, real-time object detection. In *Proceedings of the IEEE Conference on Computer Vision and Pattern Recognition*, pages 779–788, 2016.

67 Shaoqing Ren, Kaiming He, Ross Girshick, and Jian Sun. Faster R-CNN: Towards real-time object detection with region proposal networks. *Advances in Neural Information Processing Systems 28 (NIPS 2015)*, 2015.

68 Mark Schutera, Mostafa Hussein, Jochen Abhau, Ralf Mikut, and Markus Reischl. Night-to-day: Online image-to-image translation for object detection within autonomous driving by night. *IEEE Transactions on Intelligent Vehicles*, 6(3):480–489, 2021.

69 Adnan Sharif, Guangtao Zhai, Jun Jia, Xiongkuo Min, Xiangyang Zhu, and Jiahe Zhang. An accurate and efficient 1-D barcode detector for medium of deployment in IoT systems. *IEEE Internet of Things Journal*, 8(2):889–900, 2021.

70 Changyang She, Chengjian Sun, Zhouyou Gu, Yonghui Li, Chenyang Yang, H. Vincent Poor, and Branka Vucetic. A tutorial on ultrareliable and low-latency communications in 6G: Integrating domain knowledge into deep learning. *Proceedings of the IEEE*, 109(3):204–246, 2021.

71 Zixing Song, Xiangli Yang, Zenglin Xu, and Irwin King. Graph-based semi-supervised learning: A comprehensive review. *IEEE Transactions on Neural Networks and Learning Systems*, 1–21, 2022.

72 Ke Sun, Bin Xiao, Dong Liu, and Jingdong Wang. Deep high-resolution representation learning for human pose estimation. In Proceedings of the IEEE/CVF Conference on Computer Vision and Pattern Recognition, pages 5693–5703, Long Beach, CA, USA, 2019.

73 Sudeep Tanwar, Qasim Bhatia, Pruthvi Patel, Aparna Kumari, Pradeep Kumar Singh, and Wei-Chiang Hong. Machine learning adoption in blockchain-based smart applications: The challenges, and a way forward. *IEEE Access*, 8:474–488, 2019.

74 Alexander Toshev and Christian Szegedy. DeepPose: Human pose estimation via deep neural networks. In Proceedings of the IEEE Conference on Computer Vision and Pattern Recognition, pages 1653–1660, Columbus, OH, USA, 2014.

75 Godwin Brown Tunze, Thien Huynh-The, Jae-Min Lee, and Dong-Seong Kim. Sparsely connected CNN for efficient automatic modulation recognition. *IEEE Transactions on Vehicular Technology*, 69(12):15557–15568, 2020.

76 Jesper E. Van Engelen and Holger H. Hoos. A survey on semi-supervised learning. *Machine Learning*, 109(2):373–440, 2020.

77 Jinbao Wang, Shujie Tan, Xiantong Zhen, Shuo Xu, Feng Zheng, Zhenyu He, and Ling Shao. Deep 3D human pose estimation: A review. *Computer Vision and Image Understanding*, 210:103225, 2021.

78 Qiyue Wang, Wenhua Jiao, Peng Wang, and YuMing Zhang. Digital twin for human-robot interactive welding and welder behavior analysis. *IEEE/CAA Journal of Automatica Sinica*, 8(2):334–343, 2021.

79 Shupeng Wang, Shouming Sun, Xiaojie Wang, Zhaolong Ning, and Joel J. P. C. Rodrigues. Secure crowdsensing in 5G internet of vehicles: When deep reinforcement learning meets blockchain. *IEEE Consumer Electronics Magazine*, 10(5):72–81, 2021.

80 Jiasi Weng, Jian Weng, Jilian Zhang, Ming Li, Yue Zhang, and Weiqi Luo. DeepChain: Auditable and privacy-preserving deep learning with blockchain-based incentive. *IEEE Transactions on Dependable and Secure Computing*, 18(5):2438–2455, 2021.

81 Georgios N. Yannakakis and Julian Togelius. A panorama of artificial and computational intelligence in games. *IEEE Transactions on Computational Intelligence and AI in Games*, 7(4):317–335, 2015.

82 Jianpeng Zhang, Yutong Xie, Yan Wang, and Yong Xia. Inter-slice context residual learning for 3D medical image segmentation. *IEEE Transactions on Medical Imaging*, 40(2):661–672, 2021.

83 Fenghua Zhu, Yisheng Lv, Yuanyuan Chen, Xiao Wang, Gang Xiong, and Fei-Yue Wang. Parallel transportation systems: Toward IoT-enabled smart urban traffic control and management. *IEEE Transactions on Intelligent Transportation Systems*, 21(10):4063–4071, 2020.

6

Virtual/Augmented/Mixed Reality Technologies for Enabling Metaverse

Howe Yuan Zhu and Chin-Teng Lin

Computational Intelligence and Brain-Computer Interface, Australian Artificial Intelligence Institute, University of Technology Sydney, Ultimo, NSW, Australia

After reading this chapter, you should be able to:

- Identify the main challenges hindering the realization of Metaverse and the recent innovations that may enable the Metaverse.
- Explain the key technologies in Virtual/Augmented/Mixed Reality systems for the visualization and user interfacing within the Metaverse framework.
- Understand the future works and challenges of Virtual/Augmented/Mixed Reality technology within the Metaverse.

6.1 Introduction

6.1.1 Into the Metaverse: Human Consciousness in a Virtual World

The Metaverse is a framework for multimodal user access to a singular, immersive, perpetual, and social virtual world [70]. Two essential technologies enable the Metaverse. The first is the innovations in communication and networking technology such as the Internet of Things (IoT) [73], Edge Computing [104], Blockchain technology, and artificial intelligence (AI) [107]. The second is the emergence and popularity of portable virtual reality (VR), augmented reality (AR), and mixed reality (MR) technologies that enable the user to project their consciousness into a virtual environment [70]. In this chapter, we will discuss the importance and recent innovations of VR/AR/MR technologies for enabling Metaverse and further explore the emerging challenges that arise from these innovations.

Antti Renvonsuo, a cognitive neuroscientist and psychologist, proposed that dreams are a form of conscious experience. Revonsuo [86] asserted that

Metaverse Communication and Computing Networks: Applications, Technologies, and Approaches, First Edition.
Edited by Dinh Thai Hoang, Diep N. Nguyen, Cong T. Nguyen, Ekram Hossain, and Dusit Niyato.
© 2024 The Institute of Electrical and Electronics Engineers, Inc. Published 2024 by John Wiley & Sons, Inc.

human consciousness is associated with sensory input and motor output. When investigating a person's visual cortex activity and eye movement, Renvonsuo rationalized that within a dream state, the person is experiencing a sensory input and exerting a motor output (eye movement) and thus projecting their consciousness into the dream. Interestingly, Renvonsuo equated a dream to a mentally constructed "subjective virtual reality" and the human consciousness is transferable into this VR [86]. In the awake state, recreational video games are an excellent example of a person projecting their consciousness into a virtual environment [29]. With a suitable video game, a player could be completely immersed in a virtual environment and achieve a sense of consciousness through the visual and auditory stimuli of the game and the haptic motor output of the haptic user interface. Sanchez-Vives and Slater [88] furthered this discussion by identifying the user's immersion and presence as core factors to the successful transportation of consciousness through VR technology. The authors posed that a VR system convinces the user that they are present within the environment through multisensory input (visual, auditory, and haptic) and provide the user with a sense of agency and embodiment (ability to move and interact with the virtual environment).

The Metaverse framework furthers the concept of transference of human consciousness by scaling VR into a virtual world. Rather than individual and temporary virtual environments, the Metaverse is a singular virtual world where a user can experience a conscious sense of presence while maintaining the ability to switch between the virtual and real world [21]. Dionisio et al. [21] outlined immersive realism and the user's ubiquity of access as essential factors to the Metaverse. As depicted in Figure 6.1, the ubiquity of access is the user's ability to enter the

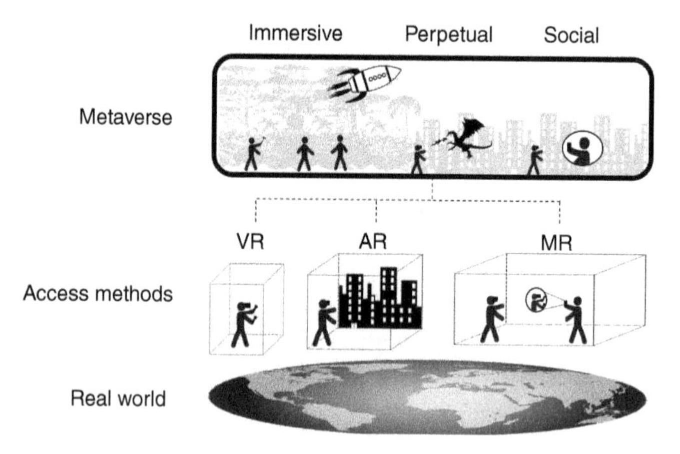

Figure 6.1 The real world and the Metaverse. An overview of the relationship between the user's in the real world and their projected consciousness in the Metaverse.

Metaverse through multiple access points such as a VR headset, AR glasses, or a handheld phone (MR) [76]. The implication is that all the access points must provide a similar level of user immersion and a sense of presence [70]. If a genuinely immersive Metaverse is realized, it could potentially bridge the digital conscious state to the real-world physical consciousness and create a continuity between the two worlds.

6.1.2 VR/AR/MR Challenges

The actualization of the Metaverse is hindered by a wide array of challenges, such as cybersecurity, mental and physical health, technological, ethics, privacy, immersion, intuitive interfacing, and inconsistent user experience [70, 76]. From the VR/AR/MR principles and challenges established by Sanchez-Vives and Slater [88] and Dionisio et al. [21], this chapter will focus on three challenges, user immersion, metaverse continuity, and user interaction, for each method of metaverse access.

The immersive realism of the Metaverse is the primary factor in the transference of user consciousness. The major challenge in the Metaverse context is for each modality (VR/AR/MR) to provide a similar level of immersion [21] to the user. From the principles asserted by Lenggenhager et al. [54] and Bowman and McMahan [10], immersion can be achieved by providing the users with realistic multisensory inputs (visual, auditory, olfactory, taste, and haptic) that are consistent with the current virtual environment and allow the users to use their motors functions (walking and tactile) to interact with the environment. Immersion is typically achieved through functional capabilities of the technology, such as the devices' visual rendering capabilities, the wearable device's comfort, the physical environment's ability to generate sensory feedback, and the intuitive design of the user interface, and Jennett et al. [41]. The ultimate goal is to provide users with a system that can convince them that they are within the virtual world.

The second challenge is maintaining a sense of continuity within the real world and the Metaverse. In video games and VR scenes, the sense of continuity is lost when the user switches between various virtual environments (changing games) or returns to the real world. For the Metaverse, the user must be able to stay within the virtual world and move to various parts of the world with continuity and singular identity of the Metaverse world [21]. The other factor is the continuity between transitioning between the virtual and the real world. The real world provides a conscious physical embodiment [78]. Likewise, the Metaverse needs to be a continuous world where the user can transition between lives through virtual embodiment. In essence, the embodiment provides a perpetual narrative within the Metaverse to be a living virtual world rather than a self-contained local space [89]. The goal of continuity is to provide a perpetual virtual world with a distinct identity that operates in parallel to the real world [103].

The third challenge is enabling the user to perform meaningful interactions with the virtual world and other users. The ability to interact with the world provides the user with a sense of agency and presence within the virtual world [88]. First, the user should be able to intuitively interact with their surrounding environment, such as picking up objects, moving scenery, or interacting with screens/user interfaces. The ability to interact with the surrounding environment can enhance the user's sense of presence and realism within the virtual world [94]. Second, the Metaverse world is a singular shared virtual space. Therefore, the users must be able to visualize and socially interact with other users both within the Metaverse (VR) and the real world (AR/MR) [109]. With the COVID-19 isolation restrictions across the globe, the online virtual conferencing market has dramatically increased in demand and popularity [82]. This increase in popularity broadens the scope of discussion for the user presence in social interactions and communications, with each technology having several benefits and drawbacks. For example, VR has dramatically aided in remote education [80] but may marginalize rural students with limited accessibility to VR devices [83]. The Metaverse's challenge is providing effective and personalized means of environmental and social interactions to the user across the VR/AR/MR platforms for ubiquitous access to the Metaverse.

In this chapter, we will outline the VR/AR/MR technologies and discuss the impact of each technology on enabling Metaverse. In each section, we will outline the definition and progressive development of the technology to the current state of the art. Each technology will be assessed by the three challenges outlined in Figure 6.2. We will discuss the ongoing research within the respective technologies and how they may overcome the current challenges to enable Metaverse.

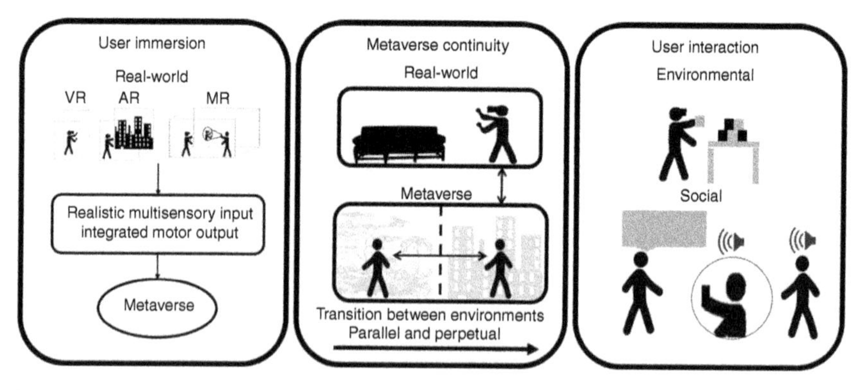

Figure 6.2 Three Challenges for VR/AR/MR in enabling Metaverse. An overview of the three challenges that will be focused upon for this chapter.

Many works featured in this chapter will be from the Computational Intelligence and Brain-Computer Interface (CIBCI) within the Australian Artificial Intelligence Institute (AAII) of the University of Technology Sydney (UTS).

6.2 Virtual Reality

6.2.1 Technology Overview

VR is the simulated experience when a person projects their consciousness from the real world to an artificially created space. One of the earliest examples of VR is the Victorian Era stereoscope. Viewers would put their eyes in front of two lenses in wooden goggles and visualize a still photograph or picture. In modern times, VR is more closely associated with a consumer's head-mounted display (HMD) that can display a virtual environment through a high-definition screen directly to the wearer's eyes [3]. VR offers the most immersive experience of the Metaverse compared to AR and MR systems [109]. With the appropriate hardware, a user could be wholly immersed in a virtual world with multisensory inputs and full body motor function usage [110]. As illustrated in Figure 6.3, VR technology has significantly progressed with multiple major innovations in the popularity and portability of HMD VR systems over the past decade [8].

The release of Oculus DK1 in 2012 was a significant milestone in VR HMD technology [20]. Even though other HMD VRs and projector-based VR (Cave VR [36], see Figure 6.3) were already on the market, the Oculus DK1 significantly increased the benchmark of VR devices with improvements in head tracking, pixel resolution, screen latency, reducing motion blur, the field of view, and stereoscopic 3D rendering [20]. Since then, the VR HMD market saw gradual improvement in rendering technology, tracking area (the physical space the wearer can move in), the field of view (FOV), tracking error, and software support [72]. Between 2012 and 2018, the trend of the VR HMD design was to create

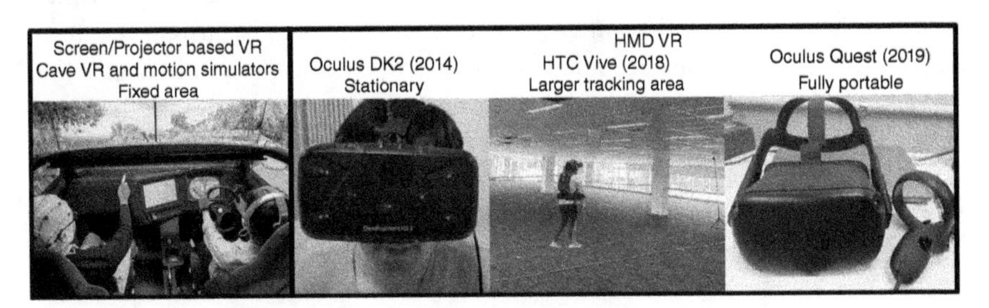

Figure 6.3 An overview of the progress of VR technology. Four examples (Cave/Motion Simulator VR, Oculus DK2, HTC Vive Pro, and Oculus Quest) of successful consumer VR systems, and how these systems have progressed over time.

highly capable devices with external optical cameras and trackers to track a user's head, body, and spatial location. Works such as Liu et al. [56–58] demonstrated that using a VR backpack and a sufficient amount HTC Vive base station (external camera) (e.g. 4 per $10\,m^2$), one could create an extremely large tracking area for walking. A common drawback of a highly capable VR HMD is the size and weight of the device. An example is the Varjo XR3 [2], which leads the market (as of 2022) in display resolution, tracking accuracy, peripheral features (eye tracking, hand tracking, and see-through mode), and FOV. However, the headset must be tethered by a cable (limited walking distance) and weighs 980 g. It can be argued that balancing the device weight can reduce the strain of the HMD weight. Studies such as Ito et al. [39], Rebenitsch and Owen [85], Yan et al. [105] and Kelly [45] have consistently agreed that overall HMD weight can contribute to user discomfort and faster muscle fatigue.

A more recent trend of VR HMD products is the shift from maximizing device capability to improving usability with a slightly lower performance. Headsets like the Oculus Quest (see Figure 6.3) [37] are designed to be completely wireless (onboard rendering), lightweight (503 g) and portable (no external sensors) through the use of Inside-Out tracking. These VR systems use integrated cameras within the HMD. Through computer vision, the cameras map the user's surrounding environment and convert it into a tracked space. This negates the need for external devices for head/spatial tracking and assists with obstacle avoidance and see-through mode [32]. Another recent feature in consumer VR systems is the inclusion of foveated rendering. Traditionally, a VR real-time rendering will render the entire virtual environment on display. Foveated rendering reasons that a person's eye gaze (or fovea) does not perceive the entire VR display, and thus only the areas that are gazed upon require high-quality rendering (peripheral vision regions are blurred) [77]. Using VR-integrated eye tracking, the VR rendering can be made more efficient by selectively rendering the center gaze region of the user instead of the entire display.

The advancements in VR HMD technology in the past decade have significantly furthered the research field toward a Metaverse world [109]. The current fidelity of display rendering, tracking accuracy, portability, and device weight indicates that the VR HMD hardware is already capable of enabling the visualization of the Metaverse. On the other hand, the challenges for VR Metaverse are primarily in user sensory input and walking in the virtual world (User Immersion), the longevity of the user to stay in the virtual world (Metaverse Continuity), and building meaningful methods of social interaction/communication (user interaction).

6.2.2 User Immersion

The ultimate goal for immersion of a VR system when entering the Metaverse is for the user to experience sensory congruence between the real-world senses and perform motor functions in the virtual world in the same manner as the real world. To achieve this, the VR system (HMD and peripheral) should aim to provide as many natural sensory stimuli to the user and improve various motor functions such as locomotion [70].

The human sensory perception is a complex system that includes visual, auditory, haptic (touch), olfactory (smell), proprioception (joint/limb/muscle sensation), temperature, and taste stimuli. The majority of VR HMDs provide the basic sensory input of visual (from the display) and auditory (onboard speakers) stimuli [8]. Previous studies have explored adding haptic feedback [111], olfactory [64], taste [46], and thermal feedback [6] to a VR system. However, one key issue is the limitation of the sensation range of each technology compared to the variety of potential environments in the Metaverse. As observed in Figure 6.4 and the study Zhu et al. [111], environmental haptic feedback can be extremely effective in providing an additional layer of realism by simulating the expected tactile sensation. On the other hand, haptic feedback requires the physical objects to be constructed and calibrated to a specific virtual environment. In the case of a Metaverse world, the users would constantly need to break their immersion to set up the haptic feedback when moving between virtual environments. Similarly, olfactory, taste, and thermal feedback hardware can only provide a specific and limited range of sensations.

Conversely, proprioception is an achievable and essential feature in a VR system. Human proprioception is a function of the somatosensory input (feeling the skin sensations) and visual perception of seeing their limbs [99]. This mechanism plays a vital role in essential motor functions such as reaching for an object or walking around an environment. Likewise, when a user accesses the Metaverse, the VR system should continue to provide stimuli for the user's proprioception. The somatosensory input remains consistent between the real world and the Metaverse. Therefore, the VR system will need to accurately visualize the user's limbs to provide proprioception [54]. This visualization can be achieved through a virtual avatar system and motion capture [30]. Generally, Motion capture requires external camera-based sensors or wearable trackers [66]. The motion capture system can calculate the user's limb positions by detecting

Figure 6.4 An example of haptic feedback with a real-time virtual avatar. The images are from a study by [111]. Zhu et al. 2021 / with permission of IEEE which explored the effects of physical elevation during virtual height exposure.

the joint positions (either wearing a tracker or computer vision) and using inverse kinematic [75]. The VR system will then use this information to visualize a virtual avatar for the user. Traditionally, joint detection is achieved through wearable markers, as shown in Figure 6.4, which uses six trackers (two hands, two feet, one waist, and one head) inverse kinematic avatar. With the improvements in markerless full body camera tracking [69], future virtual avatars may only require a single external camera instead of multiple worn trackers. In addition to the user's immersion, proprioception can also improve the efficiency of the user locomotion [62] and nonverbal communication [61].

6.2.3 Metaverse Continuity

The user's sense of continuity is crucial in maintaining the parallel between the real world and Metaverse [70]. When using a VR system, there is a distinct transfer to the Metaverse compared to AR and MR systems. Therefore, the challenge for a VR user is the longevity of remaining within the Metaverse. In the daily use of a VR system, there could be many reasons to exit the VR HMD, including walking into obstacles, eating, bathroom breaks, and fatigue. While it is not unlikely and unhealthy to never leave the Metaverse, VR technology can be used to reduce the more trivial reasons, such as physical space limitations.

One unique challenge of the Metaverse is the continuity of virtual walking. The users of a VR system would likely be located within their own homes with limited physical space. The Metaverse is intended to exist at a scale similar (or larger) to the real world; thus, the user needs to be able to walk indefinitely within the virtual world. Traditional VR games use controller-based teleportation to solve this issue [11]. Using a handheld controller, the user can point, click, and teleport to their intended location rather than using physical movement. This mode of travel may be effective in a short-term (one game/session) situation. However, previous studies [16, 17, 67] have shown that the use of teleportation can increase the risk of VR sickness. Redirected walking is a technique that directly addresses the issue of limited physical walking space [92]. The redirected walking technique subtly alters the user's walking trajectory by rotating the virtual environment on the VR HMD. The user will not consciously notice the rotation in small increments; however, they would subconsciously adjust their walking to align with the path [91]. If done correctly, the user will continuously walk in a circle in the physical space while perceiving themselves as walking in a straight line in the virtual space, as illustrated in Figure 6.5.

There are two ongoing challenges to using redirected walking for Metaverse travel. First, redirected walking is not feasible in nonlinear/circular paths. Environments that require more complex navigation or obstacle avoidance cannot be simply mapped into a circle. Researchers such as Williams et al. [100]

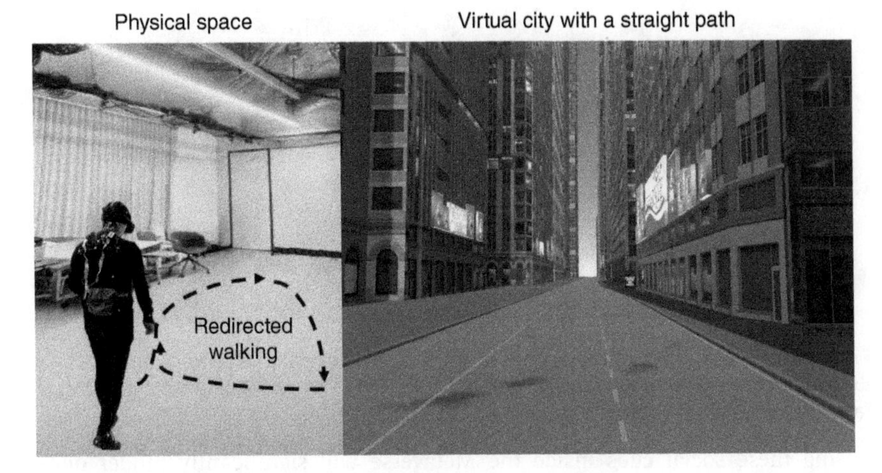

Figure 6.5 The concept of redirected walking. This figure depicts the concept of redirected walking which repaths the walking trajectory of users to allow them to travel long virtual distances while remaining in a relatively smaller physical space [92].

and Thomas and Rosenberg [97] have proposed algorithms and controllers to measure the path complexity of the virtual environment and remap the physical trajectory. However, these methods require more extensive testing to be feasible in a Metaverse context. The second challenge is the scenario of multiple VR users within one physical space [23]. The user's physical space, height, and stride length determine the trajectory during redirected walking. Collision is highly likely when two users share a physical space and perform redirected walking. Similar to the first ongoing challenge, there is no straightforward solution to this issue. This issue could potentially be solved with further research into using deep learning [53] or reinforcement learning [93] to create a more reactive redirected walking path planning system.

6.2.4 User Interaction

A core factor of the Metaverse is the singularity and cohabitation nature of the virtual world [70]. The user's ability to communicate and perform social interactions is crucial in creating a sense of agency and presence. Therefore, the VR system must provide the user with a means for verbal (speech), written (typing/text), and nonverbal (face and gesture) communication.

VR Facial tracking and visualization are essential in facilitating social interactions and communications. In the real-world, social conversations often involve listening to speech and visually observing the speaker's facial features for particular social cues such as the speaker's intentions or emotional state [96, 108].

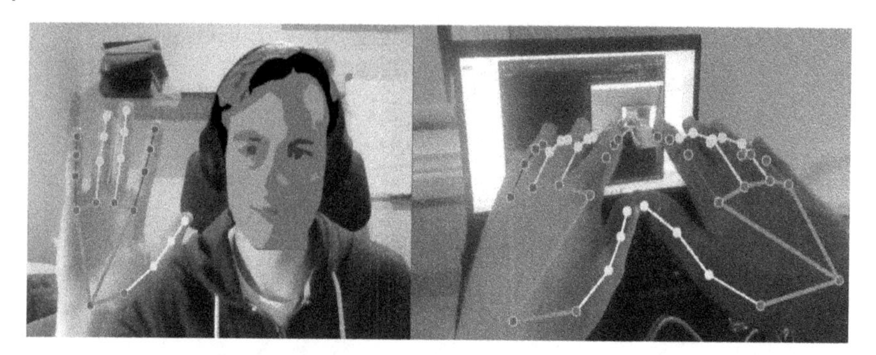

Figure 6.6 The concept of camera-based detection for hand and face tracking. An illustration of using computer vision to detect hand joints for typing and facial features to face visualization.

Losing these social cues inside the Metaverse will significantly hinder one's communication ability. To this end, certain modern VR HMDs, like the HP G2 Reverb Omnicept Edition [1], features an additional camera designed to track the facial features and mouth movements of the wearer [55]. The facial camera data provide the facial information, which can be rigged to the face of a virtual avatar and animated to match the wearer's face [1], the concept of this technique is depicted in Figure 6.6. A study by Kandalaft et al. [42] found that real-time facial rendering can improve one's ability to read social cues for verbal and nonverbal communication.

Written communication is another vital factor of VR communication. It can be used for indirect communication and relaying precise instructions to other users in the Metaverse. Finding a method of efficiently typing on a keyboard in VR is a widely explored area in ergonomics research [24]. Real world typing is a skill achieved through muscle memory and real-time visual feedback by observing the keyboard and screen. The challenge in VR is determining an effective method of typing information through a virtual keyboard. Most consumer VR systems use a controller point-and-click system for typing on a virtual keyboard [24]. This method is accurate. However, it is also slower when compared to physical keyboard typing [33]. Various works have suggested interesting alternatives, for example, using controllers like drumsticks for typing by Boletsis and Kongsvik [9], typing through gaze by Rajanna and Hansen [84], or typing through a haptic glove by Wu et al. [102]. With the advancement of hand tracking technology (as shown in Figure 6.6, researchers have begun to explore more intuitive methods of virtual keyboard typing [106]. Interestingly, recent studies by Kim and Xiong [48] found that even with hand tracking, the presence of haptic feedback significantly contributes to efficient typing. This is an ongoing open issue of finding a keyboard typing method that is intuitive and transferable between real-world skills and virtual typing.

6.2.5 Summary

In summary, VR technology has significantly progressed over the past decade. In the current state of the art, VR can enable a form of Metaverse access. However, the VR systems will need to address specific open and ongoing challenges for the user to experience a truly immersive, continuous, and meaningful Metaverse experience. Three ongoing challenges were determined in this section:

- Integration of hardware for a wider range of sensory stimuli enabling haptic, olfactory, taste, and temperature sensing.
- Explore and formulate a method for users to travel within a complex Metaverse environment in a smaller and more limited physical space.
- Finding an effective means of typing written information in the Metaverse.

6.3 Augmented Reality

6.3.1 Technology Overview

AR refers to the practice of visualizing an overlaid graphic onto the user's perception of the real world. AR retained the user's presence in the physical world with virtual additives and masks; in contrast, MR aims to blend the real and virtual world into a coexisting state [26]. As shown in Figure 6.7, AR technology is commonly divided into either a handheld interface or a wearable heads-up display (HUD) [7]. Handheld interfaces offer the benefits of accessibility as smartphones and tablets are readily available in most countries. Handheld AR devices often utilize the integrated camera on the device to capture the real-world environment and then render the appropriate visual graphics onto the device screen to create the

Figure 6.7 An overview of the various AR technologies available. The figure outlines the two types of AR devices. The first is handheld (phones and tablets) AR illustrating the use of AR for furniture shopping. The second type is wearable AR HUDs with two different models shown.

AR display. The wearable AR HUD offers the user a hands-free solution. The user will wear AR glasses with a translucent display and a projector when using an AR HUD. The projector would project a visual overlay onto the display, which creates the AR display. The current trend of wearable AR technology is to include an additional front-facing camera that can, by computer vision, detect hands, surrounding obstacles, and other people [47]. Wearable HUD ARs have succeeded in the manufacturing and assembly industry by improving worker training and efficiency [19, 79].

Unlike VR, the AR user's primary sense of presence remains within the real world. This yields unique challenges in the context of the Metaverse as the AR access point does not require a complete immersion into the virtual world [48]. Without the need for complete immersion, the challenge is for the AR system visual and convey meaningful and adaptive information for both the real world and the Metaverse (user immersion). Another challenge is maintaining the user's constant parallel embodiment in the real world and the Metaverse (Metaverse Continuity). The third challenge is cultivating Metaverse interactions while being present in the real world (User Interaction).

6.3.2 User Immersion

For the AR user's immersion, AR HUD's burden is determining the appropriate presentation of information to enhance the user's sensory perception without inhibiting or obstructing the sensory inputs from the real world. Studies by Jeffri and Rambli [40] and Woodward and Ruiz [101] have shown that certain AR HUDs could improve a user's situational awareness and cognitive workload. Conversely, administering too much information or ambiguous/erroneous can distract or hinder the user's sensory perception and adversely affect their cognitive state. In the context of the Metaverse, AR systems need to convey meaningful information on the virtual world's state without inhibiting the user's real-world functions.

One example of providing a more seamless AR HUD is the concept of displaying a Virtual Global Landmark (VGL) proposed by Singh et al. [90]. As illustrated in Figure 6.8, the authors proposed that using VGL would provide a less intrusive method of using an AR HUD to assist navigation. Later studies [56–58] found that VGL encourages and improves spatial learning while not unnecessarily increasing the user's mental workload. Other works have also explored different forms of overlaying information, such as using text annotations [5], providing building outlines/schematic information [38], and using color cues [22]. These display methodologies are crucial to Metaverse access as they allow the presentation of the Metaverse without intruding into the user's flow within the real world.

Figure 6.8 An illustration of an AR Virtual Global Landmark Concept. The figure depicts the concept proposed by Singh et al. [90] from the CIBCI lab, AAII, UTS. By using a VGL marker, the user can navigate to a target goal without the use of map HUD or other more intrusive AR indicators.

6.3.3 Metaverse Continuity

Following AR immersion, the parallel continuity between the real world and the Metaverse is another critical aspect. When using an AR system, the user is experiencing a parallel presence when they are physically not only present and active in the real world but also partially present in the Metaverse. In order to truly achieve a parallel presence between the Metaverse and the real world, the AR system will need to track the user's movements, actions, and localized positions and then mirror the digital behavior in the Metaverse.

This challenge could be addressed by creating a digital twin for an AR user. The concept of a digital twin is to digitize a parallel version of a real-world system or object [49]. The digital twin can actively mirror its real-world counterpart through sensory data. With the progress of reinforcement learning and deep learning, a digital twin can be used to stimulate or predict in real-time potential hazards, faults, or anomalous behaviours [68]. Figure 6.9 presents one example of a digital twin for a drone and the AR user. The work by Moya et al. [68] demonstrated that a digital twin for a drone is feasible and that a VR user would remotely pilot a physical drone. As outlined in Section 6.2.2, with the innovation in body, facial, gesture, and movement tracking, it is possible to generate a human digital twin that mirrors the AR user within the Metaverse. A human digital twin within the Metaverse will improve the connectivity between the real-world user and their digital counterpart

Figure 6.9 An example of using mirror AR interaction in a Virtual World. The figure illustrates an AR user controlling the drone in the real world while a collaborative user observes in VR, based on the VR work by Le et al. [52].

within the Metaverse. This functionality can also support other Metaverse-related interactions such as user collaboration, conferencing, presentations, and educational training/demonstrations [25].

6.3.4 User Interaction

The third challenge is the need for meaningful AR Metaverse interactions that can be completed in the real world. In essence, any Metaverse task in AR can be considered a dual task as the user will need to perform the task both in the context of the real and virtual world. One example was proposed by MacCallum and Parsons [59], who suggested that AR can be used for hybrid teaching (remote and face-to-face classes). In a hybrid educational setting, the teacher would utilize the AR HUD to monitor the remote-learning students and the face-to-face students. Kye et al. [50] outlined an ongoing issue with AR in education is the difficulty of social interaction and a sense of community within the classroom for remote students. The educators will likely be biased toward the physically present student compared to remote students.

Even though the AR HUD fosters the duality between the real world and the Metaverse, the real world will always be dominant. An AR system will need to provide the user with intuitive and equivalent interaction methods. The currently marketed AR devices will typically allow for hand-tracked control (gesture, pinch, and touch), controller interfacing, and voice control [14, 65]. Similar to Section 6.2.4, the implementation of features such as facial tracking can significantly improve the social interaction between AR and remote VR users by providing a similar level of information and control as in the real world. Another suggestion by Maisto et al. [60] is to include wearable haptic devices to provide the sensation of physical touch. This feature is helpful in a remote learning environment as learning is a multisensory activity with seeing, hearing, and hands-on activities MacCallum and Parsons [59].

6.3.5 Summary

In summary, AR technology is a valuable method of accessing the Metaverse in daily life in the real world. The technology can enable users in the real world to interact with virtual objects and other users in the Metaverse. The AR HUD plays a crucial role in visualizing the Metaverse features in the real world. Future developers should consider the appropriate amount of information to display to the user and use techniques such as VGL and subtle visual cues to avoid unnecessarily distracting the AR wearer from the real world. Using the sensing technology on the AR system, one can also create a human digital twin within the Metaverse. This forms a continuity and ensures the real-world actions are accurately paralleled in the Metaverse. Overall, AR technology has multiple potential applications (education, training, and industry) that can be enabled inside the Metaverse. Three ongoing challenges were determined in this section:

- Further research and development into actualizing a parallel Metaverse world for the AR system access and form a digital twin.
- The AR system needs to provide more intuitive and natural controls to reduce the user's bias toward real-world interactions.
- Further explore methods of multisensory integration to enhance the collaborative and social experience between AR users and other Metaverse users.

6.4 Mixed Reality

6.4.1 Technology Overview

MR is an emerging technology that aims to merge and blend the real and virtual world into a singular mixed presentation [18]. In contrast, AR is predominantly in the real world, and VR is within the virtual world. MR aims to equally fuse the two worlds to create a hybrid representation of both. Fundamentally, MR is a superset framework that involves using VR, AR, IoT, computer vision, and machine learning [87]. One popular method of creating an MR environment is using the pass-through VR technique to visualize the real world through the VR HMD and blend virtual elements into the presentation [51]. Pass-through VR is enabled through a high-quality integrated camera on the VR headset and high-resolution rendering to present a photorealistic image of the environment. Figure 6.10 illustrates one example of pass-through VR being used for MR shopping. The MR user would wear a portable VR headset with a pass-through feature to travel naturally in the real world. Then, at a storefront, they could virtually enter the store with a remote store clerk virtually assisting the user in their store. Examples of VR headsets that support pass-through (or XR mode) are the Varjo XR3 [2] and the Lynx-R1 [51]. Another popular MR visualization method uses projectors to

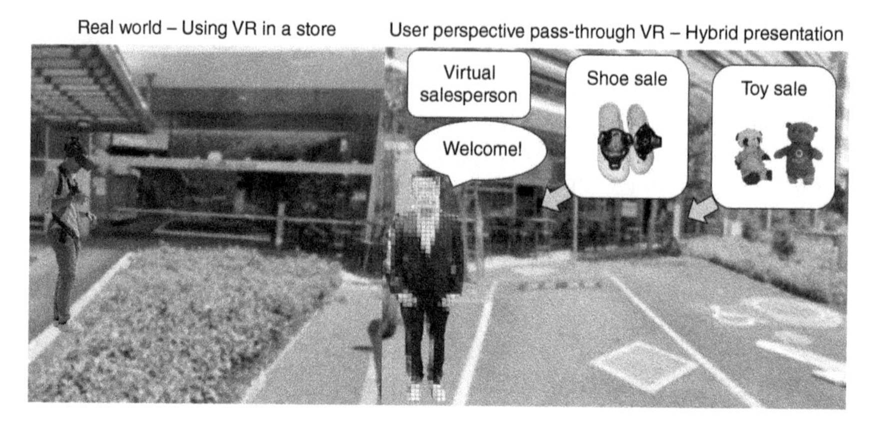

Figure 6.10 An example of MR through pass-through VR. An illustration of hybrid shopping using pass-through VR which is a popular method of presentation for MR applications.

display virtual visuals directly onto real-world surfaces. These projectors could be static [74] or dynamically moving [43]. This method will limit the visualization to a 2D display surface (until 3D holographic projectors are available) but holds the benefit of being less intrusive to the user compared to wearable VR and AR devices.

In the context of the Metaverse, the challenge with MR is to determine an appropriate method of fusing the two parallel worlds (real world and Metaverse) into a singular cohesive presentation. As the MR and Metaverse field is relatively new, many challenges must be solved before MR access to the Metaverse is possible. One challenge is the quality and method of the Metaverse visuals when merged with the real world (user immersion). Metaverse visualisation should be distinctive and blended with the user's real-world perception. Another challenge is achieving an accurate synchronization and map environment between the real and virtual worlds (metaverse continuity). The third challenge is the ambiguity of collaborative and social interactions between the real world and the Metaverse between real world (non-Metaverse) people and Metaverse users (User Interaction).

6.4.2 User Immersion

Photorealism in VR is considered the ultimate goal of the rendering/display hardware progression [12]. This is especially important when using pass-through VR as it directly renders a video stream as lower resolution quality results in poor acuity. Studies by Tirado Cortes et al. [98] and Fernandes and Feiner [27] found that the display FOV, framerate, pixel resolution, and correspondence with movement significantly contribute to the user's susceptibility to VR sickness. This poses an ongoing technical challenge to provide a display solution that has a high-quality integrated camera, high-resolution display, low latency rendering, and accurate motion tracking. The MR system for Metaverse access would not be feasible if the user is constantly experiencing VR sickness or cannot be effectively used in the real world (VR only). The Varjo XR3 [2] and the Lynx-R1. The pass-through solution by the Varjo XR3 is superior to the Lynx-R1 in display quality (higher resolution) and camera sensors (additional Lidar for depth sensing). However, the Lynx-R1 is half the weight of the Varjo XR3 and is wireless, which may be more comfortable for users. Overall, the commonality in both headsets is the use of dual cameras (one per eye), hand tracking, a high refresh rate (90 Hz), and a high-resolution display of at least (1600 by 1600 pixels per eye). It is apparent in the current state of the art that there is a trade-off between device capability and device weight. The ongoing challenge is the need for an MR device that can support a photorealistic display that is comfortable for the user (low weight and wireless).

6.4.3 Metaverse Continuity

Similar to AR displays, synchronicity between the real world and the Metaverse is important when displaying a hybrid presentation. Like the real world, the Metaverse should be considered a continuous but parallel world. A simple task such as shopping would not be feasible if the virtual catalogue items were not appropriately located within the store when using the MR system. Therefore, it is essential for the MR to accurately generate a spatial map of the surrounding real-world environment to merge the Metaverse environment accurately. This problem could be solved using VSLAM (Visual Simultaneous Localisation and Mapping) to survey the user's surrounding environment [13]. VSLAM is a technique that uses only visual data from a camera to map and construct a structure in an unknown environment and determine the user's position in that environment [95]. This technique is heavily researched in the AR field with works by Polvi et al. [81] and Gao et al. [31], finding success in detecting the environment and the position of the AR system. The MR system can be used with other sensors such as GPS, head position, and hand position to map the real world's surroundings and interactive objects and give the user agency in interacting with the environment. By actively surveying the environment, the Metaverse can continually update based on the changes in the real world (such as moving objects or entering a new room) and allow users to interact with virtual objects and users in the correct spatial location.

6.4.4 User Interaction

MR intends to provide users with the ability to interact actively with both the physical and virtual worlds. An example is shopping in AR would look like a user in the

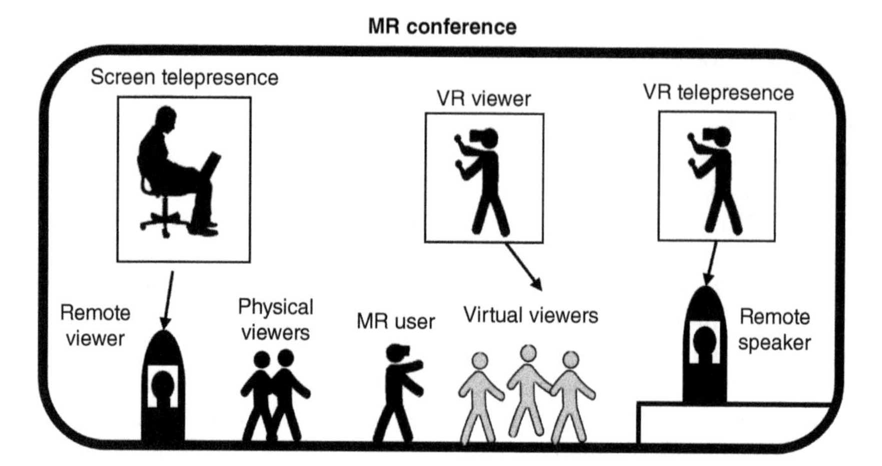

Figure 6.11 A conceptual figure of an MR Conference. This figure illustrates an example of a hybrid conference where users can use multiple methods of joining the conference and an MR user can visualize every participant.

real world visualising certain virtual shopping items in their HUD [63]. Shopping in MR would look like a customer walking into a hybrid (real and virtual) shop with distinct virtual shopping areas combined with users from the real world and Metaverse [28]. MR interaction allows users to interact simultaneously with the real world and the Metaverse.

Figure 6.11 illustrates an example of a meaningful method of using MR for conferencing. In light of the COVID-19 pandemic, many conferences have explored MR methods for remote conferencing [4]. This problem led to the concept of hybrid conferences where individuals could join through remote and face-to-face solutions. One innovation of the idea of using robots for remote telepresence [44]. By using a digital device such as a personal computer or VR HMD, a user could embody a physical robot present at the conference. The robot will provide visual and audio stimuli to the user to provide a sense of presence within the virtual world. Local Remote guests could also virtually join to interact with other virtual guests. One current limitation of hybrid conferences is the lack of interaction between physical and virtual guests [35]. This is often due to the variance and technical difficulty of creating a method for local guests to interact socially with virtual guests. Telepresence robots could address this issue. However, they are often costly (purchase and maintenance) and unreliable (collisions, falling, and breaking) [71]. MR technology could solve this problem by allowing local guests to visualize and interact with the virtual guests. Similar examples of MR technology being used for hybrid presentations can be found in the tourism [34] and the gaming industry [15].

6.4.5 Summary

In summary, MR technology is still in its early stages of conception and ambiguity in the potential forms of MR systems. The current trend is using pass-through VR technology and other wearable devices to generate a photorealistic rendering of both the real and virtual worlds to create a hybrid presentation. This form of hybrid presentation can greatly benefit Metaverse users as it provides a bridge between non-Metaverse people, VR Metaverse users, and AR/MR users. In this section, we have identified three key challenges that hinder the possibility of using MR to enable Metaverse:

- The limited availability due to the cost, weight, and device capability of the current technology.
- Further, improvements in VSLAM algorithms and technology are to be used continuously to synchronize the user's surroundings between the real world and the Metaverse.
- To build a reliable and effective infrastructure to enable the use of MR in various situations to avoid segregation of real-world and virtual-world users.

6.5 Conclusions and Future Research Directions

In this chapter, we explore the potential of using VR/AR/MR to enable users to access the Metaverse. Each technology was assessed against the three key Metaverse challenges of User Immersion, Metaverse Continuity, and User Interaction (outlined in Figure 6.2). We evaluated each technology's development progress and the steps required to enable or enhance the Metaverse.

The current state of VR technology is at a well-developed stage to enable access to the Metaverse. With recent innovations such as full body motion tracking, redirected walking, and integrated hand/face tracking, a VR wearer, can experience a fully immersive Metaverse world while naturally moving in smaller real world space (e.g. indoor room). The main challenges of VR HMD technology are to minimize the device's weight, improve the comfort of the device, and improve the virtual environment design for typing and walking within the Metaverse.

Similarly, AR technology is also well developed and in a stage to allow for a window into the Metaverse while the user remains in the real world. Through careful visualization selection, the AR display can blend Metaverse elements into the real world without a break in immersion. Additionally, incorporating human digital twin and building a more comprehensive range of AR interactions will further improve the continuity and user interaction between the real world and the Metaverse. The critical challenge of AR HUDs is to further actualize the digital twin concept and build more intuitive methods of integrating user controls.

On the other hand, MR technology is an emerging technology needing further development and clarity. The ultimate goal of MR technology is to create a multi-modal platform that blends/mixes the real world with the virtual world. MR development will likely be parallel to the Metaverse due to similarity in the end goal. The Metaverse will serve as the next step of a singular and continuous virtual world that is constantly blended with the real world. The current developments with see-through VR, VSLAM AR HUDs, and real-world MR events (e.g. conferences) provide a good indication that the Metaverse is possible with further development of MR technology.

The Metaverse can offer a new medium for human interaction in a shared virtual space. As VR/AR/MR devices improve in accessibility, portability, rendering quality, and sensory capabilities; we will likely see increased popularity for the Metaverse. There are many exciting opportunities to create novel virtual environments such as education and training centers, vendors (shopping and trying simulated products online), gyms, workplaces, and social hubs. In conclusion, the Metaverse can be realized with further development of VR/AR/MR technologies. We suspect that as the Metaverse and the surrounding develops, there will be emerging issues such as privacy, equitable accessibility (e.g. people with disabilities), addiction (gaming addiction), and mental health (withdrawal). Another factor is the emerging applications of the Metaverse in new areas such as education (virtual schools), sports (Metaverse games), industry work (telepresence and online workspace), and shared environments (e.g. walking in a virtual park). Once actualized, the Metaverse will be an unprecedented new frontier that will likely reshape how people connect and socially interact.

Acknowledgments

We would like to thank the students and alumni of the Computational Intelligence and Brain–Computer Interface at the Australian Artificial Intelligence Institute of the University of Technology Sydney. Many of their research projects are featured in this chapter. We would specifically like to acknowledge the following CIBCI members, Avinash Kumar Singh, Carlos Alfredo Tirado Cortes, Tien-Thong Nguyen Do, Jia Liu, Alka Rachel John, Daniel Leong, Braeden Knox, and Li-Ting Tsai, whose work were referenced in this chapter. We would also like to acknowledge that this work was supported in part by the Australian Research Council (ARC) under discovery grants DP180100670, DP180100656, DP210101093, and DP220100803. The research was also sponsored in part by the Australia Defence Innovation Hub under Contract No. P18-650825, Australian Cooperative Research Centers Projects (CRC-P) Round 11 CRCPXI000007, Lockheed Martin Corporation under Cooperative Agreement Number 4038,

US Office of Naval Research Global under Cooperative Agreement Number ONRG - NICOP - N62909-19-1-2058, and AFOSR - DST Australian Autonomy Initiative agreement ID10134. We also thank the New South Wales (NSW) Defence Innovation Network and the NSW State Government of Australia for financial support in part of this research through grant DINPP2019 S1-03/09 and PP21-22.03.02.

Bibliography

1 HP Reverb G2 Omnicept Edition. URL https://www.hp.com/us-en/vr/reverb-g2-vr-headset-omnicept-edition.html.

2 Varjo XR-3 - the industry's highest resolution XR headset: Varjo, July 2022. URL https://varjo.com/products/xr-3/.

3 Christoph Anthes, Rubén Jesús García-Hernández, Markus Wiedemann, and Dieter Kranzlmüller. State of the art of virtual reality technology. In *2016 IEEE Aerospace Conference*, pages 1–19, 2016. doi: 10.1109/AERO.2016.7500674.

4 Vaibhav Bajpai, Oliver Hohlfeld, Jon Crowcroft, Srinivasan Keshav, Henning Schulzrinne, Jörg Ott, Simone Ferlin, Georg Carle, Andrew Hines, and Alexander Raake. Recommendations for designing hybrid conferences. *ACM SIGCOMM Computer Communication Review*, 52(2):63–69, April 2022. doi: 10.1145/3544912.3544920.

5 Blaine Bell, Tobias Höllerer, and Steven Feiner. An annotated situation-awareness aid for augmented reality. In *Proceedings of the 15th Annual ACM Symposium on User Interface Software and Technology - UIST '02*. ACM Press, 2002. doi: 10.1145/571985.572017.

6 M. Benali-Khoudjal, M. Hafez, J.-M. Alexandre, J. Benachour, and A. Kheddar. Thermal feedback model for virtual reality. In *MHS2003. Proceedings of 2003 International Symposium on Micromechatronics and Human Science (IEEE Cat. No.03TH8717)*, pages 153–158, 2003. doi: 10.1109/MHS.2003.1249925.

7 Mark Billinghurst, Adrian Clark, and Gun Lee. A survey of augmented reality. *Foundations and Trends® in Human–Computer Interaction*, 8(2–3):73–272, 2015. doi: 10.1561/1100000049.

8 Frank Biocca. Virtual reality technology: A tutorial. *Journal of Communication*, 42(4):23–72, December 1992. doi: 10.1111/j.1460-2466.1992.tb00811.x.

9 Costas Boletsis and Stian Kongsvik. Text input in virtual reality: A preliminary evaluation of the drum-like VR keyboard. *Technologies*, 7(2):31, April 2019. doi: 10.3390/technologies7020031.

10 Doug A. Bowman and Ryan P. McMahan. Virtual reality: How much immersion is enough? *Computer*, 40(7):36–43, 2007. doi: 10.1109/MC.2007.257.

11 Evren Bozgeyikli, Andrew Raij, Srinivas Katkoori, and Rajiv Dubey. Point & teleport locomotion technique for virtual reality. In *Proceedings of the 2016 Annual Symposium on Computer-Human Interaction in Play*. ACM, October 2016. doi: 10.1145/2967934.2968105.

12 Alan Chalmers and Andrej Ferko. Levels of realism. In *Proceedings of the 24th Spring Conference on Computer Graphics - SCCG '08*. ACM Press, 2010. doi: 10.1145/1921264.1921272.

13 Denis Chekhlov, Andrew P. Gee, Andrew Calway, and Walterio Mayol-Cuevas. Ninja on a plane: Automatic discovery of physical planes for augmented reality using visual SLAM. In *2007 6th IEEE and ACM International Symposium on Mixed and Augmented Reality*. IEEE, November 2007. doi: 10.1109/ismar.2007.4538840.

14 Yunqiang Chen, Qing Wang, Hong Chen, Xiaoyu Song, Hui Tang, and Mengxiao Tian. An overview of augmented reality technology. *Journal of Physics: Conference Series*, 1237(2):022082, June 2019. doi: 10.1088/1742-6596/1237/2/022082.

15 A. D. Cheok, Anuroop Sreekumar, C. Lei, and L. N. Thang. Capture the flag: Mixed-reality social gaming with smart phones. *IEEE Pervasive Computing*, 5(2):62–69, 2006. doi: 10.1109/MPRV.2006.25.

16 Jeremy Clifton and Stephen Palmisano. Effects of steering locomotion and teleporting on cybersickness and presence in HMD-based virtual reality. *Virtual Reality*, 24(3):453–468, November 2019. doi: 10.1007/s10055-019-00407-8.

17 Jeremy Clifton and Stephen Palmisano. Comfortable locomotion in VR: Teleportation is not a complete solution. In *25th ACM Symposium on Virtual Reality Software and Technology*. ACM, November 2019. doi: 10.1145/3359996.3364722.

18 Enrico Costanza, Andreas Kunz, and Morten Fjeld. Mixed reality: A survey. In *Lecture Notes in Computer Science*, pages 47–68. Springer, Berlin, Heidelberg, 2009. doi: 10.1007/978-3-642-00437-7_3.

19 Oscar Danielsson, Magnus Holm, and Anna Syberfeldt. Augmented reality smart glasses in industrial assembly: Current status and future challenges. *Journal of Industrial Information Integration*, 20:100175, December 2020. doi: 10.1016/j.jii.2020.100175.

20 Parth Rajesh Desai, Pooja Nikhil Desai, Komal Deepak Ajmera, and Khushbu Mehta. A review paper on oculus rift-a virtual reality headset, 2014. URL https://arxiv.org/abs/1408.1173.

21 John David N. Dionisio, William G. Burns III, and Richard Gilbert. 3D Virtual worlds and the metaverse. *ACM Computing Surveys*, 45(3):1–38, June 2013. doi: 10.1145/2480741.2480751.

22 Benjamin J. Dixon, Michael J. Daly, Harley Chan, Allan D. Vescan, Ian J. Witterick, and Jonathan C. Irish. Surgeons blinded by enhanced navigation: The effect of augmented reality on attention. *Surgical Endoscopy*, 27(2):454–461, July 2012. doi: 10.1007/s00464-012-2457-3.

23 Tianyang Dong, Yue Shen, Tieqi Gao, and Jing Fan. Dynamic density-based redirected walking towards multi-user virtual environments. In *2021 IEEE Virtual Reality and 3D User Interfaces (VR)*. IEEE, March 2021. doi: 10.1109/vr50410.2021.00088.

24 Tafadzwa Joseph Dube and Ahmed Sabbir Arif. Text entry in virtual reality: A comprehensive review of the literature. In *Human-Computer Interaction. Recognition and Interaction Technologies*, pages 419–437. Springer International Publishing, 2019. doi: 10.1007/978-3-030-22643-5_33.

25 Matt Dunleavy and Chris Dede. Augmented reality teaching and learning. In *Handbook of Research on Educational Communications and Technology*, pages 735–745. Springer, New York, May 2013. doi: 10.1007/978-1-4614-3185-5_59.

26 Mana Farshid, Jeannette Paschen, Theresa Eriksson, and Jan Kietzmann. Go boldly! *Business Horizons*, 61(5):657–663, September 2018. doi: 10.1016/j.bushor.2018.05.009.

27 Ajoy S. Fernandes and Steven K. Feiner. Combating VR sickness through subtle dynamic field-of-view modification. In *2016 IEEE Symposium on 3D User Interfaces (3DUI)*. IEEE, March 2016. doi: 10.1109/3dui.2016.7460053.

28 Carlos Flavián, Sergio Ibá nez-Sánchez, and Carlos Orús. The impact of virtual, augmented and mixed reality technologies on the customer experience. *Journal of Business Research*, 100:547–560, July 2019. doi: 10.1016/j.jbusres.2018.10.050.

29 Gonzalo Frasca. Rethinking agency and immersion: Video games as a means of consciousness-raising. *Digital Creativity*, 12(3):167–174, September 2001. doi: 10.1076/digc.12.3.167.3225.

30 Guo Freeman and Divine Maloney. Body, avatar, and me: The presentation and perception of self in social virtual reality. *Proceedings of the ACM on Human-Computer Interaction*, 4(CSCW3):1–27, January 2021. doi: 10.1145/3432938.

31 Qing Hong Gao, Tao Ruan Wan, Wen Tang, Long Chen, and Kai Bing Zhang. An improved augmented reality registration method based on visual SLAM. In *E-Learning and Games*, pages 11–19. Springer International Publishing, 2017. doi: 10.1007/978-3-319-65849-0_2.

32 Michael J. Gourlay and Robert T. Held. Head-mounted-display tracking for augmented and virtual reality. *Information Display*, 33(1):6–10, January 2017. doi: 10.1002/j.2637-496x.2017.tb00962.x.

33 Jens Grubert, Lukas Witzani, Eyal Ofek, Michel Pahud, Matthias Kranz, and Per Ola Kristensson. Text entry in immersive head-mounted display-based virtual reality using standard keyboards. In *2018 IEEE Conference on Virtual Reality and 3D User Interfaces (VR)*. IEEE, March 2018. doi: 10.1109/vr.2018.8446059.

34 Tony Hall, Holger Schnädelbach, Martin Flintham, Luigina Ciolfi, Liam Bannon, Mike Fraser, Steve Benford, John Bowers, Chris Greenhalgh, Sten-Olof Hellström, et al. The visitor as virtual archaeologist. In *Proceedings of the 2001 Conference on Virtual Reality, Archeology, and Cultural Heritage - VAST '01*. ACM Press, 2001. doi: 10.1145/584993.585008.

35 Simon Hamm, Elspeth Frew, and Clare Lade. Hybrid and virtual conferencing modes versus traditional face-to-face conference delivery: A conference industry perspective. *Event Management*, 22(5):717–733, October 2018. doi: 10.3727/152599518x15299559637635.

36 Paul Havig, John McIntire, and Eric Geiselman. Virtual reality in a cave: Limitations and the need for HMDs? In Peter L. Marasco and Paul R. Havig, editors, *SPIE Proceedings*. SPIE, May 2011. doi: 10.1117/12.883855.

37 Cornel Hillmann. Comparing the gear VR, oculus go, and oculus quest. In *Unreal for Mobile and Standalone VR*, pages 141–167. Apress, 2019. doi: 10.1007/978-1-4842-4360-2_5.

38 Javier Irizarry, Masoud Gheisari, Graceline Williams, and Bruce N. Walker. InfoSPOT: A mobile augmented reality method for accessing building information through a situation awareness approach. *Automation in Construction*, 33:11–23, August 2013. doi: 10.1016/j.autcon.2012.09.002.

39 Kodai Ito, Mitsunori Tada, Hiroyasu Ujike, and Keiichiro Hyodo. Effects of the weight and balance of head-mounted displays on physical load. *Applied Sciences*, 11(15):6802, July 2021. doi: 10.3390/app11156802.

40 Nor Farzana Syaza Jeffri and Dayang Rohaya Awang Rambli. A review of augmented reality systems and their effects on mental workload and task performance. *Heliyon*, 7(3):e06277, March 2021. doi: 10.1016/j.heliyon.2021.e06277.

41 Charlene Jennett, Anna L. Cox, Paul Cairns, Samira Dhoparee, Andrew Epps, Tim Tijs, and Alison Walton. Measuring and defining the experience of immersion in games. *International Journal of Human-Computer Studies*, 66(9):641–661, September 2008. doi: 10.1016/j.ijhcs.2008.04.004.

42 Michelle R. Kandalaft, Nyaz Didehbani, Daniel C. Krawczyk, Tandra T. Allen, and Sandra B. Chapman. Virtual reality social cognition training for young adults with high-functioning autism. *Journal of Autism and Developmental Disorders*, 43(1):34–44, May 2012. doi: 10.1007/s10803-012-1544-6.

43 T. Karitsuka and K. Sato. A wearable mixed reality with an on-board projector. In *The Second IEEE and ACM International Symposium on Mixed and Augmented Reality, 2003. Proceedings*. IEEE Computer Society. doi: 10.1109/ismar.2003.1240740.

44 Lisa Keller, Oliver Gawron, Tamin Rahi, Philipp Ulsamer, and Nicholas H. Müller. Driving success: Virtual team building through telepresence robots. In *Learning and Collaboration Technologies: Games and Virtual Environments for Learning*, pages 278–291. Springer International Publishing, 2021. doi: 10.1007/978-3-030-77943-6_18.

45 Jonathan W. Kelly. Distance perception in virtual reality: A meta-analysis of the effect of head-mounted display characteristics. *IEEE Transactions on Visualization and Computer Graphics*, 1–13, 2022. doi: 10.1109/TVCG.2022.3196606.

46 Erika Kerruish. Arranging sensations: Smell and taste in augmented and virtual reality. *The Senses and Society*, 14(1):31–45, January 2019. doi: 10.1080/17458927.2018.1556952.

47 Hojoong Kim, Young-Tae Kwon, Hyo-Ryoung Lim, Jong-Hoon Kim, Yun-Soung Kim, and Woon-Hong Yeo. Recent advances in wearable sensors and integrated functional devices for virtual and augmented reality applications. *Advanced Functional Materials*, 31(39):2005692, October 2020. doi: 10.1002/adfm.202005692.

48 Woojoo Kim and Shuping Xiong. Pseudo-haptics and self-haptics for freehand mid-air text entry in VR. *Applied Ergonomics*, 104:103819, October 2022. doi: 10.1016/j.apergo.2022.103819.

49 Mareike Kritzler, Markus Funk, Florian Michahelles, and Wolfgang Rohde. The virtual twin. In *Proceedings of the Seventh International Conference on the Internet of Things*. ACM, October 2017. doi: 10.1145/3131542.3140274.

50 Bokyung Kye, Nara Han, Eunji Kim, Yeonjeong Park, and Soyoung Jo. Educational applications of metaverse: Possibilities and limitations. *Journal of Educational Evaluation for Health Professions*, 18:32, December 2021. doi: 10.3352/jeehp.2021.18.32.

51 Stan Larroque. Digital pass-through head-mounted displays for mixed reality. *Information Display*, 37(4):17–21, July 2021. doi: 10.1002/msid.1228.

52 Nguyen Thanh Trung Le, Howe Yuan Zhu, and Hsiang-Ting Chen. Remote visual line-of-sight: A remote platform for the visualisation and control of an indoor drone using virtual reality. In *Proceedings of the 27th ACM Symposium on Virtual Reality Software and Technology*. ACM, December 2021. doi: 10.1145/3489849.3489910.

53 Dong-Yong Lee, Yong-Hun Cho, and In-Kwon Lee. Real-time optimal planning for redirected walking using deep q-learning. In *2019 IEEE Conference on Virtual Reality and 3D User Interfaces (VR)*. IEEE, March 2019. doi: 10.1109/vr.2019.8798121.

54 Bigna Lenggenhager, Tej Tadi, Thomas Metzinger, and Olaf Blanke. Video ergo sum: Manipulating bodily self-consciousness. *Science*, 317(5841):1096–1099, August 2007. doi: 10.1126/science.1143439.

55 Hao Li, Laura Trutoiu, Kyle Olszewski, Lingyu Wei, Tristan Trutna, Pei-Lun Hsieh, Aaron Nicholls, and Chongyang Ma. Facial performance sensing head-mounted display. *ACM Transactions on Graphics*, 34(4):1–9, July 2015. doi: 10.1145/2766939.

56 Jia Liu, Avinash Kumar Singh, and Chin-Teng Lin. Using virtual global landmark to improve incidental spatial learning. *Scientific Reports*, 12(1):6744, April 2022. doi: 10.1038/s41598-022-10855-z.

57 Jia Liu, Avinash Kumar Singh, and Chin-Teng Lin. Predicting the quality of spatial learning via virtual global landmarks. *IEEE Transactions on Neural Systems and Rehabilitation Engineering*, 30:2418–2425, 2022. doi: 10.1109/tnsre.2022.3199713.

58 Jia Liu, Avinash Kumar Singh, Anna Wunderlich, Klaus Gramann, and Chin-Teng Lin. Redesigning navigational aids using virtual global landmarks to improve spatial knowledge retrieval. *npj Science of Learning*, 7(1):17, July 2022. doi: 10.1038/s41539-022-00132-z.

59 Kathryn MacCallum and David Parsons. Teacher perspectives on mobile augmented reality: The potential of metaverse for learning. In *Proceedings of World Conference on Mobile and Contextual Learning 2019*, pages 21–28, September 2019. URL https://www.learntechlib.org/p/210597.

60 Maurizio Maisto, Claudio Pacchierotti, Francesco Chinello, Gionata Salvietti, Alessandro De Luca, and Domenico Prattichizzo. Evaluation of wearable haptic systems for the fingers in augmented reality applications. *IEEE Transactions on Haptics*, 10(4):511–522, October 2017. doi: 10.1109/toh.2017.2691328.

61 Divine Maloney, Guo Freeman, and Donghee Yvette Wohn. "Talking without a voice": understanding non-verbal communication in social virtual reality. *Proceedings of the ACM on Human-Computer Interaction*, 4(CSCW2):1–25, October 2020. doi: 10.1145/3415246.

62 Yusuke Matsuda, Junya Nakamura, Tomohiro Amemiya, Yasushi Ikei, and Michiteru Kitazaki. Enhancing virtual walking sensation using self-avatar in first-person perspective and foot vibrations. *Frontiers in Virtual Reality*, 2:654088, April 2021. doi: 10.3389/frvir.2021.654088.

63 Graeme McLean and Alan Wilson. Shopping in the digital world: Examining customer engagement through augmented reality mobile applications.

Computers in Human Behavior, 101:210–224, December 2019. doi: 10.1016/j.chb.2019.07.002.

64 Lorenzo Micaroni, Marina Carulli, Francesco Ferrise, Alberto Gallace, and Monica Bordegoni. An olfactory display to study the integration of vision and olfaction in a virtual reality environment. *Journal of Computing and Information Science in Engineering*, 19(3):031015, July 2019. doi: 10.1115/1.4043068.

65 Mark Roman Miller, Hanseul Jun, Fernanda Herrera, Jacob Yu Villa, Greg Welch, and Jeremy N. Bailenson. Social interaction in augmented reality. *PLoS ONE*, 14(5):e0216290, May 2019. doi: 10.1371/journal.pone.0216290.

66 Thomas B. Moeslund, Adrian Hilton, and Volker Krüger. A survey of advances in vision-based human motion capture and analysis. *Computer Vision and Image Understanding*, 104(2–3):90–126, November 2006. doi: 10.1016/j.cviu.2006.08.002.

67 Kasra Moghadam, Colin Banigan, and Eric D. Ragan. Scene transitions and teleportation in virtual reality and the implications for spatial awareness and sickness. *IEEE Transactions on Visualization and Computer Graphics*, 26(6):2273–2287, June 2020. doi: 10.1109/tvcg.2018.2884468.

68 Beatriz Moya, Alberto Badías, Icíar Alfaro, Francisco Chinesta, and Elías Cueto. Digital twins that learn and correct themselves. *International Journal for Numerical Methods in Engineering*, 123(13):3034–3044, September 2020. doi: 10.1002/nme.6535.

69 Lars Mündermann, Stefano Corazza, and Thomas P. Andriacchi. The evolution of methods for the capture of human movement leading to markerless motion capture for biomechanical applications. *Journal of NeuroEngineering and Rehabilitation*, 3(1):6, March 2006. doi: 10.1186/1743-0003-3-6.

70 Stylianos Mystakidis. Metaverse. *Encyclopedia*, 2(1):486–497, February 2022. doi: 10.3390/encyclopedia2010031.

71 Carman Neustaedter, Gina Venolia, Jason Procyk, and Daniel Hawkins. To beam or not to beam. In *Proceedings of the 19th ACM Conference on Computer-Supported Cooperative Work & Social Computing*. ACM, February 2016. doi: 10.1145/2818048.2819922.

72 Diederick C. Niehorster, Li Li, and Markus Lappe. The accuracy and precision of position and orientation tracking in the HTC vive virtual reality system for scientific research. *i-Perception*, 8(3):204166951770820, May 2017. doi: 10.1177/2041669517708205.

73 Huansheng Ning, Hang Wang, Yujia Lin, Wenxi Wang, Sahraoui Dhelim, Fadi Farha, Jianguo Ding, and Mahmoud Daneshmand. A survey on metaverse: The state-of-the-art, technologies, applications, and challenges, 2021. URL https://arxiv.org/abs/2111.09673.

74 Kohei Okumura, Hiromasa Oku, and Masatoshi Ishikawa. Lumipen: Projection-based mixed reality for dynamic objects. In *2012 IEEE International Conference on Multimedia and Expo*. IEEE, July 2012. doi: 10.1109/icme.2012.34.

75 Mathias Parger, Joerg H. Mueller, Dieter Schmalstieg, and Markus Steinberger. Human upper-body inverse kinematics for increased embodiment in consumer-grade virtual reality. In *Proceedings of the 24th ACM Symposium on Virtual Reality Software and Technology*. ACM, November 2018. doi: 10.1145/3281505.3281529.

76 Sang-Min Park and Young-Gab Kim. A metaverse: Taxonomy, components, applications, and open challenges. *IEEE Access*, 10:4209–4251, 2022. doi: 10.1109/ACCESS.2021.3140175.

77 Anjul Patney, Marco Salvi, Joohwan Kim, Anton Kaplanyan, Chris Wyman, Nir Benty, David Luebke, and Aaron Lefohn. Towards foveated rendering for gaze-tracked virtual reality. *ACM Transactions on Graphics*, 35(6):1–12, November 2016. doi: 10.1145/2980179.2980246.

78 Rolf Pfeifer, Max Lungarella, and Fumiya Iida. Self-organization, embodiment, and biologically inspired robotics. *Science*, 318(5853):1088–1093, November 2007. doi: 10.1126/science.1145803.

79 Roberto Pierdicca, Emanuele Frontoni, Rama Pollini, Matteo Trani, and Lorenzo Verdini. The use of augmented reality glasses for the application in industry 4.0. In *Lecture Notes in Computer Science*, pages 389–401. Springer International Publishing, 2017. doi: 10.1007/978-3-319-60922-5_30.

80 Kevin D. Plancher, Jaya Prasad Shanmugam, and Stephanie C. Petterson. The changing face of orthopaedic education: Searching for the new reality after COVID-19. *Arthroscopy, Sports Medicine, and Rehabilitation*, 2(4):e295–e298, August 2020. doi: 10.1016/j.asmr.2020.04.007.

81 Jarkko Polvi, Takafumi Taketomi, Goshiro Yamamoto, Arindam Dey, Christian Sandor, and Hirokazu Kato. SlidAR: A 3D positioning method for SLAM-based handheld augmented reality. *Computers & Graphics*, 55:33–43, April 2016. doi: 10.1016/j.cag.2015.10.013.

82 Francesco Porpiglia, Enrico Checcucci, Riccardo Autorino, Daniele Amparore, Matthew R. Cooperberg, Vincenzo Ficarra, and Giacomo Novara. Traditional and virtual congress meetings during the COVID-19 pandemic and the post-COVID-19 era: Is it time to change the paradigm? *European Urology*, 78(3):301–303, September 2020. doi: 10.1016/j.eururo.2020.04.018.

83 M. Raja and G. G. Lakshmi Priya. Using virtual reality and augmented reality with ICT tools for enhancing quality in the changing academic environment in COVID-19 pandemic: An empirical study. In *Technologies, Artificial Intelligence and the Future of Learning Post-COVID-19*, pages 467–482. Springer International Publishing, 2022. doi: 10.1007/978-3-030-93921-2_26.

84 Vijay Rajanna and John Paulin Hansen. Gaze typing in virtual reality. In *Proceedings of the 2018 ACM Symposium on Eye Tracking Research & Applications*. ACM, June 2018. doi: 10.1145/3204493.3204541.

85 Lisa Rebenitsch and Charles Owen. Evaluating factors affecting virtual reality display. In *Lecture Notes in Computer Science*, pages 544–555. Springer International Publishing, 2017. doi: 10.1007/978-3-319-57987-0_44.

86 Antti Revonsuo. Consciousness, dreams and virtual realities. *Philosophical Psychology*, 8(1):35–58, January 1995. doi: 10.1080/09515089508573144.

87 Somaiieh Rokhsaritalemi, Abolghasem Sadeghi-Niaraki, and Soo-Mi Choi. A review on mixed reality: Current trends, challenges and prospects. *Applied Sciences*, 10(2):636, 2020. doi: 10.3390/app10020636.

88 Maria V. Sanchez-Vives and Mel Slater. From presence to consciousness through virtual reality. *Nature Reviews Neuroscience*, 6(4):332–339, April 2005. doi: 10.1038/nrn1651.

89 Donghee Shin. Empathy and embodied experience in virtual environment: To what extent can virtual reality stimulate empathy and embodied experience? *Computers in Human Behavior*, 78:64–73, January 2018. doi: 10.1016/j.chb.2017.09.012.

90 Avinash Kumar Singh, Jia Liu, Carlos A. Tirado Cortes, and Chin-Teng Lin. Virtual global landmark: An augmented reality technique to improve spatial navigation learning. In *Extended Abstracts of the 2021 CHI Conference on Human Factors in Computing Systems*. ACM, May 2021. doi: 10.1145/3411763.3451634.

91 F. Steinicke, G. Bruder, J. Jerald, H. Frenz, and M. Lappe. Estimation of detection thresholds for redirected walking techniques. *IEEE Transactions on Visualization and Computer Graphics*, 16(1):17–27, January 2010. doi: 10.1109/tvcg.2009.62.

92 Frank Steinicke, Gerd Bruder, Jason Jerald, Harald Frenz, and Markus Lappe. Analyses of human sensitivity to redirected walking. In *Proceedings of the 2008 ACM Symposium on Virtual Reality Software and Technology - VRST '08*. ACM Press, 2008. doi: 10.1145/1450579.1450611.

93 Ryan R. Strauss, Raghuram Ramanujan, Andrew Becker, and Tabitha C. Peck. A steering algorithm for redirected walking using reinforcement learning. *IEEE Transactions on Visualization and Computer Graphics*, 26(5):1955–1963, May 2020. doi: 10.1109/tvcg.2020.2973060.

94 Alistair Sutcliffe, Brian Gault, and Jae-Eun Shin. Presence, memory and interaction in virtual environments. *International Journal of Human-Computer Studies*, 62(3):307–327, March 2005. doi: 10.1016/j.ijhcs.2004.11.010.

95 Takafumi Taketomi, Hideaki Uchiyama, and Sei Ikeda. Visual SLAM algorithms: A survey from 2010 to 2016. *IPSJ Transactions on Computer Vision and Applications*, 9(1):16, June 2017. doi: 10.1186/s41074-017-0027-2.

96 Justus Thies, Michael Zollhöfer, Marc Stamminger, Christian Theobalt, and Matthias Nießner. FaceVR: Real-time facial reenactment and eye gaze control in virtual reality, 2016. URL https://arxiv.org/abs/1610.03151.

97 Jerald Thomas and Evan Suma Rosenberg. A general reactive algorithm for redirected walking using artificial potential functions. In *2019 IEEE Conference on Virtual Reality and 3D User Interfaces (VR)*. IEEE, March 2019. doi: 10.1109/vr.2019.8797983.

98 Carlos A. Tirado Cortes, Hsiang-Ting Chen, and Chin-Teng Lin. Analysis of VR sickness and gait parameters during non-isometric virtual walking with large translational gain. In *The 17th International Conference on Virtual-Reality Continuum and its Applications in Industry*. ACM, November 2019. doi: 10.1145/3359997.3365694.

99 Irene Valori, Phoebe E. McKenna-Plumley, Rena Bayramova, Claudio Zandonella Callegher, Gianmarco Altoè, and Teresa Farroni. Proprioceptive accuracy in immersive virtual reality: A developmental perspective. *PLoS ONE*, 15(1):e0222253, January 2020. doi: 10.1371/journal.pone.0222253.

100 Niall L. Williams, Aniket Bera, and Dinesh Manocha. ARC: Alignment-based redirection controller for redirected walking in complex environments. *IEEE Transactions on Visualization and Computer Graphics*, 27(5):2535–2544, May 2021. doi: 10.1109/tvcg.2021.3067781.

101 Julia Woodward and Jaime Ruiz. Analytic review of using augmented reality for situational awareness. *IEEE Transactions on Visualization and Computer Graphics*, 29(4):2166–2183, 2022. doi: 10.1109/tvcg.2022.3141585.

102 Chien-Min Wu, Chih-Wen Hsu, Tzu-Kuei Lee, and Shana Smith. A virtual reality keyboard with realistic haptic feedback in a fully immersive virtual environment. *Virtual Reality*, 21(1):19–29, September 2016. doi: 10.1007/s10055-016-0296-6.

103 Jiangnan Xu, Konstantinos Papangelis, John Dunham, Jorge Goncalves, Nicolas James LaLone, Alan Chamberlain, Ioanna Lykourentzou, Federica L. Vinella, and David I. Schwartz. Metaverse: The vision for the future. In *CHI Conference on Human Factors in Computing Systems Extended Abstracts*. ACM, April 2022. doi: 10.1145/3491101.3516399.

104 Minrui Xu, Wei Chong Ng, Wei Yang Bryan Lim, Jiawen Kang, Zehui Xiong, Dusit Niyato, Qiang Yang, Xuemin Sherman Shen, and Chunyan Miao. A full dive into realizing the edge-enabled metaverse: Visions, enabling technologies, and challenges, 2022. URL https://arxiv.org/abs/2203.05471.

105 Yan Yan, Ke Chen, Yu Xie, Yiming Song, and Yonghong Liu. The effects of weight on comfort of virtual reality devices. In *Advances in Ergonomics* *in Design*, pages 239–248. Springer International Publishing, June 2018. doi: 10.1007/978-3-319-94706-8_27.

106 Zhen Yang, Cheng Chen, Yuqing Lin, Duming Wang, Hongting Li, and Weidan Xu. Effect of spatial enhancement technology on input through the keyboard in virtual reality environment. *Applied Ergonomics*, 78:164–175, July 2019. doi: 10.1016/j.apergo.2019.03.006.

107 Qinglin Yang, Yetong Zhao, Huawei Huang, Zehui Xiong, Jiawen Kang, and Zibin Zheng. Fusing blockchain and ai with metaverse: A survey. *IEEE Open Journal of the Computer Society*, 3:122–136, 2022. doi: 10.1109/OJCS.2022.3188249.

108 Yuanhang Zhang, Shuang Yang, Jingyun Xiao, Shiguang Shan, and Xilin Chen. Can we read speech beyond the lips? rethinking RoI selection for deep visual speech recognition. In *2020 15th IEEE International Conference on Automatic Face and Gesture Recognition (FG 2020)*. IEEE, November 2020. doi: 10.1109/fg47880.2020.00134.

109 Yuheng Zhao, Jinjing Jiang, Yi Chen, Richen Liu, Yalong Yang, Xiangyang Xue, and Siming Chen. Metaverse: Perspectives from graphics, interactions and visualization. *Visual Informatics*, 6(1):56–67, March 2022. doi: 10.1016/j.visinf.2022.03.002.

110 Howe Yuan Zhu, Hsiang-Ting Chen, and Chin-Teng Lin. The effects of a stressful physical environment during virtual reality height exposure. In *2021 IEEE Conference on Virtual Reality and 3D User Interfaces Abstracts and Workshops (VRW)*. IEEE, March 2021. doi: 10.1109/vrw52623.2021.00116.

111 Howe Yuan Zhu, Hsiang-Ting Chen, and Chin-Teng Lin. The effects of virtual and physical elevation on physiological stress during virtual reality height exposure. *IEEE Transactions on Visualization and Computer Graphics*, 29(4):1937–1950, 2021. doi: 10.1109/tvcg.2021.3134412.

7

Blockchain for the Metaverse: State-of-the-Art and Applications

Pawan Kumar Hegde[1], Rajeswari Chengoden[1], Nancy Victor[1], Thien Huynh The[2], Sweta Bhattacharya[1], Praveen Kumar Reddy Maddikunta[1], Thippa Reddy Gadekallu[1,3], and Quoc-Viet Pham[4]

[1]*School of Information Technology and Engineering, Vellore Institute of Technology, Tamil Nadu, India*
[2]*Department of Computer and Communication Engineering, Ho Chi Minh City University of Technology and Education, Ho Chi Minh city, Vietnam*
[3]*Department of Electrical and Computer Engineering, Lebanese American University, Byblos, Lebanon*
[4]*School of Computer Science and Statistics, Trinity College Dublin, Dublin, Ireland*

After reading this chapter you should be able to:

- Understand the role of blockchain in resolving the issues of the Metaverse considering various aspects, including data acquisition, data interoperability, data privacy, and data sharing.
- Describe various use cases to demonstrate the significance of incorporating blockchain in the Metaverse.
- Explore an overview of existing real-time blockchain-based projects for the implementation of the Metaverse.

7.1 Introduction

Various innovations in computer science and engineering have revolutionized the processes involved in human interaction and social transactions. The Metaverse is one such technology that creates a virtual world replicating all the opportunities similar to the real world. In simplified terms, the Metaverse is a digital virtual space developed using three-dimensional (3D) technologies to mimic the real world through virtual reality (VR) technologies [45]. The relevance of the Metaverse has increased steadily with the rapid growth of the Internet, smartphones, and VR goggles, wherein most companies incline to develop

Metaverse Communication and Computing Networks: Applications, Technologies, and Approaches, First Edition.
Edited by Dinh Thai Hoang, Diep N. Nguyen, Cong T. Nguyen, Ekram Hossain, and Dusit Niyato.
© 2024 The Institute of Electrical and Electronics Engineers, Inc. Published 2024 by John Wiley & Sons, Inc.

their Metaverse ecosystems. The word "Metaverse" was first outlined by Neal Stephenson in his novel "Snow Crash" written in 1992. However, it recently gained popularity when Facebook was renamed Meta with the primary focus on utilizing the Metaverse for enhancing business connectivity, business efficiency, community discovery, and growth of business [22]. The science fiction novel shows that imaginary characters can escape a virtual world. "Metaverse" from the virtual world with digital avatars became a reality with the advent of innovative technologies like augmented reality (AR), VR, artificial intelligence (AI), machine learning (ML), blockchain, and several others [68]. The Metaverse is thus a virtual digital 3D universe created by integrating various virtual spaces. Users are allowed to enter into this digital universe through the authentication of their virtual identity forming digital avatars which can explore the Metaverse space for performing social interactions, communications, and transactions similar to the real world. The Metaverse enables users to gain an immersive experience and perform activities meant for isolated environments to be done within their comfort zones [57].

The operation of the Metaverse requires the implementation of multiple cutting-edge technologies, namely VR, AR, AI, ML, blockchain, cloud/edge computing, the Internet of Things, 5G and beyond, spatial technologies, head-mounted displays (HMDs), and 3D constructions. Using such avant-garde technologies need to be supported by various hardware resources, software tools, apps, and contents created by users [45, 65]. The aforementioned cutting-edge technologies enable an immersive experience that attracts the users to an imaginary space, allowing them to relate and manipulate their environment, blending visualization, sound, and technology to deliver an amalgamation of an extraordinary and engaging world [42]. VR technologies help in capturing the VR environment digitally by 360° photography and videography and submerge the user into the digital replication of the physical world using virtual headsets. AR integrates the real world and digital elements using smartphone displays and headsets. It enhances the virtual elements enabling users to use the digital features in the physical world. Mixed Reality (MR) integrates AR and VR to create an immersive experience using a fusion of virtual and holographic images wherein the user wears customized glasses to visualize. The peripheral sight of the users is blocked, which channelizes them to focus on the digital revelations being projected through the virtual headset. The physical noises can be eliminated when the user wears the VR headset, and the touch sensation is enhanced through VR accessories enabling them to concentrate completely on the virtual world, thereby providing haptic feedback for the users. The 3D constructions create the virtual spaces and provide a realistic look and feel to the Metaverse forming a digital ecosystem. The advanced 3D cameras accurately depict objects, buildings, and physical locations, making them 3D photorealistic. These captured HD pictures are computationally processed, wherein the 3D spatial data are used to generate digital twins of the real world. The blockchain validates all the credentials, data transfer, and data storage within the Metaverse [42, 66].

The Metaverse has tremendous potential to improve human sustainability benefiting the entire ecosystem. For example, in the education sector, students would be allowed to perform various scientific tests and operations without pondering the waste of resources or risks associated with chemical hazards or health damages. The Metaverse classrooms will provide a real-time and detailed view of all live demonstrations through the creation of digital twins, holographic images, and various other enabling technologies [26]. In the retail consumer-branding sector, the Metaverse is all set to offer new options to firms in terms of website construction, e-commerce, social media interactions, and live broadcasting. The virtual customized avatars enable customers to get a mix of in-store and online experiences allowing them to interact better with the brands [4]. The Metaverse can bring forth the wave of revolution in the healthcare sector by making medical services open, accessible, and interoperable. The healthcare Metaverse will gain more momentum if the data and services that constitute the virtual world have easier mobility between networks and platforms [17]. In the financial sector, the incorporation of nonfungible tokens (NFTs) into cryptocurrencies has great potential to strengthen the digital economy, ensuring users have complete control over their digital assets [58]. Blockchain is vital in providing immutable ownership confirmation in the virtual world, wherein users can perform transactions safely. Blockchain can also mitigate risks that could be incurred from potential losses [47].

Blockchain is a technology that records transactions permanently in a decentralized and public database known as a ledger. Bitcoin is the most popular and predominantly used blockchain-based cryptocurrency. Whenever a bitcoin is purchased, the transaction gets recorded in the blockchain, thus enabling the record to be distributed to thousands of computers across the world. The decentralized recording system is secured, and the transactions pertaining to public blockchains in the form of bitcoin and ethereum are highly transparent in comparison to traditional banking systems. Ethereum is similar to bitcoin, which is programmable through smart contracts wherein blockchain-based software routines get executed automatically when certain conditions are met. Smart contracts in blockchain could be used to establish ownership of digital objects, also known as crypto assets, in currencies, securities, or artworks. Thus, the role of blockchain is essential when it comes to creating, owning, and monetizing decentralized assets. The primary application of blockchain is in the case of NFTs in the Metaverse. Items that are unique and nonreplaceable are known as nonfungible. In the Metaverse, users intend to mimic the physical world by replicating physical world commodities into the digital world, and NFTs ensure that these items are unique and have proof of ownership. NFTs can be used in digital art

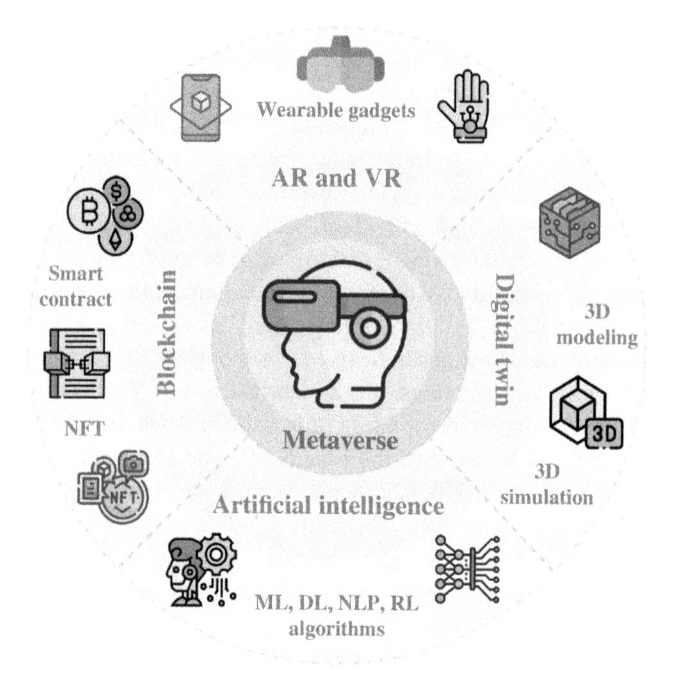

Figure 7.1 Blockchain implementation areas in the Metaverse.

exchange, purchasing of gaming assets, unique user avatars, virtual real estate, and various other commodities. Blockchain has many applications in the case of peer-to-peer (P2P) transactions. For example, cryptocurrencies enable individuals in the Metaverse to perform a direct transaction without the mediator's involvement. Crypto-powered P2P transactions save money and offer convenience, and the essential identity verification process using blockchain-based business certificates in the Metaverse eliminates the possibility of unauthorized access and related malpractices. The Metaverse eventually will become more decentralized and autonomous compared with the physical world wherein decentralized autonomous organizations (DAOs) and virtual organizations enable people to join and control the organizational operations [24, 44].

Although blockchain and the Metaverse are two distinct and different concepts, utilizing the potential of their integration can provide an exceptional experience to the users. Some of the applications of blockchain implementation in the Metaverse are depicted in Figure 7.1 and discussed below:

- **Virtual Currencies:** One of the most common applications of blockchain in the Metaverse is financial transactions. This will enable customers to shop in virtual stores as consumption needs are ever-increasing, and thus, retail is slowly progressing toward retail businesses. Million-dollar transaction dealings are being completed through games in Decentraland. Decentralized Finance (DeFi) is a term in the crypto world which is a protocol that does not require intermediaries to perform transactions. Blockchain in DeFi eliminates the need for banks, brokers, and stock exchanges to perform the transactions [20, 67].

- **NFTs:** NFTs are predominantly used in the trading of digital arts as they provide the ability to show proof of ownership. In the Metaverse ecosystem, any individual can clone the assets using NFTs; however, blockchain makes it necessary to maintain original ownership of the NFT and create virtual NFT marketplaces in that environment [18].
- **Self-Identity Authentication:** The blockchain technology helps securing user data, stops the manipulation of identity data, and prevents illegal activities. The user details are saved in the blockchain to ensure the reliability and transparency of the data [60].
- **Real-Estate:** The major challenges in real estate involving the Metaverse are the assessment and regulation of digital properties. The blockchain helps the mediation of all such activities, which include the creation, purchasing, selling, transfer, and disposal of digital lands [69].

In this chapter, we first present the fundamentals of blockchain and discuss how blockchain can be used to solve major challenges in the Metaverse in Section 7.2. Then, we provide various use cases of blockchain in the Metaverse, including privacy, security, traceability, decentralization, ownership, governance, and trust accountability, in Section 7.3. Next, in Section 7.4, we demonstrate the use of blockchain in the Metaverse via several real projects. Finally, we conclude this chapter in Section 7.5.

7.2 Background

7.2.1 Blockchain

The adoption of modern technologies in the past few years has transformed physical data into digital data. Modern technology has changed the way we live and work. Most companies and industries have changed business logic by adopting technologies leading to enormous digital data. The rapid digitalization in various sectors requires new forms of records, electronic signatures, and digital identities that are to be managed efficiently without compromising the privacy and security of data in a decentralized environment [23, 28, 34, 49]. Recent research has proved that the decentralized distributed ledger is a promising technology for securely maintaining digital data. Blockchain was developed as an underlying technology for Bitcoin in 2008 by Satoshi Nakamoto [43]. However, due to great features, such as decentralization, immutability, transparency, auditability, and secure cryptographic solutions, blockchain has been adopted in various applications beyond Bitcoin, such as healthcare, supply chain, digital media transfer, and remote services delivery. Blockchain can be briefly described as a public, trusted, shared ledger that organizes all committed transactions in blocks in a peer-to-peer network system. Blockchain encompasses core technologies, such as digital signature, cryptographic hash, and distributed consensus algorithm [40]. Ever since the invention of blockchain, many researchers and organizations have contributed to the development of three different stages Blockchain 1.0, 2.0, and 3.0 [36]. Blockchain 1.0 is mainly applied to digital currency, where Bitcoin is the most dominating one, and its application is limited to the storage and transfer of value. However, in the second stage of blockchain, which supports the creation of advanced smart contracts, extending the doors for new application areas, enabling different industries to collaborate through applications, and promoting automation of resource allocation by resolving the mutual trust and identity among the participants. The next generation Blockchain 3.0 has emerged as application-centric by allowing blockchain to confirm the property rights of the information values on the Internet. Hence, blockchain can be used to track and control assets while trading. Apart from that, blockchain has also extended its scope in various domains, such as health, identity certification, logistics, and voting [16, 29].

Blockchain is an interconnection of blocks structured in chronological order, and each block is chained to another using a cryptographic hash of the previous block, ensuring the infeasibility of illegal tampering. However, the individual block in the blockchain consists of a block header with a hash of the previous block, nonce, Merkle root of transactions, timestamp, and block body consisting of transaction data. The structure of blocks in a blockchain is depicted in Figure 7.2. Nodes in the blockchain network will have a copy of the ledger, and all transactions need to be validated by the individual node of the network and contribute to the block creation by achieving the consensus mechanism among the various nodes [19, 55, 64]. The consensus mechanism, one of the core components of blockchain, resolves the ambiguity among the nodes by maintaining a common agreement among the nodes to maintain the network and finally choosing a node responsible for creating the block [25]. The Proof of Work (PoW) is the most popular consensus algorithm which is used in Bitcoin, Ethereum, and many crypto-currencies where the miner nodes will compete among themselves by solving a large puzzle until the target is achieved and the miner who has achieved the target will get a chance to create a block, and this process continues for the subsequent blocks in the blockchain [70]. There are several other consensus algorithms, such as Proof of Elapsed time (PoET), Proof of Stake (PoS), Delegated Proof of Stake (DPoS), Practical Byzantine Fault Tolerance (PBFT), Raft Tendermint, Ripple, Proof of

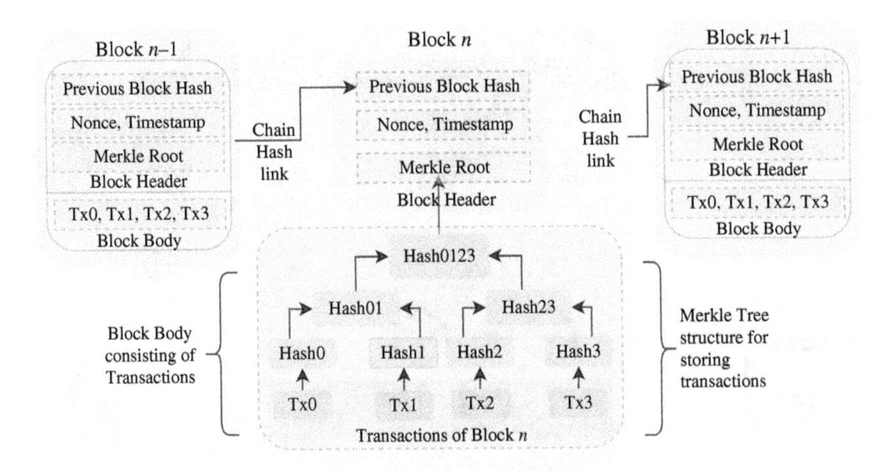

Figure 7.2 Structure of block in blockchain.

Burn (PoB), Proof of Capacity (PoC), and Proof of Authority, are used in the later versions of blockchain to achieve a particular task in an optimized and efficient manner [31, 63].

Blockchain is a buzzword in the current market, and more industries and organizations have started adopting it for their daily operations. The main reason is the blend of various technologies such as cryptography, distributed data storage, networking, and consensus mechanism. Due to these modern technologies, blockchain is capable of exhibiting various properties. First, there will not be any centralized authority to manage the nodes, and their operations and information are automatically shared among the peer-to-peer nodes in the network. Second, every transaction is shared among all the network nodes for verification and validation. However, blocks containing valid transactions are chained through a cryptographic hash value, thus making it a tamperproof system. Third, blockchain allows organizations and individuals to audit and trace transaction history because every transaction executed in the blockchain is recorded in a ledger and validated by a digital timestamp. Finally, each user can interact with the blockchain network by generating an address; hence, the identity of the individual or user is not exposed [72]. In a blockchain, every request will be considered as a transaction, and the execution or processing of transactions is carried out by multiple nodes in the blockchain network [54]. The overall processing of transactions in a secure blockchain is depicted in Figure 7.3. If any client initiates a transaction request in the blockchain network, then the transaction is transferred to all the nodes for the verification process once the transaction is verified and validated by all the nodes, a block is created with the verified transactions. Finally, the newly created block will be added to the

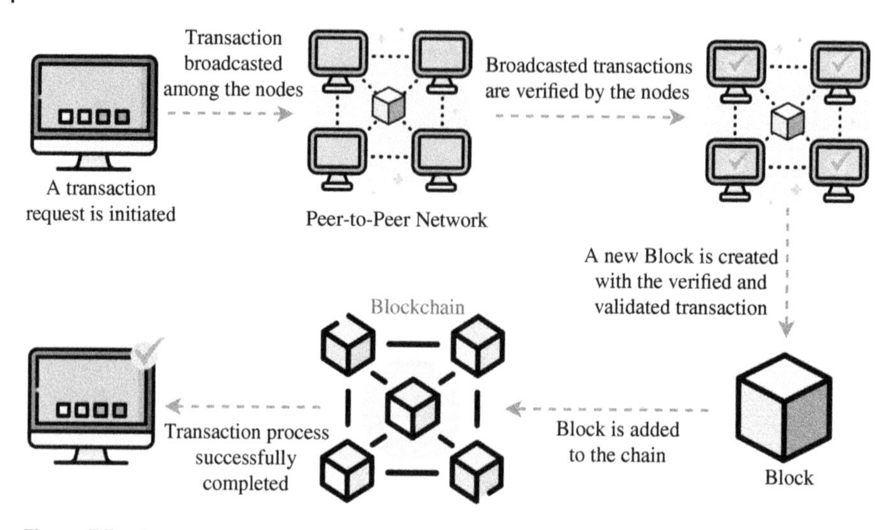

Figure 7.3 Transaction processing in blockchain.

blockchain, ensuring the completion of the transaction process [9]. Performance efficiency, cost-effectiveness, and trust among multiple parties without any risk are the major advantages of the blockchain, as reflected in various fields, such as healthcare, cryptocurrency, finance market, data provenance, the Metaverse, and beyond 5G networks.

7.2.2 Blockchain-Based Solutions for the Metaverse

The Metaverse is a blend of two words "Meta" and "Universe," which refers to a hypothetical virtual world connected to the physical world with the aid of enabling technologies such as VR, AR, AI, cloud edge computing, and blockchain. The phrase "Metaverse" was first coined by Neal Stephenson in his 1992 science fiction novel Snow Crash, in which people-controlled avatars are used to interact and form a relationship with other electronic agents in a VR world called Metaverse [11]. Since we are mostly dependent on technology for our day-to-day activities, the modernization of existing technology to minimize human intervention and the optimization of the resources in various tasks have been considered to be a predominant job in the last few decades. However, the subsequent evolutions in Internet technologies have reduced the separation gap between the digital and physical world, thus, motivated toward the development of the Metaverse. The Metaverse is a scalable and persistent network of interconnected virtual worlds mainly focusing on real-time interactions where people can work, socially interact, transact, and play. The Metaverse is a virtual, digital 3D space built by combining various kinds of virtual spaces, where users can enter the space using their virtual identities in the form of avatars to perform various activities in the Metaverse space, similar to the real world [47]. The Metaverse is not owned by a single entity or organization, but it is an ecosystem with multiple virtual spaces and technologies interconnected through different layers to develop shared experiences [52]. The Metaverse is still in its evolution stage, the modern technologies, such as VR, AR, AI, digital twins, 6G, blockchain, and IoT will become key enabling technologies to strengthen the interconnected virtual world in various aspects [15]. Incorporating various modern technologies in the Metaverse helps exhibit various characteristics, such as persistent, synchronous, immersive, decentralized, and accessible.

Even though the Metaverse and blockchain are two different concepts, these two technologies have the potential to offer a new experience to users when combined. At present, many blockchain-based platforms use NFTs and cryptocurrency, providing an ecosystem for creating, owning, and monetizing decentralized digital assets. Blockchain can work globally as a digital source in a decentralized manner, restricting the Metaverse from adopting the capabilities of the traditional Internet. The Metaverse will be incomplete without the aid of blockchain technology, and blockchain-based Metaverse provides access to any digital space without the interference of a centralized institution [48]. The blockchain-based solutions for the major challenges faced by the Metaverse are listed below:

- **Data Acquisition:** Since the Metaverse is an amalgamation of multiple technologies and users with various activities, an enormous amount of unstructured, real-time transaction data has to be collected from the source devices. The reliability and integrity of services offered by the Metaverse are determined based on the quality of collected data. The enormous amount of data will also burden the data acquisition system. During the acquisition process, duplicate and inaccurate data may also be acquired, which will change the quality of the data. Hence, ensuring the quality of gathered data is a challenging task in the Metaverse ecosystem [50].
  Adopting blockchain in the Metaverse will ensure authentic data acquisition using distributed ledger technology by validating transaction records and tracing the Metaverse data. The acquired data will be stored in the blocks consisting of a cryptographic hash of the previous block, thus making it secure and tamper-resistant. A specific validation procedure in the blockchain will prevent the alteration and duplication of data during the acquisition process; hence, the blockchain-enabled acquisition system will develop a most reliable Metaverse ecosystem [24].
- **Data Interoperability:** Metaverse is a blend of numerous virtual digital spaces; however, the transfer of users' digital assets like NFT and avatars among the virtual spaces is restricted due to the lack of openness, which makes it difficult

to relocate in virtual spaces. The interoperability of the Metaverse is measured based on the number of interactions managed between the virtual spaces. However, establishing interactions among the virtual worlds irrespective of their locations, technology, and digital applications in an appropriate manner is a major limitation of the traditional approach [50].
A cross-chain protocol allows the transfer of users' digital assets like NFT and avatars among the virtual spaces by ensuring interoperability among the virtual spaces in the Metaverse. Thus, adopting cross-blockchain technology can eliminate the intermediaries required for achieving interoperability in Metaverse [24].

- **Data Privacy:** In the Metaverse, users are not aware of whom they are interacting with, and users may admit they are dealing with a real person. Hence, attackers might steal the user's sensitive or personal information by tricking the users. Thus, maintaining the confidentiality of personally identifiable information needs to be addressed in Metaverse. However, incorporating valid information in the Metaverse will enormously increase the data volume and consequently increase the complexity of data management [50].
The blockchain-based solution for preserving users' sensitive information confidentiality is the adoption of public and private keys, thus enabling users to control their personal data. Blockchain also enables the data owners to decide when and how the third party can access the data, which can prevent the misuse or acquisition of sensitive data by third-party intermediaries. The adoption of zero-knowledge proof on blockchain permits the users to access and identify essential data by protecting the privacy and maintaining ownership over their possessions [24].
- **Data Sharing:** The Metaverse generates an enormous amount of real-time digital data, but sharing data in a centralized exchange platform will expose the sensitive and personal information of the owners, thus creating a huge risk for data owners in the Metaverse ecosystem. The secure data sharing of VR and AR data is very much necessary in the Metaverse ecosystem for smooth operation; however, data flexibility turns out to be an issue as the demand for real-time data increases in the traditional data-sharing environment.
Blockchain technology aids in the gathering of information from reliable sources and provides Metaverse users with complete control over their data. Blockchain technology ensures all the transactions that arise in crypto exchange, and other applications are transparent and precise in the Metaverse ecosystem. However, each and every stakeholder involved in the application can have a copy of the generated immutable transaction records, thus making it more transparent. Smart contracts in blockchain improve the flexibility of data sharing by automating the execution of agreements without the involvement of any intermediaries [24].

7.3 Use Cases of Blockchain for the Metaverse

This section discusses the technical and non-technical aspects of adopting blockchain for the Metaverse. Figure 7.4 depicts various use cases of blockchain for the Metaverse.

Consider a healthcare application in which the blockchain plays a significant role. The combination of modern technologies such as AR, VR, Web3.0, AI, Blockchain with Metaverse has changed the traditional approach of consultation, treatments in healthcare domain. The adoption of Metaverse has changed perspectives of providing therapies to telemedicine and telehealth, and patient's interactions with the physicians will happen through virtual 3D clinics. Apart from that, the digital workouts and digital activities have become a part of our daily routine. However, the use of decentralized distributed technology in Metaverse will strengthen the Healthcare applications such as Patient Monitoring, Medical record Management, and Medical Diagnosis. Since the Metaverse is in its evolution phase and the applications that are designed using Metaverse are more prone to vulnerabilities, it requires an aid of technologies such as blockchain, 6G to strengthen the Metaverse-based applications in various aspects such as

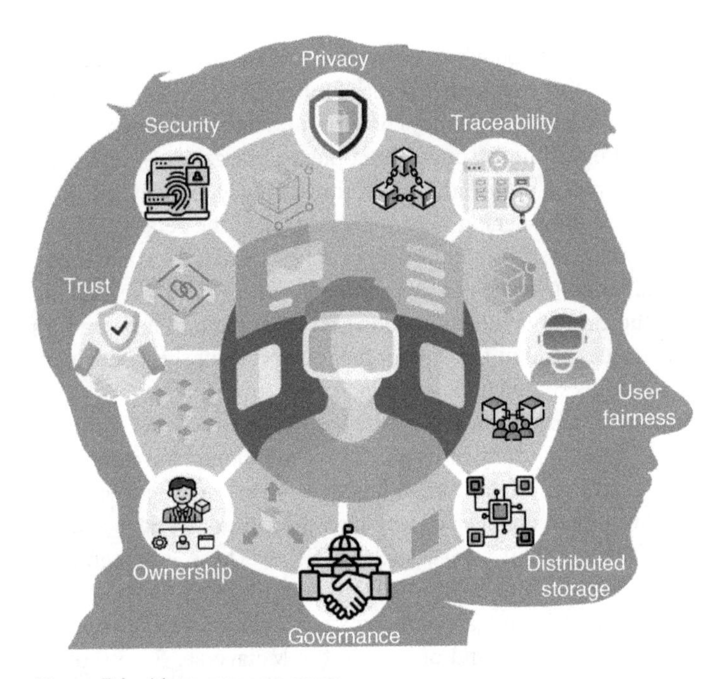

Figure 7.4 Metaverse use cases.

data security, data privacy, trust, data exchange facilities. The decentralized, immutable nature of the Blockchain will improve the privacy of the Patient's sensitive data. Cryptographic techniques are used for processing and validating every transaction that are carried by the patient or the medical service provider. Also, every valid transaction pertaining to an individual patient is permanently stored in a cryptographically linked chain of blocks, resulting in tamper-proof model; hence, the overall security of Healthcare application is achieved. Blockchain also manages the interactions that are happing between physical and virtual environment with improved security. The blockchain consensus algorithms will facilitate to maintain the integrity of the data in a distributed environment by maintaining the mutual trust among the peer. Every patient or the medical authorities in the blockchain-based Metaverse ecosystem will have a copy of the ledger with all the valid transactions that are executed in the system; however, the patients or the medical authorities will have the access to the history of medical reports as well as the current status via shared ledger. This helps the doctor to understand medical history of the patient. Only the authorized, validated owners or members will have the access to the blockchain as well as the sensitive information present in the Block.

7.3.1 Privacy

Digital assets play a major role in today's economy by enabling anything and everything to be stored digitally and can further be used for realizing value. The work in [33] conducted a detailed study on privacy concerns and how the same could be regulated in the second life, within and out of the context of the game. Even though the concept seems to be based on openness and transparency, privacy infringements can occur due to the third-person view of the avatars, as the details could be observed by attackers or hackers without the user being aware of that. Maintaining anonymity is quite challenging in such an environment. Privacy in the second life is as important as in the case of real life due to the fact that digital life resembles actual life. Even the devices used in the environment can cause privacy infringements if the data collected is shared with a common platform.

Privacy in the Metaverse corresponds to privacy at the sensory level, communication and behavior, and the safety of users. With the unique convergence of blockchain technology with the Metaverse, cohesive services can be provided to the customers by making them change the way how they interact and exchange cryptocurrencies and other digital assets, such as NFTs. These cryptographic assets have unique identification codes as well. However, buying digital spaces in the Metaverse may also raise privacy concerns. The work in [60] surveyed the fundamental aspects of security and privacy in the Metaverse. Even though decentralization allows for secure digital transactions, user authentication still plays a significant role, as fake crypto domains can be created by attackers. NFTs play a significant role in the Metaverse, as they are unique and are linked to virtual assets. Any kind of attack can further lead to the loss of data. As the network is decentralized, and the data are stored in a distributed manner, the blockchain-enabled Metaverse offers better privacy guarantees. Metaverse alone can be a threat to privacy, whereas blockchain-enabled Metaverse will offer privacy and security, thereby enabling secure access to the virtual world created.

7.3.2 Security

The Metaverse-enabled applications, such as gaming, healthcare, and real estate mimic the real world to give the user an immersive experience and hence collect numerous amounts of data from brainwaves, biometrics, health information, preferences, and more. The data collected from several Metaverse applications can be misused by malicious users, posing significant challenges in securing the user data in the Metaverse. The technologies used to track the information on the web can also be used to track the information in the Metaverse, but with more intensity. However, it is essential to protect the various forms of data generated from the Metaverse applications. In the virtual world, the users will use avatars as an identity to represent themselves. Users can choose the preference of having multiple user accounts, and they can enable the data flow between their accounts. Data through the Metaverse can lead to identity theft. For instance, by hacking a VR glass, hackers can steal personal information. One can no longer apply the same practices that he/she does on the web to protect the data in the Metaverse. Hence, providing security to the applications of the Metaverse is of paramount importance.

Blockchain-enabled Metaverse is a key solution to this problem as it provides inherent security to the application. The significant character of blockchain is that the transactions are stored in every node of the system. The data generated from every participating node and user are circulated among all other nodes in the Metaverse application. This blockchain property will ensure that if any node is hacked, the integrity of the original data will not be compromised. The work in [56] proposed to use blockchain in the Metaverse that introduces asymmetric cryptography to secure transactions between users, which have the potential to generate two different keys for security purposes. The keys are random values generated by the algorithm; hence, intruders cannot easily hack the blockchain-enabled Metaverse applications.

7.3.3 Traceability

The Metaverse has the potential to serve as a platform for education, virtual interactions, gaming, and shopping as well. Hence, the Metaverse has become the next inevitable phase of our digital lives. Traceability is an important requirement for several applications in the Metaverse. One of the applications that need traceability is shopping enabled by the Metaverse. When we do shopping in the real world, we visit a store, browse through shelves, find appropriate products, and walk out of the store buying the desired products. The same could be replicated in the virtual world. The purchasing style could be simply a voice command or walking your virtual self out of the store with the product. In order to support such a seamless process and advanced experience to the user, the retailers had to take the additional responsibility of offering online experience to the customers in the physical world. Retailers should offer an immersive environment in the Metaverse to their consumers so as to walk into the virtual store make, shop with 3D and holographic experience, and purchase at their convenience. The retailers of these Metaverse-enabled stores could be under pressure more than ever to fast-track the immediacy of delivery, offer more delivery and pick-up options, and enable the consumers to know exactly where the products are and when they will have them. Hence, tracing and tracking these tasks are a very important requirement in this use case. Example applications of retailers in this space include Nike and Gucci. The US Drug Supply Chain Security Act of 2013 also requires pharmaceutical companies to identify and trace the drugs so as protect consumers from harmful products [7]. Traceability in the Metaverse can be realized through blockchain, which leverages the use of smart contracts in traceability in supply chain management. Hence, traceability can enable end-to-end visibility throughout the supply chain in real time.

7.3.4 Decentralization

The concept of decentralization has its root in political science, decision science, and economics over a period of more than 200 years, and has proved to be immensely successful since then [13]. Instead of a centralized mechanism where all the decisions are carried out by a single authority or an organization, decentralization allows the same to be carried out by different parties. There are two types of Metaverse platforms: centralized and decentralized. Online games have caught the attention of both young and adult populations equally, and the Metaverse is contributing significantly to this trend. The centralized Metaverse does not adapt blockchain and operates on a centralized system, as in the case of games (e.g. Roblox), whereas Axie Infinity is an example of a decentralized Metaverse [30]. However, the users will have the utmost control over the project when the system is decentralized. The adoption of blockchain technology makes decentralization a reality in the Metaverse environment. The "Bored Ape Business Model" proposed by Lee [32] focuses on the significance of decentralized collaboration using blockchain. The Metaverse-blockchain integration enhances user engagement and interaction and helps create and earn digital assets.

The work in [53] proposed the distributed Metaverse for MR applications. The study particularly aims to archive, recycle, and share virtual spaces among multiple applications, thereby reducing the computational load to a great extent. The proposed model uses a decentralized blockchain to store the information about spaces, where each space is represented as a block containing essential information (e.g. timestamps), links to photospheric imagery, and other relevant attributes. Such a mechanism has a wide range of applications, including navigation, as it can create an "independent and self-sustaining" archive of geo-tagged URLs. A service management model using blockchain for the Metaverse is presented in [37]. A "multi-flow synchronized service provisioning" is proposed here using blockchain technology, that enables better means of sensing and visualization in the Metaverse. An interesting study by Hopkins [27] detailed how blockchain-enabled Metaverse can contribute to virtual commerce. Extensive research is being carried out in the blockchain-based Metaverse systems and is a hot research topic in the current scenario.

7.3.5 Ownership

Ownership is a significant aspect in the Metaverse that requires legal policies to understand the ownership associated with the virtual properties and to what extent the property rights are given to the owners. The Metaverse proves to be advantageous in having a virtual presence in even sales and marketing. However, there is always the question of ownership in the Metaverse. Integration of blockchain technologies with the Metaverse aids in enabling digital ownership of properties. The unhackability and immutability of blockchain help in the realization of ownership of digital assets in the Metaverse ecosystem. NFTs are unique cryptographic tokens in a public blockchain that can serve as a record of ownership. These digital assets are nonreplaceable and nonexchangeable. Even though anyone can "view" the item, only the buyer can have the ownership and access to its status.

The work in [6] details the NFTs, Web3, and cryptocurrencies concerning possessions and ownership in the digital world. Similarly, the work in [14] explores the opportunities and challenges of using NFTs as an ownership mechanism. A discussion on whether these NFTs can be considered as the optimal digital asset for the Metaverse was carried out in [12]. NFTs have been proven useful for owning artworks, in-game assets, and even mechanic horses in horse races. Ownership in the Metaverse should be made possible in such a way that equal rights are provided for every user in terms of usage and creation. However, the Matthew effect [51] of "rich getting richer and poor getting poorer" shows up significantly in the Metaverse environment as well. A novel value system for the Metaverse with NFTs allows high-value owners to lock their tokens with smart contracts, enabling the other users to participate. Ownership determination thus becomes a concern in this case. "NFTing the NFT" is the idea proposed in [18], which can be accomplished in two ways. The first is done by NFTing the NFTs with a single and consistent representation, while the latter is done by NFTing the unique NFTs. Dividing the NFTs enables the decentralization of ownership rights and incentivizes the users. The work in [41] presented an exposition on the issues pertaining to the copyrights associated with the NFTs. Ownership in the Metaverse is still an open issue that needs to be solved with proper standardization and governance mechanisms.

7.3.6 Governance

The Metaverse can be regarded as a parallel digital world created for real, physical life. Such a system generates massive data since everything in the Metaverse creates a data file, and it continues to grow according to the different activities performed. However, generating, storing, maintaining, and sharing such data poses significant challenges as well. Hence, governance is considered as one of the crucial and multifaceted challenges to be addressed in the Metaverse environment. This is primarily due to the trade-offs between various factors, such as privacy, utility, security, and interoperability. Also, all the users in the Metaverse have to be treated fairly. An interesting example that can explain the importance of governance in the Metaverse is online gaming, where the digital assets of other players can be stolen by the other party. Automated tools are widely used to control user behavior in the current online social media platforms. However, significant measures are required for governance in the Metaverse. A recent initiative on the Metaverse governance by "The World Economic Forum" focuses on defining the parameters of an "economically viable, interoperable, safe and inclusive Metaverse" [1].

Blockchain helps in overcoming the challenges related to the governance of the Metaverse by ensuring data integrity, user authentication, and trust in information exchange and other transactions. Global organization and coordination can be provided by the "decentralized autonomous organizations (DAO)," as these are generally based on smart contracts [59]. The rules governing management and operations are typically encoded in blockchain as smart contracts and are operated autonomously in a decentralized manner. A modular framework that considers the ethical aspects of the Metaverse is proposed in [21]. Similarly, Lin et al. [35]

proposed a collaborative governance system for Metaverse manufacturing using blockchain. The system not only focuses on cost reduction and value increase but also on improving trust, thereby promoting the consensus on manufacturing collaboration in the Metaverse. Blockchain-enabled Metaverse proves to be important in almost all aspects of the future digital world, where anything and everything will be represented as digital avatars of the real-world entity and strict governance mechanisms are therefore necessary for the successful deployment of the system.

7.3.7 Trust and Accountability

The Metaverse is considered to be a bridge between real world and the virtual world. Alike, applications provide fully immersive experiences to the users. The Metaverse ecosystem is realized by seamlessly collecting data from various participating devices, storing the collected data, performing analysis and computation, and communicating between the participating devices and the users. Embracing all these activities will increase the amount of trust and accountability needed among the participating nodes in the Metaverse environment. Trust and accountability is an important challenge in many virtual applications such as gaming, healthcare, architecture planning, and building smart cities. Industrial Metaverse is an example of interest considered to demonstrate how the Metaverse can be harnessed from the dimensions of trust and accountability. In the industrial Metaverse, the designs are created in a shared virtual space that supports multiusers to collaborate and create the 3D designs, where there is a likelihood of design models being accessed by malicious users. In order to address this potential challenge of the Metaverse, one of the possible solutions is the users need to know that the avatars they interact with are verified and need to trust these are authentic nodes. Every avatar created on behalf of the user and the participating devices must communicate only within its zone and has to consider other avatars or devices as malicious. An idealized framework that includes zero-trust mechanism and cross chain technology of blockchain is proposed by the authors in [8]. This framework addresses the issue by allowing trusted interactions without the need for a third-party verification platform. In this way, the consensus algorithm achieves trust between the participating nodes in the Metaverse ecosystem. The proposed framework not only creates trust but further promotes tracking of the information from end-to-end throughout the process. This feature of blockchain leads to accountability in the Metaverse environment. Assigning accountability in a Metaverse platform is difficult as the technology extends notably without geographical boundaries. Blockchain-enabled Metaverse facilitates the participants by recording all of the modifications made to the document in real-time and making the changes transparent.

7.4 Projects

This section presents the most important blockchain projects for the Metaverse, which are essential for the Metaverse implementation.

7.4.1 Axie Infinity

Sky Mavis, a Vietnamese studio, developed Axie Infinity, a gaming application in the Metaverse. It is an NFT-based online video game.[1] Axie Infinity was launched in 2017 by Nguyen Thanh Trung, Tu Doan, Aleksander Larsen, Jeffrey Zirlin, and Andy Ho. The game crashed in February 2022, losing more than 99% of its value. Sky Mavis tried to sustain the price by introducing new game features; all of the new features made the game more popular and increased the company's profits [38]. Axie Infinity match runs on the Ethereum blockchain, which is integrated with Ronin, a side chain that reduces the fees and transaction delays [5]. Smooth Love Potion (SLP) and Axie Infinity Shards (AXS) are the two main tokens associated with Axie Infinity. It is one of the world's largest games and one of the biggest and most popular NFT projects, allowing users to own their virtual assets and reward the players.

The game's configuration is loaded with Axies, where players collect pets. The goal of the game is to fight, breed, gather, increase, and build kingdoms for their Axies. The game has many features in which players can purchase, sell, and exchange resources earned in-game. Sky Mavis promoted the game using a "play-to-earn" concept, in which players can earn an Ethereum-based cryptocurrency by playing in the virtual environment. Every 14 days, Axie Infinity users can cash out their reward points. The challenge of the game is to destroy all enemies in the Axies. In every move, a player should tactically play cards to increase their chances of winning. Every fight gradually strengthens a player's Axies. In various modes (a player-versus-environment (PvE), player-versus-player (PvP)), players with larger and more powerful Axies have a better chance of winning [46].

7.4.2 Decentraland MANA

Decentraland is a 3D Virtual Online Platform that runs on the web, basically, a Metaverse where you can buy and sell lands. It is the first ever fully decentralized virtual space developed by Ari Meilich and Esteban Ordano in 2015 and launched for public access in February 2020. Many companies began investing in Decentraland Metaverse crypto projects and started investing in virtual properties in the years 2021 and 2022. Decentraland hosted a Metaverse Fashion Week,

1 https://www.techinasia.com/vietnamese-developer-axie-infinity

which included a Metaverse auction some famous brands, such as Adidas, Nike, Samsung, Atari, and Sotheby's participated.

As the name suggests Decentraland is an amalgamation of two words, decentral and land. An exclusive platform allowing users to create, trade, and monetize their digital content and virtual real estate properties in a decentralized and virtual world of Metaverse. The Decetraland software is built on the Ethereum Blockchain, where users can acquire virtual properties as NFTs using the MANA cryptocurrency. The software leverages the Ethereum blockchain to track ownership of digital land, and it requires users to hold its MANA token within an Ethereum wallet [3]. The main tokens used in the Decentraland are as follows:

- LAND – A NFT used to define the ownership of digital real estate.
- MANA – A cryptocurrency that facilitates purchases of LAND, as well as virtual goods like avatars, names, wearables, and services used in Decentraland.

Changes to the Decentraland software are done through a collection of blockchain-based smart contacts, which allow participants who own MANA to vote on policy updates, land auctions, and subsidies for new developments. However, the trading in Metaverse takes place in the Decentraland Marketplace, a one-stop-shop for everything related to land, estates avatars, and various commodities required to build a world [39].

7.4.3 The Sandbox (SAND)

The Sandbox is an Ethereum-based application that runs a decentralized virtual entertainment world enabling users to create their digital avatars and new game assets using the various tools provided in the virtual environment. Later, they can be converted to NFT Tokens for trading. The Sandbox allows players to purchase virtual plots of land, sale of in-game and virtual land, and create virtual real-time experiences. In addition to the NFT LAND token, the Sandbox has introduced two other tokens that guide the operational processes in its Metaverse [71].

- SAND tokens – guide users while performing various transactions such as purchasing virtual land, interacting with other users, and staking it to participate in the governance of a DAO [62].
- ASSET tokens – provide real-time equipment for avatars and facilitate the requirements for creating LAND [62].

Sandbox software leverages the Ethereum blockchain to track the ownership of the digital LAND and NFT ASSETS on its application. Users participating in the Sandbox ecosystem have to hold their SAND tokens in the Ethereum wallets. VoxEdit, Game Maker, and Sandbox Marketplace are the three main products of Sandbox Metaverse that provide a user-friendly experience using User-Generated

Content and also bring promising opportunities for gamers, creators, and designers. However, the Sandbox gives players control over their in-game experiences and also allows them to monetize their digital assets [61].

7.4.4 Enjin (ENJ)

Enjin is an online gaming platform built on the Ethereum blockchain. Enjin Coin was introduced in July 2017 and went live on the Ethereum mainnet in June 2018 by Maxim Blagov. Enjin's goal is to improve the playing experience for both players and developers by providing crypto-backed value and tools, such as software development kits (SDKs), wallets, game plugins, and online payment services. Game players and content providers can efficiently manage, distribute, and exchange digital products using the Enjin platform (NFTs). To begin investments across communities, users must first convert their online products into NFTs through minting. Minting is the process of converting a digital product into a unique token. Users can build their own NFTs on Enjin by linking them to smart contracts. The main objective of Enjin is to provide a safe and secure space to store virtual goods that are used in various games and also to provide complete control over in-game assets to the content developer and game players [10]. The major features of Enjin are as follows [2]:

- **Smart Wallets:** The primary blockchain wallet combines various games and applications by providing dedicated subscription services and enables server owners to run tile-limited services. The smart wallets also validate the connections between every user account on every reliable platform.
- **Public API Accessibility:** Ethereum developers and game developers can use public APIs developed on the JOSN-RPC platform to help users check their account balance without any hindrance and also provide easy access to the smart contract.
- **Escrow-Based Marketplace:** Enji supports an Escrow-based Marketplace for managing multiple transactions of in-game assets and crypto coins, ensuring enhanced user experience by facilitating player-to-player exchanges in a few steps and high-security.

7.5 Conclusions and Future Research Directions

In this chapter, we have presented the use of blockchain in the Metaverse. In particular, we have focused on the following points:

- Demonstrate the underlying mechanisms of blockchain and the Metaverse
- Explore the use cases of blockchain implementation in the Metaverse

- Understand the blockchain-based solutions for dealing with the challenges posed by the Metaverse
- Gain a deep insight into the blockchain projects for the Metaverse

There are some potential research directions:

- Federated learning can be integrated with blockchain-based Metaverse to enhance security and privacy.
- Explainable AI can be integrated with blockchain-based Metaverse as it can create more explainable models that provide a better understanding for Metaverse users and providers.

Bibliography

1 Defining and building the metaverse. URL https://initiatives.weforum.org/defining-and-building-the-metaverse/home. Accessed: 2022-11-01.

2 Ayushi Abrol. What is Enjin (ENJ), and how does it work? 2022. URL https://www.blockchain-council.org/cryptocurrency/enjin/.

3 Ayushi Abrol. Decentraland Metaverse — A Complete Guide, 2022. URL https://www.blockchain-council.org/metaverse/decentraland-metaverse/#::text=Decentraland%20is%20an%20Ethereum%2Dbased,and%20apps%20or%20a%20marketplace.

4 Donald Adams. Virtual retail in the metaverse: Customer behavior analytics, extended reality technologies, and immersive visualization systems. *Linguistic & Philosophical Investigations*, (21):73–88, 2022.

5 Axie. Axie Infinity, 2022. URL https://en.wikipedia.org/wiki/Axie_Infinity#Development_and_history.

6 Russell Belk, Mariam Humayun, and Myriam Brouard. Money, possessions, and ownership in the metaverse: NFTs, cryptocurrencies, Web3 and Wild Markets. *Journal of Business Research*, 153:198–205, 2022.

7 I. B. Bernstein. Drug supply chain security act. *(Eds.): 'Book Drug Supply Chain Security Act'(2017, edn.)*, 2013.

8 Pronaya Bhattacharya, Deepti Saraswat, Amit Dave, Mohak Acharya, Sudeep Tanwar, Gulshan Sharma, and Innocent E. Davidson. Coalition of 6G and blockchain in AR/VR space: Challenges and future directions. *IEEE Access*, 9:168455–168484, 2021.

9 Bharat Bhushan, Preeti Sinha, K. Martin Sagayam, and J. Andrew. Untangling blockchain technology: A survey on state of the art, security threats, privacy services, applications and future research directions. *Computers & Electrical Engineering*, 90:106897, 2021.

10 Maxim Blagov. Enjin NFTs for everyone, 2022. URL https://enjin.io/company.

11 Rachel Breia. Metaverse meaning: Definition, origin and opportunities, 2022. URL https://sensoriumxr.com/articles/metaverse-meaning.

12 Rodney Brown Sr., Soo Il Shin, and Joo Baek Kim. Will NFTs be the best digital asset for the metaverse? In *SAIS 2022 Proceedings*, 2022.

13 Longbing Cao. Decentralized AI: Edge intelligence and smart blockchain, metaverse, Web3, and DeSci. *IEEE Intelligent Systems*, 37(3):6–19, 2022.

14 Dominic Chalmers, Christian Fisch, Russell Matthews, William Quinn, and Jan Recker. Beyond the bubble: Will NFTs and digital proof of ownership empower creative industry entrepreneurs? *Journal of Business Venturing Insights*, 17:e00309, 2022.

15 Luyi Chang, Zhe Zhang, Pei Li, Shan Xi, Wei Guo, Yukang Shen, Zehui Xiong, Jiawen Kang, Dusit Niyato, Xiuquan Qiao, et al. 6G-enabled edge AI for metaverse: Challenges, methods, and future research directions. *arXiv preprint arXiv:2204.06192*, 2022.

16 Yourong Chen, Hao Chen, Yang Zhang, Meng Han, Madhuri Siddula, and Zhipeng Cai. A survey on blockchain systems: Attacks, defenses, and privacy preservation. *High-Confidence Computing*, 2(2):100048, 2022.

17 Rajeswari Chengoden, Nancy Victor, Thien Huynh-The, Gokul Yenduri, Rutvij H. Jhaveri, Mamoun Alazab, Sweta Bhattacharya, Pawan Hegde, Praveen Kumar Reddy Maddikunta, and Thippa Reddy Gadekallu. Metaverse for healthcare: A survey on potential applications, challenges and future directions. *arXiv preprint arXiv:2209.04160*, 2022.

18 LEE David and Low Swee Won. NFT of NFT: Is our imagination the only limitation of the metaverse? *The Journal of The British Blockchain Association*, 5(2):36444, 2022.

19 Natarajan Deepa, Quoc-Viet Pham, Dinh C. Nguyen, Sweta Bhattacharya, B. Prabadevi, Thippa Reddy Gadekallu, Praveen Kumar Reddy Maddikunta, Fang Fang, and Pubudu N. Pathirana. A survey on blockchain for big data: Approaches, opportunities, and future directions. *Future Generation Computer Systems*, 131:209–226, 2022.

20 Haihan Duan, Jiaye Li, Sizheng Fan, Zhonghao Lin, Xiao Wu, and Wei Cai. Metaverse for social good: A university campus prototype. In *Proceedings of the 29th ACM International Conference on Multimedia*, pages 153–161, 2021.

21 Carlos Bermejo Fernandez and Pan Hui. Life, the metaverse and everything: An overview of privacy, ethics, and governance in metaverse. *arXiv preprint arXiv:2204.01480*, 2022.

22 David Frederick. The beauty metaverse: Virtual worlds through digital innovation. *Global Cosmetic Industry*, 177(7):30–33, 2009.

23 Thippa Reddy Gadekallu, Quoc-Viet Pham, Dinh C. Nguyen, Praveen Kumar Reddy Maddikunta, Natarajan Deepa, B. Prabadevi, Pubudu N. Pathirana, Jun Zhao, and Won-Joo Hwang. Blockchain for edge of things: Applications, opportunities, and challenges. *IEEE Internet of Things Journal*, 9(2):964–988, 2021.

24 Thippa Reddy Gadekallu, Thien Huynh-The, Weizheng Wang, Gokul Yenduri, Pasika Ranaweera, Quoc-Viet Pham, Daniel Benevides da Costa, and Madhusanka Liyanage. Blockchain for the metaverse: A review. *arXiv preprint arXiv:2203.09738*, 2022.

25 Arunima Ghosh, Shashank Gupta, Amit Dua, and Neeraj Kumar. Security of cryptocurrencies in blockchain technology: State-of-art, challenges and future prospects. *Journal of Network and Computer Applications*, 163:102635, 2020.

26 Emily Hedrick, Michael Harper, Eric Oliver, and Daniel Hatch. Teaching & learning in virtual reality: Metaverse classroom exploration. In *2022 Intermountain Engineering, Technology and Computing (IETC)*, pages 1–5. IEEE, 2022.

27 Emily Hopkins. Virtual commerce in a decentralized blockchain-based metaverse: Immersive technologies, computer vision algorithms, and retail business analytics. *Linguistic & Philosophical Investigations*, (21):203–218, 2022.

28 Abdul Rehman Javed, Muhammad Abul Hassan, Faisal Shahzad, Waqas Ahmed, Saurabh Singh, Thar Baker, and Thippa Reddy Gadekallu. Integration of blockchain technology and federated learning in vehicular (IoT) networks: A comprehensive survey. *Sensors*, 22(12):4394, 2022.

29 Abdullah Ayub Khan, Asif Ali Laghari, Thippa Reddy Gadekallu, Zaffar Ahmed Shaikh, Abdul Rehman Javed, Mamoon Rashid, Vania V. Estrela, and Alexey Mikhaylov. A drone-based data management and optimization using metaheuristic algorithms and blockchain smart contracts in a secure fog environment. *Computers and Electrical Engineering*, 102:108234, 2022.

30 Liew Voon Kiong. *Metaverse Made Easy: A Beginner's Guide to the Metaverse: Everything you need to know about Metaverse, NFT and GameFi*. Liew Voon Kiong, 2022.

31 Randhir Kumar, Rakesh Tripathi, Ningrinla Marchang, Gautam Srivastava, Thippa Reddy Gadekallu, and Neal N. Xiong. A secured distributed detection system based on IPFS and blockchain for industrial image and video data security. *Journal of Parallel and Distributed Computing*, 152:128–143, 2021.

32 Edward Lee. The bored ape business model: Decentralized collaboration via blockchain and NFTs. *Available at SSRN 3963881*, 2021.

33 Ronald Leenes. Privacy in the metaverse. In *IFIP International Summer School on the Future of Identity in the Information Society*, pages 95–112. Springer, 2007.

34 Devan Leos. How blockchain technology is changing the world from the metaverse to NFTs, 2022. URL https://www.entrepreneur.com/science-technology/how-blockchain-technology-is-changing-the-world-from-the/426489.

35 Zhiyu Lin, Peng Xiangli, Zhi Li, Fuhe Liang, and Aofei Li. Towards metaverse manufacturing: A blockchain-based trusted collaborative governance system. In *The 2022 4th International Conference on Blockchain Technology*, pages 171–177, 2022.

36 Yang Lu. The blockchain: State-of-the-art and research challenges. *Journal of Industrial Information Integration*, 15:80–90, 2019.

37 Taras Maksymyuk, Juraj Gazda, Gabriel Bugár, Vladimír Gazda, Madhusanka Liyanage, and Mischa Dohler. Blockchain-empowered service management for the decentralized metaverse of things. *IEEE Access*, 10:99025–99037, 2022.

38 Sky Mavis. The Dawn of P2E giants? An inside look at axie infinity, 2022. URL https://nftnow.com/guides/inside-axie-infinity-a-deep-dive-into-the-worlds-largest-p2e-game/.

39 Ari Meilich. Decentraland, 2022. URL https://decentraland.org/.

40 Ahmed Afif Monrat, Olov Schelén, and Karl Andersson. A survey of blockchain from the perspectives of applications, challenges, and opportunities. *IEEE Access*, 7:117134–117151, 2019.

41 Michael D. Murray. NFT ownership and copyrights. *Available at SSRN 4152468*, 2022.

42 Stylianos Mystakidis. Metaverse. *Encyclopedia*, 2(1):486–497, 2022.

43 Satoshi Nakamoto. Bitcoin: A peer-to-peer electronic cash system. *Decentralized Business Review*, page 21260, 2008.

44 Cong T. Nguyen, Dinh Thai Hoang, Diep N. Nguyen, and Eryk Dutkiewicz. Metachain: A novel blockchain-based framework for metaverse applications. In *2022 IEEE 95th Vehicular Technology Conference:(VTC2022-Spring)*, pages 1–5. IEEE, 2022.

45 Huansheng Ning, Hang Wang, Yujia Lin, Wenxi Wang, Sahraoui Dhelim, Fadi Farha, Jianguo Ding, and Mahmoud Daneshmand. A survey on metaverse: The state-of-the-art, technologies, applications, and challenges. *arXiv preprint arXiv:2111.09673*, 2021.

46 Alexandra Pankratyeva. Crypto games: Play-to-earn phenomenon, 2022. URL https://capital.com/crypto-games-play-to-earn-phenomenon.

47 Sang-Min Park and Young-Gab Kim. A metaverse: Taxonomy, components, applications, and open challenges. *IEEE Access*, 10:4209–4251, 2022.

48 PixelPlex Team. Decentralized economy – the role of blockchain in the metaverse, 2022. URL https://pixelplex.io/blog/importance-of-blockchain-in-metaverse/.

49 B. Prabadevi, Natarajan Deepa, Quoc-Viet Pham, Dinh C. Nguyen, Praveen Kumar Reddy M, Thippa Reddy, Pubudu N. Pathirana, and Octavia Dobre. Toward blockchain for edge-of-things: A new paradigm, opportunities, and future directions. *IEEE Internet of Things Magazine*, 4(2):102–108, 2021.

50 A. R. Rahul. The inevitable role of blockchain in the metaverse, 2022. URL https://blog.accubits.com/the-inevitable-role-of-blockchain-in-the-metaverse//.

51 Daniel Rigney. *The Matthew Effect: How Advantage Begets Further Advantage.* Columbia University Press, 2010.

52 David Roman. Understanding the metaverse: What is it and how will it shape our future? 2022. URL https://wearebrain.com/blog/innovation-and-transformation-strategy/understanding-the-metaverse-what-is-it/.

53 Bektur Ryskeldiev, Yoichi Ochiai, Michael Cohen, and Jens Herder. Distributed metaverse: Creating decentralized blockchain-based model for peer-to-peer sharing of virtual spaces for mixed reality applications. In *Proceedings of the 9th Augmented Human International Conference*, pages 1–3, 2018.

54 Abdurrashid Ibrahim Sanka and Ray C. C. Cheung. A systematic review of blockchain scalability: Issues, solutions, analysis and future research. *Journal of Network and Computer Applications*, 195:103232, 2021.

55 Bela Shrimali and Hiren B. Patel. Blockchain state-of-the-art: Architecture, use cases, consensus, challenges and opportunities. *Journal of King Saud University-Computer and Information Sciences*, 34(9):6793–6807, 2022.

56 Fengxiao Tang, Xuehan Chen, Ming Zhao, and Nei Kato. The roadmap of communication and networking in 6G for the metaverse. *IEEE Wireless Communications*, 1–15, 2022.

57 Ashwani Kumar Upadhyay and Komal Khandelwal. Metaverse: The future of immersive training. *Strategic HR Review*, 21(3):83–86, 2022.

58 David Vidal-Tomás. The new crypto niche: NFTs, play-to-earn, and metaverse tokens. *Finance Research Letters*, 47(Part B):102742, 2022.

59 Shuai Wang, Wenwen Ding, Juanjuan Li, Yong Yuan, Liwei Ouyang, and Fei-Yue Wang. Decentralized autonomous organizations: Concept, model, and applications. *IEEE Transactions on Computational Social Systems*, 6(5):870–878, 2019.

60 Yuntao Wang, Zhou Su, Ning Zhang, Rui Xing, Dongxiao Liu, Tom H. Luan, and Xuemin Shen. A survey on metaverse: Fundamentals, security, and privacy. *IEEE Communications Surveys & Tutorials*, 25(1):319–352, 2022.

61 Georgia Weston. Know everything about Sandbox metaverse, 2022. URL https://101blockchains.com/sandbox-metaverse/.

62 Georgia Weston. Know everything about Sandbox metaverse, 2022. URL https://www.kraken.com/en-gb/learn/what-is-sandbox-sand.

63 Yang Xiao, Ning Zhang, Wenjing Lou, and Y Thomas Hou. A survey of distributed consensus protocols for blockchain networks. *IEEE Communications Surveys & Tutorials*, 22(2):1432–1465, 2020.

64 Hu Xiong, Chuanjie Jin, Mamoun Alazab, Kuo-Hui Yeh, Hanxiao Wang, Thippa Reddy Gadekallu, Weizheng Wang, and Chunhua Su. On the design of blockchain-based ECDSA with fault-tolerant batch verification protocol for blockchain-enabled IoMT. *IEEE Journal of Biomedical and Health Informatics*, 26(5):1977–1986, 2021.

65 Minrui Xu, Wei Chong Ng, Wei Yang Bryan Lim, Jiawen Kang, Zehui Xiong, Dusit Niyato, Qiang Yang, Xuemin Sherman Shen, and Chunyan Miao. A full dive into realizing the edge-enabled metaverse: Visions, enabling technologies, and challenges. *arXiv preprint arXiv:2203.05471*, 2022.

66 Jeong Ok Yang and Jook Sook Lee. Utilization exercise rehabilitation using metaverse (VR· AR· MR· XR). *Korean Journal of Sport Biomechanics*, 31(4):249–258, 2021.

67 Qinglin Yang, Yetong Zhao, Huawei Huang, Zehui Xiong, Jiawen Kang, and Zibin Zheng. Fusing blockchain and AI with metaverse: A survey. *IEEE Open Journal of the Computer Society*, 3:122–136, 2022.

68 Yin Yang, Keng Siau, Wen Xie, and Yan Sun. Smart health: Intelligent healthcare systems in the metaverse, artificial intelligence, and data science era. *Journal of Organizational and End User Computing (JOEUC)*, 34(1):1–14, 2022.

69 Jongyoung Yoo. A study on transaction service of virtual real estate based on metaverse. *The Journal of the Institute of Internet, Broadcasting and Communication*, 22(2):83–88, 2022.

70 Rui Zhang, Rui Xue, and Ling Liu. Security and privacy on blockchain. *ACM Computing Surveys (CSUR)*, 52(3):1–34, 2019.

71 Steven Zhao. The Sandbox is a community-driven platform, 2022. URL https://www.sandbox.game/en/about/sand/.

72 Zibin Zheng, Shaoan Xie, Hong-Ning Dai, Xiangping Chen, and Huaimin Wang. Blockchain challenges and opportunities: A survey. *International Journal of Web and Grid Services*, 14(4):352–375, 2018.

8

Edge Computing Technologies for Metaverse

Minrui Xu and Dusit Niyato

School of Computer Science and Engineering, Nanyang Technological University, Singapore

After reading this chapter, you should be able to:

- Gain an understanding of how edge-computing technologies provide the fundamental infrastructure for characteristic services in the Metaverse, such as high-dimensional data processing, 3D virtual world rendering, and virtual avatar computing from communications, networking, and computing.
- Establishing a concept of real-time rendering and streaming of virtual reality services in the Metaverse and understanding how the bidding, asking, and transaction are completed in the virtual reality (VR) service market.
- Obtain an insight that, in the bidirectional synchronization of physical and virtual worlds in the Metaverse, the synchronizing entities in physical and virtual worlds have a mutual impact on the continuous synchronization.

Edge-computing technologies are considered to be one of the fundamental infrastructures for the realization of the full-dive Metaverse. Edge computing refers to the deployment of edge servers with cloud computing capabilities at the edge of networks that can be accessed by end devices via wireless connections. On the one hand, edge servers provide Metaverse users with extremely low-latency access to edge servers for computing physical rules, rendering 3D environments, and training and inference of artificial intelligence (AI). On the other hand, edge servers provide real-time orchestration for immersion content, such as augmented reality (AR) and virtual reality (VR), streaming, and social interactions among Metaverse users.

In this chapter, we first provide an overview of the Metaverse enabled by edge-computing technologies, i.e. the edge-enabled Metaverse. Specifically, we discuss how the edge-enabled Metaverse can be realized from the perspectives

Metaverse Communication and Computing Networks: Applications, Technologies, and Approaches, First Edition.
Edited by Dinh Thai Hoang, Diep N. Nguyen, Cong T. Nguyen, Ekram Hossain, and Dusit Niyato.
© 2024 The Institute of Electrical and Electronics Engineers, Inc. Published 2024 by John Wiley & Sons, Inc.

of communication and computation, respectively. In contrast to traditional AR/VR, the *immersive* Metaverse is characterized by massive user interaction and differentiated services. We focus on next-generation communication architectures such as human-in-the-loop communication and real-time synchronization between the physical and virtual worlds. This allows readers to develop a better understanding of how future communication systems can potentially play their roles in edge computing technologies for the Metaverse. To realize the *ubiquitous* Metaverse, we leverage edge computing for AR/VR rendering, avatar computing, and AI training and inference. Enabled by edge-computing technologies, the Metaverse can be accessed by users and providers anytime and anywhere, especially via resource-constrained edge devices. Then, investigation and implementation opportunities and challenges brought by the edge-enabled Metaverse are highlighted in this chapter.

Finally, to concretize the concept and opportunities of edge-enabled Metaverse, we carefully select two typical examples of the edge-enabled Metaverse, i.e. remote rendering for nonpanoramic VR in edge-enabled Metaverse and physical–virtual synchronization in vehicular Metaverse. In nonpanoramic VR, VR users in the wireless edge-enabled Metaverse can immerse in the virtual worlds through head-mounted displays (HMD) that is the access to VR services offered by different providers. However, VR applications are computation and communication-intensive. Therefore, VR service providers have to optimize their VR service delivery efficiently and economically, given the limited communication and computation resources at the edge. Meanwhile, in the vehicular Metaverse, accessing the Metaverse via vehicles, drivers, and passengers can immerse in and interact with 3D virtual objects overlaying the view of physical streets on head-up displays (HUDs) via AR. The seamless, immersive, and interactive experience rather relies on real-time multidimensional data synchronization between physical entities, i.e. vehicles, and virtual entities, i.e. roadside recommenders.

8.1 An Overview of Edge-enabled Metaverse

Edge-computing technologies will provide immersive and ubiquitous services to Metaverse users at the place close to where the data of users is generated, i.e. the edge-enabled Metaverse [39]. We discuss the realization from the communication and computing perspectives about edge-computing technologies for the Metaverse, respectively. In detail, ubiquitous edge devices and sensors access the edge-enabled Metaverse through radio access networks. Therefore, efficient communication and networking coverage with high-speed, low-delay, and wide-area wireless accessibility has to be provided so that users can immerse themselves in the Metaverse and have a seamless and real-time experience.

Beyond reliable and low-latency communication systems, the resource-intensive Metaverse services required to connect to the 3D virtual worlds will require costly computation resources to support [9]. For example, the implementation of the Metaverse is based on the execution of the following computation tasks:

- **High-Dimensional Data Processing:** The physical and virtual worlds will generate vast quantities of high-dimensional data, such as spatiotemporal data [37], that allows users to experience realism in the Metaverse. For example, when a user drops a virtual object, its behavior should be governed by the laws of physics for realism. The virtual worlds with physical properties can be built upon physics game engines, such as the recently developed real-time ray tracing technology to provide the ultimate visual quality by calculating more bounces in each light ray in a scene [24]. During these activities, this generated high-dimensional data will be processed and stored in the databases of the Metaverse.
- **3D Virtual World Rendering:** Rendering is the process of converting raw data of virtual worlds into displayable 3D objects. When users are immersed in the Metaverse, all the images/videos/3D objects require a rendering process to display on AR/VR devices for users to watch. For example, Meta operates research facilities that set up collaborations with museums to transform 2D paintings into 3D [23]. In addition, the Metaverse is expected to produce and render 3D objects in an intelligent way. For example, 3D objects, e.g. clouds, islands, trees, picnic blankets, tables, stereos, drinks, and even sounds, can be generated by AI (i.e. AI-generated content [AIGC]) through audio inputs.
- **Avatar Computing:** Avatars are digital representations of users in the virtual worlds that require ubiquitous computation and intelligence in avatar generation and interaction. On the one hand, avatar generation is based on the machine learning (ML) technologies, such as computer vision (CV) and natural language processing (NLP). On the other hand, the real-time interaction between avatars and users is computationally intensive to determine and predict the interaction results. For example, AI can synergize with AR applications to provide avatar services such as retrieving text information from the users' view, text amplification on paper text, and using hand gestures to copy and paste files between computers [30].

8.1.1 Communication and Networking

The level of fidelity, reliability, and latency requirements for AR/VR and haptic immersive streaming ensures that there is no break-in-present of Metaverse services of users [25]. Data traffic growth in the edge-enabled Metaverse is expected to be exponential, which may cause the constrained edge computing capacity to be alleviated by the high connectivity of edge networks. As a consequence, the B5G/6G communication infrastructure is crucial to alleviating bandwidth and latency constraints by efficiently managing edge resources as the ubiquitous connectivity grow dramatically.

To address these issues, cutting-edge communication and networking solutions should be provided for the immersive experience in the edge-enabled Metaverse. By seamlessly delivering 3D multimedia services, users are able to immerse themselves in the Metaverse. As a result, the virtual worlds and physical entities can synchronize seamlessly through VR streaming and AR adaptation. The edge communication and networking to support the Metaverse should prioritize user-centric considerations in the content delivery of communication and networking support for the Metaverse. AR/VR and the tactile Internet are context-aware and personalized content-based services that are included by the Metaverse. A paradigm shift in the focus of classical information theory is also necessary due to the explosive growth of data traffic of multimedia services and limited communication resources at the edge. In response, the semantic/goal-oriented communication solutions [41] open the gate toward Metaverse-native communications that can serve to alleviate the spectrum scarcity for next-generation multimedia services. For the Metaverse to be constructed, it is essential to have bidirectional physical–virtual synchronization in real time. It is necessary to leverage intelligent communication infrastructures like reconfigurable intelligence surfaces (RIS) and unmanned aerial vehicles (UAV) to integrate physical and virtual worlds.

8.1.1.1 Rate-Reliability-Latency 3D Multimedia Networks

With embodied experience enabled by AR/VR in the Metaverse, wireless communication and network infrastructure must provide seamless, reliable, and low-latency communication and networking services [26]. The high volume of data exchanged between virtual worlds and physical worlds requires a holistic edge networking and communication infrastructure to balance the trade-off between rate, reliability, and latency for 3D multimedia services of users to traverse between the physical and virtual worlds.

It is important to consider the adequate transmission rate to support round-trip interactions between the virtual and the physical worlds (e.g. for the user and IoT devices engaging in the physical–virtual synchronization) [15, 21, 22]. The second challenge is the interaction latency, which hinders users from experiencing realism in immersive services [43]. Players, for instance, can play massively multiplayer online games with ultra-low latency thanks to mobile edge networks. Due to latency, players receive information about their status in virtual worlds and send their responses to other players quickly. Third, the reliability of physical network services refers to the frequency of break-in-present and its impact on users immersing in the Metaverse [34]. Moreover, virtual service providers and physical service providers are challenged with allocating edge resources for AR/VR services based on the dynamic reliability requirements of AR/VR applications and users.

8.1.1.2 Human-in-the-loop Communication

In the Metaverse, there will be massive, AI-supported communication within a human-centric 3D virtual space. As wireless networks become more human-centric, we can expect to see some innovative applications, such as holograms, AR/VR, and the tactile Internet, supporting various Metaverse services [10, 33]. Simulating human perception, cognition, and physiology to formulate haptic interactions is essential in the creation of highly simulated mixed-reality (XR) systems [2, 16, 31]. During haptic communication, haptics refers to both kinesthetic perceptions (information about forces, torques, positions, and velocity sensed by muscles, joints, and tendons) as well as tactile perceptions (the sensation of surface friction and texture that the human body receives through different types of mechanoreceptors). To assess the quality of services in the Metaverse, new metrics, such as the quality of physical experience (QoPE) [26], are proposed to evaluate the quality of human-in-the-loop communication. Traditionally, quality of service (QoS) (i.e. latency, rate, and reliability) and quality of experience (QoE) (i.e. average opinion score) are applied in measuring the physiological and psychological factors of individual users separately instead of blending. In addition to cognition, physiology, and gesture, several human-related activities can be influenced by the QoPE.

8.1.1.3 Real-time Physical-Virtual Synchronization

The virtual and physical worlds are blurred by the Metaverse, which is an unprecedented medium [9] for synchronization among physical and virtual entities. The Metaverse, for instance, can enhance the performance of edge computing technologies through offline simulation and then online decision-making by transforming the physical wireless environment into virtual worlds. A widely distributed coverage of edge devices and edge servers is required to maintain bidirectional synchronization between the Metaverse and the physical world. As shown in Figure 8.1, one of the possible solutions for such virtual services is the DT [14], i.e. digital replications of real-world entities in the Metaverse, e.g. city twins and copies of smart factories. Based on the modeling, simulation, and prediction for physical entities, the DT can provide data from historical digital replications. In the Metaverse, edge networks can be made more efficient through intratwin communication and intertwin communication. A particular application of the Metaverse for edge computing technologies is monitoring the DTs of edge devices and infrastructure, such as RIS, UAV, and space–air–ground integrated

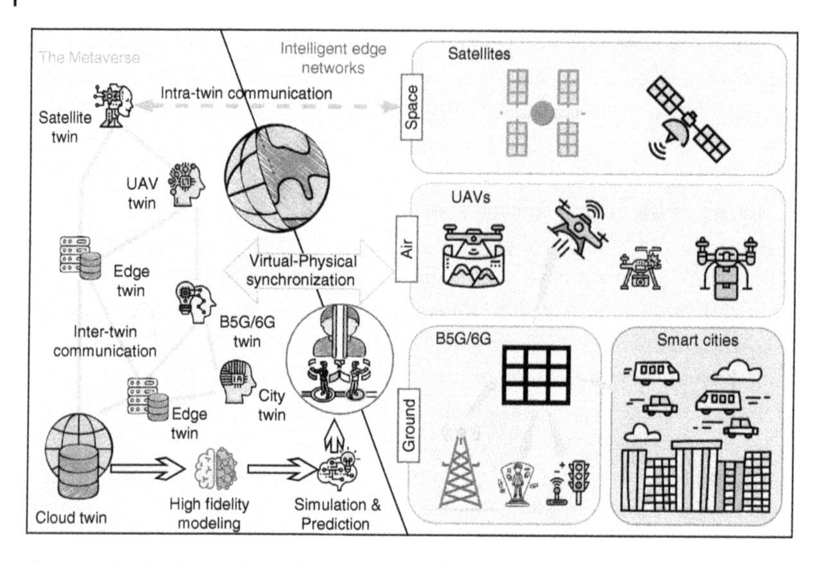

Figure 8.1 An illustration of real-time physical–virtual synchronization between the Metaverse and Intelligent edge networks.

networks (SAGIN) [7, 8, 40], in which physical systems can be instantly calibrated through the Metaverse.

8.1.2 Computation at the Edge

There are currently more than one million players online at the same time playing massively-multiplayer online (MMO) games, which requires high-performance GPUs for real-time rendering and interaction orchestration. Despite the potential of virtual reality massively multiplayer online (VRMMO) games, their appearance in the Metaverse remains scarce in the game industry. VRMMO games need to render immersive virtual worlds and interact with hundreds of players, and thus user devices such as HMDs have to connect to powerful edge/cloud servers for remote rendering. The Metaverse will be accessible to everyone through cloud-edge-end computation, as shown in Figure 8.2, that provides ubiquitous computing and intelligence for users and service providers in the Metaverse. An avatar's position and movement can be determined by the physics engine via local computations on edge devices. This way, the overhead and the latency can be reduced by edge servers to execute expensive foreground render, which requires less graphical details but tend to be faster [13]. It is also possible to execute cloud-based tasks that are more computation-intensive but less sensitive to delay, such as background rendering of the Metaverse scenes. Additionally, with the benefits of AI techniques, such as model pruning, compression, and distributed learning, the edge networks can be relieved of much of the burden.

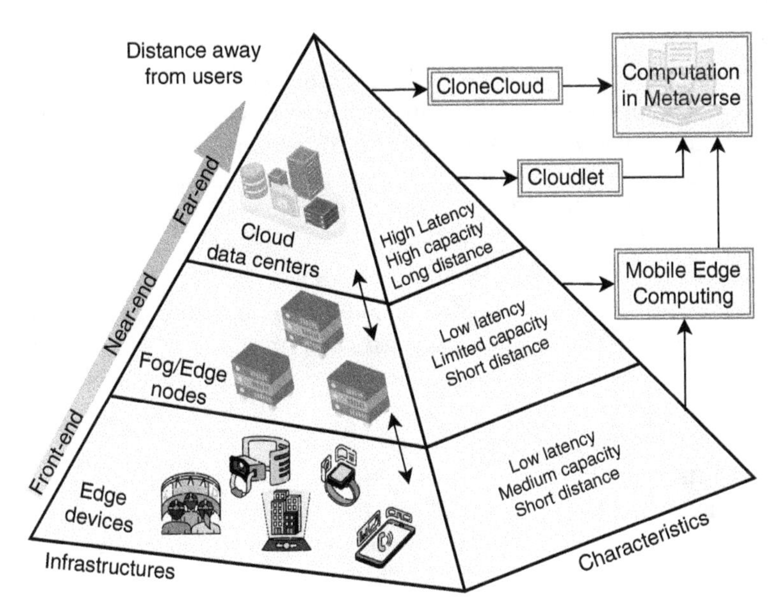

Figure 8.2 Various types of computing infrastructure to support the computation in the Metaverse and their characteristics.

8.1.2.1 Efficient AR/VR Cloud-Edge-End Rendering

In addition to wired devices, the edge-enabled Metaverse offers mobile users in the physical world the ability to experience virtual worlds via AR/VR by using VR HMDs and AR goggles. On the one hand, HMDs generate sensory images that constitute the virtual worlds [9]. For the Metaverse to operate smoothly, users need to be able to render the sensory pictures in real time so that continuous flows can be created instead of discrete events. VR can enhance lesson delivery in the virtual world, for instance, by incorporating education platforms. On the other hand, AR employs digitally modified or augmented physical reality to interact with the user in various ways. Metaverse users benefit most from AR applications when they are physically present in the real world, such as while on the job or working with a computer [17]. However, it is important to note that users' devices have limited computational capacity, memory storage, and battery life. These devices cannot support the intense AR/VR applications needed for the immersive Metaverse. Users' devices accessing the Metaverse can leverage ubiquitous computing resources to render and offload tasks remotely via cloud-edge-end collaborative computing to overcome these challenges [5].

The cloud-edge-end collaborative computing architecture in edge-computing technologies enables mobile users to perform computations, such as analytics, rendering, and avatar computing, where the data are created [32]. A vehicle can, for instance, send the data to nearby vehicles or roadside units rather than offloading it to the cloud when the user interacts with avatars in the Metaverse to perform task computations, thereby dramatically reducing the end-to-end latency. Collaboration between the cloud and mobile edge networks is a promising solution to reduce latency for mobile users [3] utilizing mobile edge networks' computing capabilities [1]. The Metaverse should be accessible no matter where mobile users are located. By shifting computations away from the core networks, mobile edge computing also reduces network traffic. An edge server provides computing resources to mobile edge users at mobile edge networks [3]. The authors in [42] define binary and minimum offloading operations for mobile users based on criteria such as partitioning tasks into smaller subtasks. This will improve the QoS of the users by transmitting VR content to nearby edge servers for immediate rendering.

8.1.2.2 Scalable AI Model Training

By providing intelligent and personalized virtual services, such as AR/VR recommender systems and cognitive avatars, the Metaverse can deliver a better user experience. For this vision to be realized, AI will have to be integrated, such as speech recognition and content generation. Avatars in the Metaverse can even be emulated with AI for intelligent decision-making by nonplayer characters (NPCs). Metaverse users can create their personalized avatars with tools provided by the Metaverse platforms. The complexity of today's deep learning models makes them computationally and storage-intensive to train and predict. Edge devices with limited resources, such as VR HMDs and AR goggles, are especially vulnerable to this problem. In addition to the computation offloading solutions discussed above, deep learning (DL) model compression and acceleration can also help improve the computation efficiency of AI models for edge devices. By compressing AI models, the processing speeds can be improved as well as storage and communication costs can be largely reduced [6]. In this way, it is possible to generate and deploy large-scale AI models locally on edge devices with limited resources.

8.1.2.3 Computational Privacy and Security

As AR/VR emerges in the Metaverse, it will enhance immersive content delivery. However, new accessing devices will also allow for multidimensional data collection in new ways. For example, users' eye-tracking data may be captured by HMDs rather than smartphones since HMDs are used instead of smartphones. This data can be leveraged to help improve VR rendering efficiency as well as allow companies to better sell their products by determining how long people are willing to spend on them [44]. However, it is possible to devise a risk assessment framework to predict the likelihood of any privacy or security intrusion occurring [12]. By utilizing the framework, administrators can prioritize threats and have a better understanding of the vulnerability of AR/VR system components.

Metaverse applications, such as AR/VR, can transform several industries and alter our lifestyles while providing new privacy and security issues for edge-computing technologies. As the Metaverse becomes more widely adopted, it is crucial to implement policies that protect users. In addition, users are in danger of being impacted in new ways due to the value of security and data protection. To ensure the security of edge devices, AR applications should be restricted from accessing sensor data [18]. Users upload images or video streams to deliver AR content, which can be shared with third-party servers and thus allow sensitive information to be leaked. The user should not be exposed to any unnecessary data, such as gestures and voice commands, in order to maintain an immersive experience while protecting their data. The correct inputs must, however, be verified, and unwanted content must be blocked. An attacker can access the user's data by copying voice commands and physical gestures without authentication. In order to protect voice access, it is possible to use a voice-spoofing defense system to determine whether the voice command is coming from the environment or is being generated by the person using the device [29].

In the Metaverse, federated learning and adversarial machine learning can both be used to enhance the privacy and security of AR/VR technologies. Our discussion will focus on trusted execution environments (TEEs), federated learning, and adversarial machine learning techniques. However, they will add extra overhead and degrade performance during the training and inference at edge devices and servers.

8.2 Opportunities and Challenges in the Edge-enabled Metaverse

8.2.1 Opportunities and Challenges in Edge Communication

Implementing the Metaverse with edge-computing technologies will initially enable efficient immersive streaming and social interaction for network edge users. To achieve an efficient allocation of communication resources, AI technologies will first transform the traditional edge network architecture into intelligent edge networks. Based on different user preferences and interaction contexts of Metaverse users, the edge-enabled Metaverse will deliver the content in advance to the edge servers near the users based on their preferences during the immersive content streaming. In addition, ubiquitous physical–virtual synchronization will autonomously form a sustainable control loop for the edge networks and the Metaverse, as the resources required for physical–virtual synchronization of the Metaverse can be provided by the virtual-to-physical optimization from the Metaverse. Finally, decentralized wireless access and incentives also enable the

Metaverse to achieve security and privacy protection over the communication access infrastructure.

Efficient Immersive Streaming and Interaction: In the Metaverse, content is delivered in multimodal formats using VR, AR, and tactile Internet technologies. A massive number of users will interact and coexist in the virtual worlds within the Metaverse, unlike traditional AR/VR, which tends to focus on single-user scenarios. A massive number of users at the edge of the network require efficient resource allocation, hence optimizing service delivery efficiency. Additionally, users will have to interact with real-time 3D FoVs and perform calculations related to real-time communication networks and rendering.

AI for the Intelligent Edge Communication: To efficiently manage communication resources in the complex and dynamic environment of networks, AI methods are required. "AI for edge" approaches utilize AI to allocate resources efficiently at the edge, e.g. for task offloading and bandwidth allocation. By using AI, traditional optimization tools can be used in combination, for example, to reduce communication costs and speed up the convergence of auctions for pricing physical service provider services. However, training and storing such AI models on devices with limited resources are computationally expensive.

Context-Aware Immersive Content Delivery: Based on current FoVs, view angles, and interacting objects, Metaverse virtual service providers deploy a cache of immersive content to physical service providers. Therefore, the user perceives lower latency, and the bandwidth needed for immersive streaming and interactions is reduced.

Self-sustainable Physical–Virtual Synchronization: It is self-sustaining to synchronize the physical and virtual worlds. When the physical–virtual synchronization is going on, the physical world maintains DTs in the virtual world. This requires a significant amount of communication resources. However, the DKs in virtual worlds can transform physical environments into smart wireless networks using physical–virtual synchronization to reduce communication resources.

Decentralized Incentive Mechanism: Wireless sensors in the physical world provide vast amounts of data to the Metaverse. In general, physical service providers and virtual service providers tend to be self-interested and do not wish to share communication resources or energy with other organizations. The long-term sustainability of the Metaverse depends on the right incentives for wireless sensors.

8.2.2 Opportunities and Challenges in Edge Computing

To provide ubiquitous access to Metaverse devices and infrastructure, edge computing must provide adaptive AR/VR cloud-edge co-rendering for Metaverse users' devices. In addition, to support ubiquitous intelligence in the Metaverse, edge AI trainers should leverage on-demand and generalized model compression to perform AI training and inference on heterogeneous edge devices. User-centric computing is important because during Metaverse virtual–real synchronization, Metaverse user data are used to optimize Metaverse services, which also risks exposing private information in the data. Finally, secure interoperable computation between Metaverses is also required so that users can seamlessly switch between different Metaverses.

Adaptive AR/VR Cloud-Edge-End Rendering: To begin with, users can offload AR/VR tasks to nodes in a network that have the necessary computing power for rendering by leveraging ubiquitous computing and intelligence in mobile edge networks. In the cloud-edge-end collaborative computing paradigm, stragglers at edge networks, heterogeneous tasks, stochastic demand, and network conditions, and stragglers at edge networks are among the performance bottlenecks in AR/VR rendering that can be adaptively overcome.

On-demand and Generalized Model Compression: To access the Metaverse, local devices can benefit from model compression. Despite these limitations, some devices will still have storage, processing power, and energy constraints. Further compression of the model must be implemented on-demand so that the model can be compressed even further. To achieve a balanced on-demand model compression, it should be studied as the trade-off between accuracy and user quality of experience. The Metaverse's computational interoperability can also be enhanced by using a generalized model compression technique to determine the most suitable compression level or technique for heterogeneous tasks.

User-Centric Computing: For the Metaverse to be trusted and operated without compromising user privacy, user-centric computing must enforce user privacy responsibly. The virtual and physical worlds can be protected, i.e. both simultaneously. Applications in virtual worlds must protect sensitive information when they perform computation on edge devices. GPS locations, voice recordings, and eye movements captured by an external gadget, such as an AR/VR device, should also be protected in the real world. The data minimum principle is essential to ensuring that users' privacy is protected, while data-hungry AI algorithms need a lot of such input.

Secure Interoperable Computing: Finally, it is also possible, however, for users' data to be stored on different edge/cloud servers in practice. As a result, edge or cloud offloading can be complicated due to different development frameworks being used for virtual worlds. Hence, distributed storage solutions are important for users to be able to move between physical and virtual environments.

8.3 Edge-Enabled Metaverse: Release the Ubiquitous Computing and Intelligence at the Edge

8.3.1 VR Remote Rendering via Edge Computing Technologies

8.3.1.1 Background

In the Metaverse, human players can interact with immersive applications, e.g. VR, as digital avatars in the 3D virtual worlds [20]. As described in the famous film *Ready Player One*, human players can become avatars and interact with the Metaverse with the immersive experience. AR goggles and VR HMDs are the accessing devices that provide users with immersive virtual experiences in the Metaverse [4]. As envisioned in *Ready Player One*, advanced technologies today allow users to build the Metaverse, including DT that model the real world, AR/VR for immersive experiences via HMDs, as well as blockchains that run distributed ledgers and decentralized autonomous organizations. Metaverse users are able to access the 3D virtual worlds anywhere and anytime, thanks to 5G and beyond wireless networks and edge computing/intelligence technologies. Several Metaverse applications, such as VRMMO video games and virtual concerts, can be experienced seamlessly immersive in wireless edge-enabled Metaverse applications with ultralow latency and high reliability [9].

8.3.1.2 Motivation

It is still necessary to further investigate the economic system design of the edge-enabled Metaverse (especially the incentive mechanisms) for motivating and incentivizing both users and service providers to participate in the Metaverse interaction, even though the wireless edge-enabled Metaverse has many advantages. To provide a quality gaming experience, VR users require extensive communication and computation resources from service providers. It is important to recognize that in designing economic mechanisms in the wireless edge-enabled Metaverse, three main challenges exist. The first advantage of wirelessly edge-enabled Metaverse is that it reduces the communication and computation costs of wireless devices by rendering VR without a panoramic view [19]. This highlights the need to rethink the Metaverse's utility for VR users and service providers by emphasizing the quality of perceptual experience. As a result, VR users and service providers interact in a bilateral manner in the Metaverse service market. In addition, the Metaverse service market, unlike continuous trading markets, is a call market requiring efficient matching and pricing between VR users and service providers.

A promising solution to these problems is the double Dutch Auction [11], which generalizes Dutch auctions to the double-sided market. Two Dutch clocks are used by the auctioneer in the double Dutch auction, a blockchain-based decentralized autonomous organization, to match and determine transactions between buyers (VR users) and sellers (VR service providers). Based on the current price of VR services, the Buyer clock shows the current cost of VR services, and the Seller clock shows the current cost of VR services, starting from low and ascending over time. Iterative and asynchronous bidding are characteristics of the double Dutch auction process. Each time the auctioneer holds an auction round, he broadcasts the current auction clock to buyers and sellers. As long as the market price is at least as favorable as the prices of the clocks they want to buy and sell, the buyers and sellers are willing to deal. A clock is automatically restarted at the same value as when it was stopped during an auction. The same clock price can be applied to multiple units. A successful auction ends when the two clocks cross, and the auctioneer clears the market at the price that the winning buyers and sellers agreed to at the crossing point of the two clocks. A deep reinforcement learning (DRL) algorithm allows an auctioneer to adjust the clock stepsize dynamically during an auction without having prior knowledge about auctions. The learning-based auctioneer can therefore increase auction efficiency while guaranteeing social welfare.

8.3.1.3 Contribution

As shown in Figure 8.3, an incentive mechanism framework based on DRL [27] is proposed here for managing and allocating nonpanoramic VR services between service providers and users in wireless edge-powered Metaverses. The researchers can use the double Dutch auction call market, in which the auction is completed in a finite amount of time, to efficiently determine allocation and pricing rules for VR services. It is possible for auctioneers acting as learning agents in DRL to learn the near-optimal policy through interaction with the environment without prior auction knowledge, thus increasing the efficiency of the double Dutch auction and

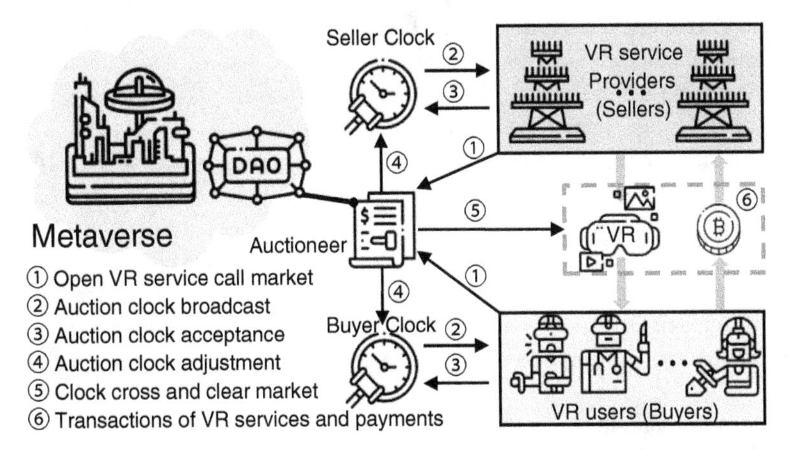

Figure 8.3 A typical framework of the VR service market in the wireless edge-empowered Metaverse.

thus reducing auction information exchange costs within the VR service market. In detail, the researchers can present an efficient learning-based incentive mechanism framework for wireless edge-enabled Metaverse to evaluate and reinforce the seamless experience of immersion of VR users. In detail, the researchers present an incentive mechanism to improve seamless immersion for VR users using wireless edge-enabled Metaverse services. As a dynamic way of matching and pricing VR services for users and VR service providers, the researchers propose a double Dutch auction instead of the existing double auctions for continuous trading markets. As long as incentive rationality, truthfulness, and budget balance are ensured, service providers and users can trade asynchronously while offering buy-bids and sell-bids simultaneously. In order to maximize social welfare and communicate efficiently, the researchers designed a DRL-based double Dutch auction. To reduce auction information exchange costs and learn the efficient auction policy without prior knowledge, i.e. the auctioneer will act as the DRL agent by interacting with the VR service market.

8.3.2 Edge-Enabled Physical–Virtual Synchronization

8.3.2.1 Background

It is expected that the vehicular Metaverse will lead to the evolution of intelligent transportation systems, with ubiquitous physical and virtual entities, such as vehicles, roadside recommenders, and roadside units interconnecting [38]. In the vehicular Metaverse, the physical–virtual synchronization system is maintained from two directions, as shown in Figure 8.4. By sensing their surrounding environments (e.g. landmarks and landscapes) with internal and external sensors, vehicles update their DTs in the virtual world continuously during the physical-to-virtual (P2V) synchronization [28, 35]. In the virtual world, they can also execute DT tasks, such as real-time vehicular status, historical sensing data, and passenger bio-data. Then, vehicles share the DTs with road-side receivers (RSRs) via RSUs the *preference caches*.[1] Lastly, RSRs send personalized AR recommendations and advertisements through the virtual-to-physical (V2P) synchronization process to vehicles and display them on their HUDs [36]. Drivers and passengers can benefit greatly from this bidirectional synchronization. For vehicular Metaverse real-time and low latency physical–virtual synchronization, both vehicles and RSRs must execute and update DT tasks and render and display AR recommendations and advertisements. A synchronization service can be provided to vehicles and RSRs by RSUs since they have high-capacity

1 Preference caches store intersections between historical sensing data and passengers' context-aware preferences and needs during a voyage.

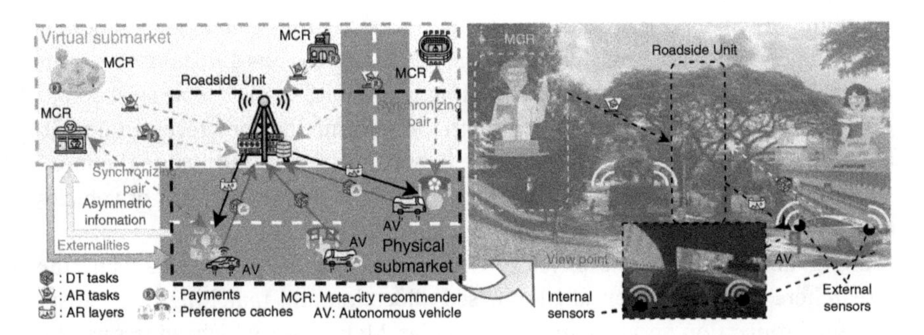

Figure 8.4 The physical–virtual synchronization system in the vehicular Metaverse.

communication and computing resources. RSUs can perform resource-intensive tasks such as DT and AR by offloading them from vehicles to RSRs when vehicles and RSRs are synchronized [39]. While maintaining synchronization services will consume the limited resources of RSUs, compensation is expected to be required. As a result, the vehicular Metaverse should be designed to allow vehicles and RSR pairs to synchronize in real time and charge each other, thus motivating RSUs to share their resources.

8.3.2.2 Motivation

A lot of research has been dedicated to developing auction-based mechanisms to handle private information in synchronization services without introducing excessive latency. Physical–virtual synchronization can be challenging due to the limitations and the unsuitability of existing solutions. The reason they are not able to synchronize physical and virtual entities is that they allocate them separately. Consequently, they introduce unnecessary delays during the physical–virtual synchronization of vehicles and RSRs via RSUs, resulting in a detrimental experience for passengers. Second, they assume that the synchronizing virtual entities will not have any external effects (or externalities) during the allocation of the physical entities in the P2V synchronization procedure. RSRs, for example, may introduce externalities to the allocation of vehicles because the match qualities and rendering delays of their recommendations and advertisements vary. As a result of the unpredictable characteristics and delays, vehicles are exposed to *adverse selection*[2] that the vehicles demand RSUs to set a prefixed threshold of synchronization delays to minimize externalities. There is yet another concern. The existing solutions are mostly based on the assumption that virtual entities

2 The adverse selection occurs when participants with asymmetric information will only pay the average market price for a vehicle. This might introduce an inefficient allocation outcome in the market.

are symmetrical observers of reality. During V2P synchronization, every RSR may not be able to measure how accurately the recommendations and advertisements match. The asymmetric information regimes expose RSRs to adverse selection as they cannot adjust their bids according to the information they receive immediately. This may result in them losing the opportunities that they value the most. In the synchronization market, this results in the adverse selection of vehicles and RSRs and the deterioration of allocation efficiency.

Using an auction-based physical–virtual synchronization mechanism, the researchers can propose an auction-based allocation and matching mechanism for synchronization services in the vehicular Metaverse, which addresses the problem of adverse selection and improves allocation efficiency. Vehicles and RSRs that are synchronizing in the physical and virtual submarkets are determined by the physical–virtual synchronization mechanism. The researchers' proposal for a synchronization-scoring rule incorporates information from RSRs in the virtual submarket to address the adverse selection problems of vehicles in the physical submarket. In accordance with vehicle preferences, RSUs can adjust processing deadlines according to efficient synchronization scoring rules. To allocate and price the synchronizing RSRs with high-quality recommendations and advertisements, enhanced allocation and pricing rules with a price-scaling factor are proposed in the virtual submarket for V2P synchronization. The surplus in the virtual submarket increases as a result of the alleviation of asymmetric information among RSUs. We first demonstrate the allocation efficiency by comparing it with an omniscient mechanism and a second-price auction-based mechanism. As compared to the omniscient benchmark, the proposed mechanisms have a lower bound on their allocation efficiency.

8.3.2.3 Contribution

A new physical–virtual synchronization system for the vehicular Metaverse is proposed, in which RSUs perform DT tasks to help vehicles and RSRs synchronize with the virtual world while rendering AR recommendations and advertisements to help RSRs synchronize with the physical world. In the proposed system, the researchers design a real-time synchronization service market where service providers, i.e. RSUs, allocate and match the synchronizing pairs of vehicles and RSRs, whose surpluses are positively correlated in the physical and the virtual submarkets. Vehicles and RSRs for RSUs are matched and allocated simultaneously to solve the adverse selection problem in existing synchronization mechanisms. A scoring rule for synchronizing vehicle allocation in the physical world is proposed to minimize the external effects of the virtual world. RSRs' recommendations and advertisements are matched by a pricing factor in the virtual world to mitigate asymmetric information. An extensive analysis and experiment are conducted to determine the effectiveness of the proposed mechanism.

8.4 Conclusions and Future Research Directions

In this chapter, we describe edge-computing technologies that can support the implementation of the Metaverse in both communication and computation. First, we introduce an important concept of the edge-enabled Metaverse, namely the location where the Metaverse data is generated, i.e. the edge of the network where the users of the Metaverse are located. On the one hand, in the edge-enabled Metaverse, the existing edge network infrastructure will deliver immersive content in a trade-off of rate, reliability, and latency, and the future edge network architecture will provide the infrastructure to enable human-in-the-loop communication and physical–virtual real-time synchronization, and provide the necessary technology and theory to achieve human-in-the-loop communication and physical real-time synchronization. On the other hand, ubiquitous computing and intelligence at the edge of the network enable resource-constrained user devices to offload compute-intensive tasks to edge servers for execution. Edge intelligence is seen as the cornerstone of the edge computing-enabled Metaverse and the ultimate goal of the Metaverse for a realistic edge network implementation. The scalability, security, and privacy of edge intelligence, therefore, depend on advanced edge-computing techniques, such as model pruning, knowledge distillation, and federated learning. To illustrate the concept of edge-enabled Metaverse, we use two typical examples: edge remote rendering for immersive services and edge-assisted physical–virtual synchronization in real-time. In the edge remote rendering example, the VR streaming system is proposed and demonstrates the detailed workflow for matching and pricing among VR service providers and users. In the physical–virtual synchronization example, the simultaneous allocating and pricing models are discussed for synchronizing entities in the physical and the virtual worlds, respectively.

Future research directions in edge-computing technologies for Metaverse can be elaborated based on three dimensions: embodied user experience, sustainable and harmonious edge networks, and comprehensive edge intelligence. First, Metaverse is aiming at providing services with an immersive and human-centered experience, leading to the development of multisensory multimedia networks and human-centered communications. Unlike the traditional 2D Internet, users in Metaverse can use multisensory multimedia services such as AR/VR, the tactile Internet, and hologram streaming to immerse themselves in the world. To effectively deliver these services, edge-computing technologies must be equipped with the necessary technology to process holographic services in real-time, including AR/VR and the tactile Internet. Second, the development of Metaverse requires that edge-computing technologies provide green computing and communication services through real-time physical–virtual integration and collaboration among digital edge networks. To promote sustainability in Metaverse, there are three possible solutions that could be implemented for edge-computing technologies. These include developing new architectures focused on sustainability, implementing energy-efficient resource allocation strategies, and integrating cloud/fog/edge networking and computing with other emerging green technologies. Finally, with the support of robust edge intelligence and blockchain technology, virtual and physical environments can be seamlessly exchanged in Metaverse without compromising security and privacy.

Bibliography

1 Nasir Abbas, Yan Zhang, Amir Taherkordi, and Tor Skeie. Mobile edge computing: A survey. *IEEE Internet of Things Journal*, 5(1):450–465, September 2017.

2 Konstantinos Antonakoglou, Xiao Xu, Eckehard Steinbach, Toktam Mahmoodi, and Mischa Dohler. Toward haptic communications over the 5G tactile internet. *IEEE Communications Surveys & Tutorials*, 20(4):3034–3059, January 2018.

3 Michael Till Beck, Martin Werner, Sebastian Feld, and S. Schimper. Mobile edge computing: A taxonomy. In *Proceedings of the Sixth International Conference on Advances in Future Internet*, pages 48–55, Lisbon, Portugal, November 2014.

4 Mingzhe Chen, Walid Saad, and Changchuan Yin. Virtual reality over wireless networks: Quality-of-service model and learning-based resource management. *IEEE Transactions on Communications*, 66(11):5621–5635, January 2018.

5 Ying Chen, Ning Zhang, Yongchao Zhang, Xin Chen, Wen Wu, and Xuemin Shen. Energy efficient dynamic offloading in mobile edge computing for Internet of Things. *IEEE Transactions on Cloud Computing*, 9(3):1050–1060, February 2019.

6 Yu Cheng, Duo Wang, Pan Zhou, and Tao Zhang. A survey of model compression and acceleration for deep neural networks. *arXiv preprint arXiv:1710.09282*, October 2017. URL https://arxiv.org/abs/1710.09282.

7 Nan Cheng, Wenchao Xu, Weisen Shi, Yi Zhou, Ning Lu, Haibo Zhou, and Xuemin Shen. Air-ground integrated mobile edge networks: Architecture, challenges, and opportunities. *IEEE Communications Magazine*, 56(8):26–32, August 2018.

8 Nan Cheng, Feng Lyu, Wei Quan, Conghao Zhou, Hongli He, Weisen Shi, and Xuemin Shen. Space/aerial-assisted computing offloading for IoT applications: A learning-based approach. *IEEE Journal on Selected Areas in Communications*, 37(5):1117–1129, March 2019.

9 Haihan Duan, Jiaye Li, Sizheng Fan, Zhonghao Lin, Xiao Wu, and Wei Cai. Metaverse for social good: A university campus prototype. In *Proceedings of the 29th ACM International Conference on Multimedia*, pages 153–161, Virtual Event, China, October 2021.

10 Gerhard P. Fettweis. The tactile internet: Applications and challenges. *IEEE Vehicular Technology Magazine*, 9(1):64–70, March 2014. doi: 10.1109/MVT.2013.2295069.

11 Daniel Friedman. *The Double Auction Market: Institutions, Theories, and Evidence*. Routledge, 2018.

12 Aniket Gulhane, Akhil Vyas, Reshmi Mitra, Roland Oruche, Gabriela Hoefer, Samaikya Valluripally, Prasad Calyam, and Khaza Anuarul Hoque. Security, privacy and safety risk assessment for virtual reality learning environment applications. In *2019 16th IEEE Annual Consumer Communications Networking Conference (CCNC)*, pages 1–9, Las Vegas, NV, USA, January 2019. doi: 10.1109/CCNC.2019.8651847.

13 Fengxian Guo, F. Richard Yu, Heli Zhang, Hong Ji, Victor C. M. Leung, and Xi Li. An adaptive wireless virtual reality framework in future wireless networks: A distributed learning approach. *IEEE Transactions on Vehicular Technology*, 69(8):8514–8528, May 2020.

14 Yue Han, Dusit Niyato, Cyril Leung, Dong In Kim, Kun Zhu, Shaohan Feng, Sherman Xuemin Shen, and Chunyan Miao. *arXiv preprint arXiv:2203.03969*, March 2022. URL https://arxiv.org/abs/2203.03969.

15 Nguyen Quang Hieu, Diep N. Nguyen, Dinh Thai Hoang, and Eryk Dutkiewicz. When virtual reality meets rate splitting multiple access: A joint communication and computation approach. *arXiv e-prints*, pages arXiv–2207, July 2022. URL https://arxiv.org/abs/2207.12114.

16 Sandra Hirche and Martin Buss. Human-oriented control for haptic teleoperation. *Proceedings of the IEEE*, 100(3):623–647, January 2012.

17 Honeywell. How augmented reality is revolutionizing job training. Accessed: 2022-01-03. URL https://www.honeywell.com/us/en/news/2018/02/how-ar-and-vr-are-revolutionizing-job-training?utm_source=TW.

18 Suman Jana, David Molnar, Alexander Moshchuk, Alan Dunn, Benjamin Livshits, Helen J. Wang, and Eyal Ofek. Enabling fine-grained permissions for augmented reality applications with recognizers. In *Proceedings of the 22nd USENIX Security Symposium*, pages 415–430, Washington, DC, USA, August 2013.

19 Viktor Kelkkanen, Markus Fiedler, and David Lindero. Bitrate requirements of non-panoramic VR remote rendering. In *Proceedings of the 28th ACM International Conference on Multimedia*, MM '20, page 3624–3631, New York, NY, USA, 2020. Association for Computing Machinery. ISBN 9781450379885. doi: 10.1145/3394171.3413681.

20 Paul Lee, Tristan Braud, Pengyuan Zhou, Lin Wang, Dianlei Xu, Zijun Lin, Abhishek Kumar, Carlos Bermejo, and Pan Hui. All one needs to know about Metaverse: A complete survey on technological singularity, virtual ecosystem, and research agenda. October 2021. doi: 10.13140/RG.2.2.11200.05124/6.

21 Khaled B. Letaief, Wei Chen, Yuanming Shi, Jun Zhang, and Ying-Jun Angela Zhang. The roadmap to 6G: AI empowered wireless networks. *IEEE Communications Magazine*, 57(8):84–90, August 2019.

22 Yijie Mao, Onur Dizdar, Bruno Clerckx, Robert Schober, Petar Popovski, and H. Vincent Poor. Rate-splitting multiple access: Fundamentals, survey, and future research trends. *IEEE Communications Surveys & Tutorials*, 24(4):2073–2126, July 2022.

23 Meta. Reality labs. Accessed: 2022-01-03 2022. URL https://tech.fb.com/ar-vr/.

24 Meta. Ray tracing - what does it mean to you? Accessed: 2022-02-16. URL https://blog.unity.com/manufacturing/ray-tracing-what-does-it-mean-to-you.

25 Jihong Park and Mehdi Bennis. URLLC-eMBB slicing to support VR multi-modal perceptions over wireless cellular systems. In *Proceedings of IEEE Global Communications Conference (GLOBECOM)*, pages 1–7, Abu Dhabi, UAE, December 2018.

26 Walid Saad, Mehdi Bennis, and Mingzhe Chen. A vision of 6G wireless systems: Applications, trends, technologies, and open research problems. *IEEE Network*, 34(3):134–142, October 2019.

27 John Schulman, Filip Wolski, Prafulla Dhariwal, Alec Radford, and Oleg Klimov. Proximal policy optimization algorithms. *arXiv preprint arXiv:1707.06347*, 2017.

28 Chris Schwarz and Ziran Wang. The role of digital twins in connected and automated vehicles. *IEEE Intelligent Transportation Systems Magazine*, 14(6):41–51, January 2022. doi: 10.1109/MITS.2021.3129524.

29 Jiacheng Shang and Jie Wu. Enabling secure voice input on augmented reality headsets using internal body voice. In *Proceedings of the 16th Annual IEEE International Conference on Sensing, Communication, and Networking (SECON)*, pages 1–9, Boston, MA, uSA, June 2019. doi: 10.1109/SAHCN.2019.8824980.

30 Bowen Shi, Ji Yang, Zhanpeng Huang, and Pan Hui. Offloading guidelines for augmented reality applications on wearable devices. In *Proceedings of the 23rd ACM International Conference on Multimedia*, pages 1271–1274, Brisbane, Australia, October 2015.

31 Meryem Simsek, Adnan Aijaz, Mischa Dohler, Joachim Sachs, and Gerhard Fettweis. 5G-enabled tactile internet. *IEEE Journal on Selected Areas in Communications*, 34(3):460–473, February 2016.

32 Ali Sunyaev. Fog and edge computing. In *Internet Computing*, pages 237–264. Springer, 2020.

33 Ming Tang, Lin Gao, and Jianwei Huang. Enabling edge cooperation in tactile internet via 3C resource sharing. *IEEE Journal on Selected Areas in Communications*, 36(11):2444–2454, November 2018. doi: 10.1109/JSAC.2018.2874123.

34 Harsh Tataria, Mansoor Shafi, Andreas F. Molisch, Mischa Dohler, Henrik Sjöland, and Fredrik Tufvesson. 6G wireless systems: Vision, requirements, challenges, insights, and opportunities. *Proceedings of the IEEE*, 109(7):1166–1199, March 2021.

35 Haoxin Wang, Tingting Liu, BaekGyu Kim, Chung-Wei Lin, Shinichi Shiraishi, Jiang Xie, and Zhu Han. Architectural design alternatives based on cloud/edge/fog computing for connected vehicles. *IEEE Communications Surveys & Tutorials*, 22(4):2349–2377, September 2020.

36 Ziran Wang, Kyungtae Han, and Prashant Tiwari. Augmented reality-based advanced driver-assistance system for connected vehicles. In *Proceedings of IEEE International Conference on Systems, Man, and Cybernetics (SMC)*, pages 752–759, Toronto, Canada, December 2020.

37 Yuntao Wang, Zhou Su, Ning Zhang, Dongxiao Liu, Rui Xing, Tom H. Luan, and Xuemin Shen. A survey on metaverse: Fundamentals, security, and privacy. *arXiv preprint arXiv:2203.02662*, March 2022. URL https://arxiv.org/abs/2203.02662.

38 Chaocan Xiang, Yaoyu Li, Yanlin Zhou, Suining He, Yuben Qu, Zhenhua Li, Liangyi Gong, and Chao Chen. A comparative approach to resurrecting the market of mod vehicular crowdsensing. In *Proceedings of IEEE INFOCOM*, pages 1479–1488, London, UK, June 2022.

39 Minrui Xu, Wei Chong Ng, Wei Yang Bryan Lim, Jiawen Kang, Zehui Xiong, Dusit Niyato, Qiang Yang, Xuemin Sherman Shen, and Chunyan Miao. A full dive into realizing the edge-enabled metaverse: Visions, enabling technologies, and challenges. *arXiv preprint arXiv:2203.05471*, March 2022. URL https://arxiv.org/abs/2203.05471.

40 Minrui Xu, Dusit Niyato, Zehui Xiong, Jiawen Kang, Xianbin Cao, Xuemin Sherman Shen, and Chunyan Miao. Quantum-secured space-air-ground integrated networks: Concept, framework, and case study. *arXiv preprint arXiv:2204.08673*, April 2022. URL https://arxiv.org/abs/2204.08673.

41 Wanting Yang, Zi Qin Liew, Wei Yang Bryan Lim, Zehui Xiong, Dusit Niyato, Xuefen Chi, Xianbin Cao, and Khaled B. Letaief. Semantic communication meets edge intelligence. *arXiv preprint arXiv:2202.06471*, February 2022. URL https://arxiv.org/abs/2202.06471.

42 Changsheng You, Kaibin Huang, Hyukjin Chae, and Byoung-Hoon Kim. Energy-efficient resource allocation for mobile-edge computation offloading. *IEEE Transactions on Wireless Communications*, 16(3):1397–1411, December 2017. doi: 10.1109/TWC.2016.2633522.

43 Xiaohu You, Cheng-Xiang Wang, Jie Huang, Xiqi Gao, Zaichen Zhang, Mao Wang, Yongming Huang, Chuan Zhang, Yanxiang Jiang, Jiaheng Wang, et al. Towards 6G wireless communication networks: Vision, enabling technologies, and new paradigm shifts. *Science China Information Sciences*, 64(1):1–74, November 2021.

44 Xiaojin Zhang, Hanlin Gu, Lixin Fan, Kai Chen, and Qiang Yang. No free lunch theorem for security and utility in federated learning. *arXiv preprint arXiv:2203.05816*, March 2022. URL https://arxiv.org/abs/2203.05816.

9

Security Issues in Metaverse

Yuntao Wang[1], *Zhou Su*[1], *Ning Zhang*[2], *Dongxiao Liu*[3], *Rui Xing*[1], *Tom H. Luan*[1], *and Xuemin Shen*[3]

[1] *School of Cyber Science and Engineering, Xi'an Jiaotong University, Xi'an, Shaanxi, China*
[2] *Department of Electrical and Computer Engineering, University of Windsor, Windsor, ON, Canada*
[3] *Department of Electrical and Computer Engineering, University of Waterloo, Waterloo, ON, Canada*

After reading this chapter, you should be able to:

- Aware of the main security and privacy threats in the Metaverse and critical challenges to address them.
- Better understand how these security/privacy threats could arise and be prevented in the Metaverse.
- Know state-of-the-art security and privacy countermeasures in both academic and industry and their feasibility in the Metaverse.

9.1 Overview of Security and Privacy Threats in Metaverse

In spite of the promising sign of Metaverse, security and privacy issues are the prime concerns that hinder its further development. A wide range of security breaches and privacy invasions may arise in the Metaverse from the management of massive data streams, pervasive user profiling activities, unfair outcomes of artificial intelligence (AI) algorithms, to the safety of physical infrastructures and human bodies.

First, since Metaverse integrates a variety of latest technologies and systems built on them as its basis, their vulnerabilities and intrinsic flaws may also be inherited by the Metaverse. There have been incidents of emerging technologies, such as hijacking of wearable devices or cloud storage, theft of

Metaverse Communication and Computing Networks: Applications, Technologies, and Approaches, First Edition.
Edited by Dinh Thai Hoang, Diep N. Nguyen, Cong T. Nguyen, Ekram Hossain, and Dusit Niyato.
© 2024 The Institute of Electrical and Electronics Engineers, Inc. Published 2024 by John Wiley & Sons, Inc.

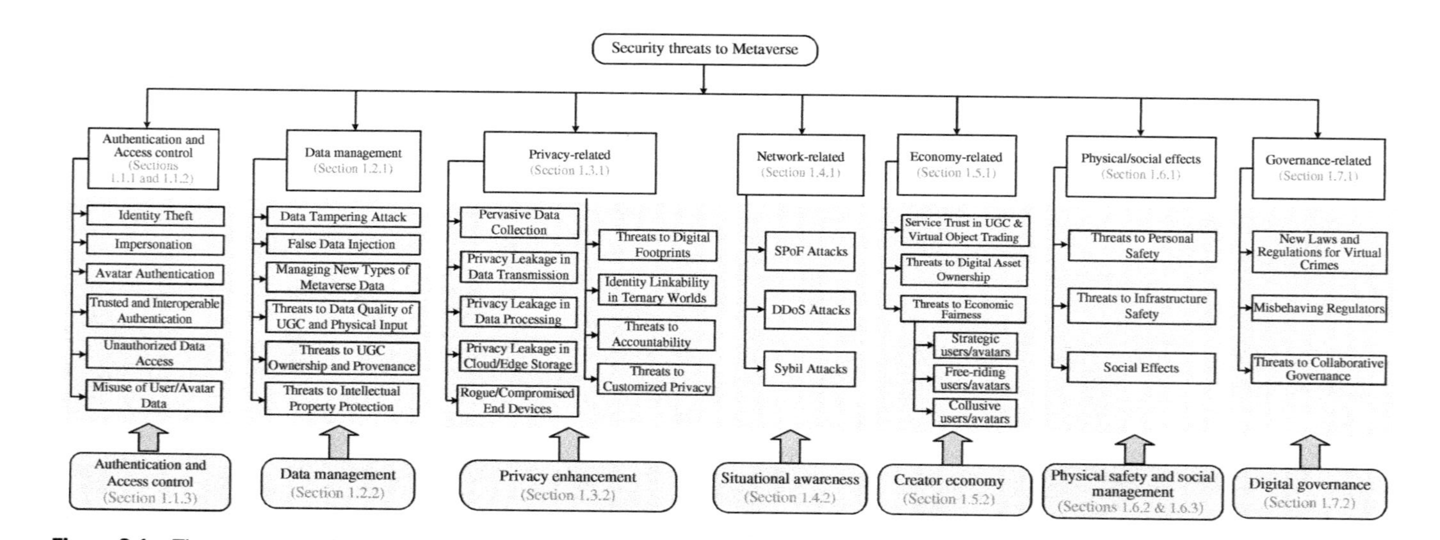

Figure 9.1 The taxonomy of security threats and corresponding security countermeasures in the Metaverse.

virtual currencies, and the misconduct of AI to produce fake news. Second, driven by the interweaving of various technologies, the effects of existing threats can be amplified and become more severe in virtual worlds, while new threats nonexistent in physical and cyber spaces can breed such as virtual stalking and virtual spying. Particularly, the personal data involved in the Metaverse can be more granular and unprecedentedly ubiquitous to build a digital copy of the real world, which opens new horizons for crimes on private big data [16]. For example, to build a virtual scene using AI algorithms, users will inevitably wear wearable augmented reality (AR) and virtual reality (VR) devices with built-in sensors to comprehensively collect brain wave patterns, facial expressions, eye movements, hand movements, speech and biometric features, as well as the surrounding environment. Lastly, hackers can exploit system vulnerabilities and compromise devices as entry points to invade real-world equipment such as household appliances to threaten personal safety, and even threaten critical infrastructures such as power grid systems, high-speed rail systems, and water supply systems via advanced persistent threat (APT) attacks [21].

This chapter summarizes the key challenges and existing/potential solutions to build the secure and privacy-preserving Metaverse from the following seven aspects (i.e. authentication and access control, data management, privacy, network, economy, governance, and physical/social effects). Figure 9.1 depicts the proposed taxonomy of security threats and the corresponding security countermeasures in the Metaverse.

9.2 Threats and Countermeasures to Authentication and Access Control in Metaverse

In the Metaverse, identity authentication and access control play a vital role for massive users/avatars in Metaverse service offering.

9.2.1 Threats to Authentication in Metaverse

The identities of users/avatars in the Metaverse can be illegally stolen, impersonated, and interoperability issues can be encountered in authentication across virtual worlds.

(1) Identity Theft: If the identity of a user is stolen in the Metaverse, his/her avatars, digital assets, social relationships, and even the digital life can be leaked and lost, which can be more severe than that in traditional information systems. For example, hackers can steal users' personal information (e.g. full names, secret keys of digital assets, and banking details) in Roblox through hacked personal VR glasses, phishing e-mail scams, and authentication loopholes to commit fraud and crimes (e.g. steal the victim's avatar and digital assets) in Roblox. For example, in 2022, the accounts of 17 users in the Opensea nonfungible token (NFT) marketplace were hacked due to smart contract flaws and phishing attacks, causing a loss of US$1.7 million [5].

(2) Impersonation Attack: An attacker can carry out the impersonation attack by pretending to be another authorized entity to gain access to a service or a system in the Metaverse [21]. For example, hackers can invade the Oculus helmet and exploit the stolen behavioral and biological data gathered by the in-built motion-tracking system to create digital replicas of the user and impersonate the victim to facilitate social engineering attacks. The hackers can also create a fake avatar using digital replicas of the victim to deceive, fraud, and even commit a crime against the victim's friends in the Metaverse. Another example is that attackers can exploit Bluetooth impersonation threats [7] to impersonate trusted endpoints and illegally access Metaverse services by inserting rogue wearable devices into the established Bluetooth pairing.

(3) Avatar Authentication Issue: Compared with real-world identity authentication, the authentication of avatars (e.g. the verification of their friends' avatars) for users in the Metaverse can be more challenging through verifying facial features, voice, video footage, and so on. Besides, adversaries can create multiple AI bots (i.e. digital humans), which appear, hear, and behave identical to user's real avatar, in the virtual world (e.g. Roblox) by imitating user's appearance, voice, and behaviors [16]. As a consequence, more additional personal information might be required as evidence to ensure secure avatar authentication, which may also open new privacy breach issues.

(4) Trusted and Interoperable Authentication: For users/avatars in the Metaverse, it is fundamental to ensure fast, efficient, and trusted cross-platform and cross-domain identity authentication, i.e. across various service domains and virtual worlds (built on distinct platforms such as blockchains). For example, the trust-free and interoperable asset exchange and avatar transfer between Roblox and Fortnite, as well as among distinct administrative domains for offering different services in Roblox.

9.2.2 Threats to Access Control in Metaverse

(1) Unauthorized Data Access: Complex Metaverse services will generate new types of personal profiling data (e.g. biometric information, daily routine, and user habits). To deliver seamless personalized services (e.g. customized avatar appearance) in the Metaverse, different virtual service providers (VSPs) in distinct sub-Metaverses need to access real-time user/avatar profiling activities. Malicious VSPs may carry out attacks for unauthorized data access to earn benefits.

An example is that malicious VSPs may illegally elevate their rights in data access via attacks such as buffer overflow and tampering access control lists [67].

(2) Misuse of User/Avatar Data: In the life-cycle of data services in the Metaverse, user/avatar-related data can be disclosed intentionally by attackers or unintentionally by VSPs for user profiling and targeted advertising activities. Besides, due to the potential noninteroperability of certain sub-Metaverses, it is hard to trace the data misuse activities in the large-scale Metaverse.

9.2.3 Security Countermeasures to Metaverse Authentication and Access Control

For the Metaverse, secure and efficient identity management is the basis for user/avatar interaction and service provisioning. Generally, digital identities can be classified into the following three kinds:

- **Centralized Identity:** Centralized identity refers to the digital identity authenticated and managed by a single institution, such as the Gmail account.
- **Federated Identity [23]:** Federated identity refers to the digital identity managed by multiple institutions or federations. It can reduce the administrative cost in identity authentication for cross-platform and cross-domain operations, and alleviate the cumbersome process of typing personal information repeatedly for users.
- **Self-Sovereign Identity (SSI) [44]:** SSI refers to the digital identity which is fully controlled by individual users. It allows users to autonomously share and associate different personal information (e.g. username, education information, and career information) in performing cross-domain operations to enable identity interoperability with users' consent.

In the Metaverse, centralized identity systems can be prone to single point of failure (SPoF) risks and suffer potential leakage risks. Federated identity systems are semicentralized, and the management of identities is controlled by a few institutions or federations, which may also suffer potential centralization risks. The identity systems built on SSIs will be dominant in future Metaverse construction. Generally, identity management schemes in the Metaverse should follow the following design principles: (i) *scalability* to massive users/avatars, (ii) *resilience* to node damage, and (iii) *interoperability* across various sub-Metaverse during authentication.

Figure 9.2 compares the hardware terminals for entering the web, mobile Internet, the Metaverse. In the Metaverse, empowered by extended reality (XR) and human–computer interaction (HCI) technologies, wearable devices such as helmet-mounted display (HMD) and brain–computer interface (BCI) enable user/avatar interactions and are expected as the major terminal to enter the Metaverse. Besides, the Metaverse usually includes various administrative domains.

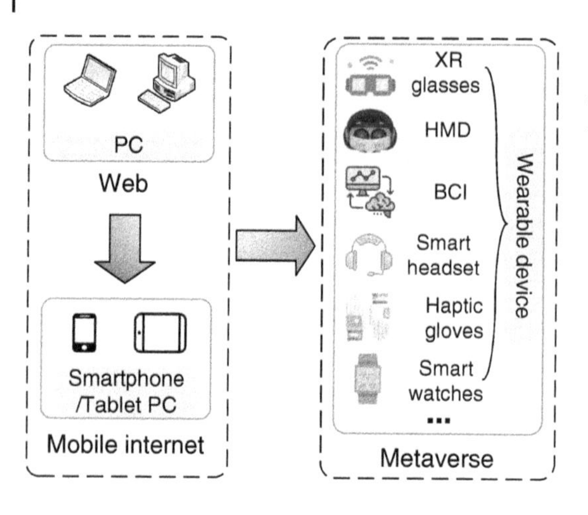

Figure 9.2 Comparison of hardware terminals for entering the web, mobile Internet, and the Metaverse.

9.2.3.1 Key Management for Wearable Devices

Wearable devices such as Oculus helmets and HoloLen headsets are anticipated to be the major terminal to enter the Metaverse. Key management (including generation, negotiation, distribution, update, revocation, and recovery) is essential for wearable devices to establish secure communication, deliver sensory data, receive immersive service, etc.

In the literature, works [30, 54] take the intrinsic features of distinct wearable devices (e.g. wireless channel and gait signal) into account in designing efficient key management schemes, which can be beneficial for future Metaverse construction. For example, Li et al. [30] design an innovative key establishment approach by leveraging the received signal strength (RSS) trajectories of two moving wearables to construct the secret key by moving or shaking the wearable devices. Experimental results validate its practicability for wearables with short-range communications and frequent movements. Besides, Sun et al. [54] exploit the gait-based biometric cryptography to design a group key generation and distribution scheme for wearable devices based on signed sliding window coding and fuzzy vault. Simulations prove that it can pass both the National Institute of Standards and Technology (NIST) and Dieharder statistical tests.

9.2.3.2 Identity Authentication for Wearable Devices

Identity authentication for wearable devices to guarantee device/user authenticity is also a promising topic in the Metaverse. To adapt to wearable devices with extremely low computing/storage capacity, Srinivas et al. [51] present a cloud-based mutual authentication model with low system cost for wearable medical devices to prevent device impersonation in healthcare monitoring systems with password change and smart card revocation functions. Rigorous security analysis proves the security of created session key against active and passive attacks. However, the one-time authentication in [51] may cause friction such as unauthorized privileges. To resolve this issue, Zhao et al. [71] propose a novel continuous authentication model to support seamless device authentication at a low cost. In [71], unique cardiac biometrics are extracted from photoplethysmography (PPG) sensors (embedded in wrist-worn wearables) for user authentication. Experimental results show that their proposed system obtains a high average continuous authentication accuracy rate of 90.73%.

9.2.3.3 Cross-Domain Identity Authentication

The Metaverse typically contains various administrative security domains created by distinct operators/standards. Identity authentication across distinct administrative domains (e.g. VR/AR services run by distinct VSPs) in the Metaverse is critical to deliver seamless Metaverse services for users/avatars. Traditional cross-domain authentication mechanisms mainly rely on a trusted intermediary and bring heavy overhead in key management. To address this issue, Shen et al. [49] employ blockchain technology to design a decentralized and transparent cross-domain authentication scheme for industrial Internet of Things (IoT) devices in different domains (e.g. factories). In their work, a consortium blockchain is employed to establish trust among distinct domains, and an anonymous authentication protocol with identity revocation capability is proposed to remedy the drawback of identity-based encryption (IBE) in terms of identity revocation. Further efforts are required to design lightweight approaches for battery-powered wearable devices with compact battery size and limited computational capacity.

9.2.3.4 Fine-Grained Access Control and Usage Audit for Wearables and UGCs

The massive personally identifiable information (PII) handled by wearables can pose a huge risk of unauthorized exposure. To address this issue, Ometov et al. [40] propose a novel delegation-of-use mode for wearable devices with privacy guarantees, where owners can lend their personal devices to others for temporary use. However, the associated attacks along with scalability and efficiency issues still need more investigations in real-world implementation.

The naive content creation (e.g. user-generated contents [UGCs]) produced by avatars is essential to maintain the creativity and sustainability of the Metaverse. As UGCs inevitably contain sensitive and private user information, efficient UGC access control and usage audit schemes should be designed. Nevertheless, the authorized entities may become traitors to illegally redistribute UGCs to the public, i.e. *illegal UGC redistribution*. To address this realistic threat, Zhang et al. [70] propose a novel secure encrypted user-generated media content (UGMC) sharing scheme with traitor tracing in the cloud via the proxy re-encryption mechanism (for secure UGMC sharing) and watermarking mechanism (for traitor tracing). Apart from access control of UGCs, the usage control (i.e. shared UGCs can be only used for intended purposes) is also essential. To bridge this gap, Wang et al. [62] propose a novel data processing-as-a-service (DPaaS) mode to complement the current data sharing ecosystem and exploit blockchain technologies for fine-grained data usage policymaking on the user's side, policy execution atop smart contracts, and policy audit on transparent ledgers.

9.3 Threats and Countermeasures to Data Management in Metaverse

9.3.1 Threats to Data Management in Metaverse

The data collected or generated by wearable devices and users/avatars may suffer from threats in terms of data tampering, false data injection, low-quality UGC, ownership/provenance tracing, and intellectual property violation in the Metaverse.

(1) Data Tampering Attack: Integrity features ensure effective checking and detection of any modification during data communication among the ternary worlds and various sub-Metaverses. Adversaries may modify, forge, replace, and remove the raw data throughout the life-cycle of Metaverse data services to interfere with the normal activities of users, avatars, or physical entities [53]. Besides, adversaries may remain undetected by falsifying corresponding log files or message-digest results to hide their criminal traces in the virtual space.

(2) False Data Injection Attack: Attackers can inject falsified information such as false messages and wrong instructions to mislead Metaverse systems [34]. For example, AI-aided content creation can help improve user immersiveness in the early stage of the Metaverse, and adversaries can inject adversary-training samples or poisoned gradients during centralized or distributed AI training, respectively, to generate biased AI models. The returned wrong feedbacks or instructions may also threaten the safety of physical equipment and even personal safety. For example, falsified feedbacks such as excessive voltage can cause damage and malfunction of wearable XR devices. Another example is that the tampered hundredfold magnifications of bodily pain in being shot in Fortnite (a Metaverse game) may cause the death of human user.

(3) Issues in Managing New Types of Metaverse Data: Compared with the current Internet, the Metaverse requires new hardware and devices to gather various new types of data (e.g. eye movement, facial expression, and head movement), which are previously uncollected, to make fully immersive user experiences [25]. Besides, end-devices in the Metaverse (e.g. VR glasses and haptic gloves) can be capable of capturing iris biometrics, fingerprints, or other user-sensitive biometric information. Consequently, it raises new challenges in collecting, managing, and storing these enormous user-sensitive Metaverse data, as well as the cyber/physical security of Metaverse devices.

(4) Threats to Data Quality of UGC and Physical Input: In Metaverse, selfish users/avatars may contribute low-quality contents under the UGC mode to save their costs, thereby undermining user experience such as unreal experience in the synthesized environment. For example, they may share unaligned and severe nonindependent identically distributed (Non-IID) data during the collaborative training process of the content recommendation model in the Metaverse, causing inaccurate content recommendation. Another example is that uncalibrated wearable sensors can generate inaccurate and even erroneous sensory data to mislead the creation of digital twins in the Metaverse, causing poor user experience.

(5) Threats to UGC Ownership and Provenance: Different from the asset registration procedure supervised by the government in the real world, the Metaverse is an open and fully autonomous space and there exists no centralized authority. Due to the lack of authority, it is hard to trace the ownership and provenance of various UGCs produced by massive avatars under different virtual worlds in the Metaverse, as well as turn UGCs into protected assets [33]. Besides, UGCs can be shared in real time within the virtual world or across various virtual worlds and unlimitedly replicated due to the digital attributes, making it harder for efficient provenance and ownership tracing.

(6) Threats to Intellectual Property Protection: Different from the actual world, the definition of intellectual property in the Metaverse should be adapted to enforce licensing boundaries and usage rights for the owners with the evolvement and expanding scale of the Metaverse [20]. Moreover, severe challenges may arise in defining and protecting intellectual property (e.g. avatars and UGCs) in the new Metaverse ecology, as the geographic boundaries of countries are broken down in the Metaverse. For example, there have already been disputes owing to the use of celebrity lookalikes in video games. Given the commercial value created by avatars, such kinds of disputes may spike exponentially in the future Metaverse.

9.3.2 Security Countermeasures to Metaverse Data Management

The Metaverse is a digital world built on digital copies of the physical environment and avatars' digital creations. Analogy to the value created by human activities in the real world, digital twins and UGCs as well as avatars' behaviors (e.g. chat records and browsing records) will produce certain value in the Metaverse. Information security is an important prerequisite for the development and prosperity of the Metaverse. In the following, we discuss the data security in Metaverse in terms of data reliability, data quality, and provenance.

9.3.2.1 Data Reliability of AIGC, Digital Twin, and Physical Input

In the Metaverse, AI such as generative adversarial network (GAN) can help not only generate high-quality AI-generated contents (AIGCs), such as dynamic game scenarios and context images, but also poses security threats such as adversarial and poisoned samples which is hard to detect for humans. In the literature, by taking adversarial samples as part of training data, various efforts have been done to resist adversarial samples via virtual adversarial learning, adversarial representation learning, adversarial reinforcement learning, adversarial transfer learning, and so on, which can be beneficial to resist adversarial threats in the construction of the Metaverse.

To address the trustworthiness of data collected from disparate data silos in the Metaverse, as shown in Figure 9.3, Liao et al. [35] leverage permissioned blockchain technology for trusted digital twin (DT) service transactions between VSPs and service requesters in intelligent transportation systems (ITS). A DT-DPoS (delegated proof of stake) consensus protocol is devised to improve consensus efficiency by using distributed DT servers to form the validator committee. Besides, to facilitate users' customized DT services, an on-demand DT-as-a-service (DTaaS) architecture is presented for fast response to meet diverse DT requirements in ITS.

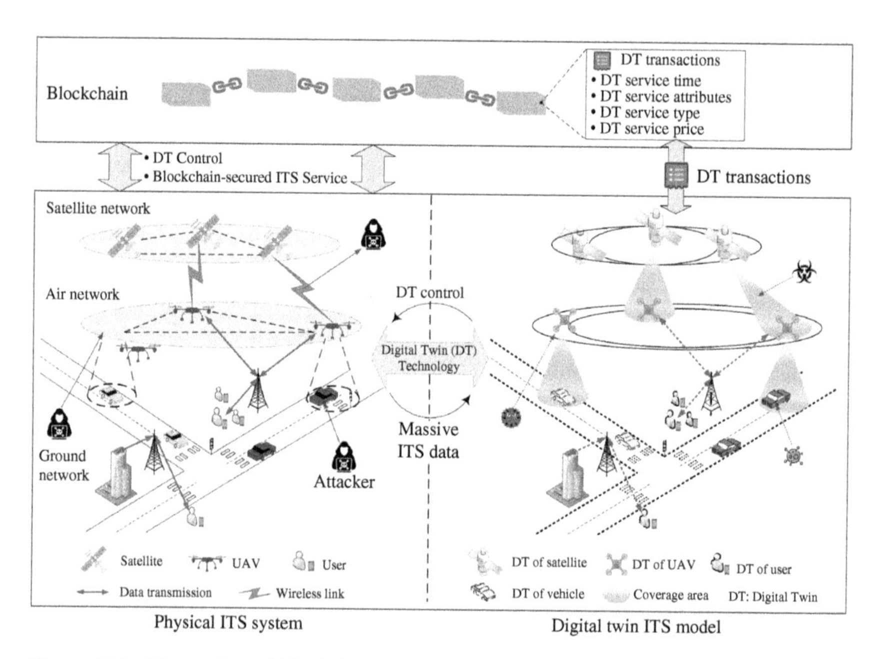

Figure 9.3 Illustration of blockchain-enabled digital twin (DT)-as-a-service (DTaaS) in intelligent transportation systems (ITS) [35].

Zimmermann and Ke [73] investigate parametric audio rendering to match and improve the visual experience in 3D virtual worlds. The work [73] presents an interactive audio streaming mechanism with high scalability based on peer-to-peer (P2P) topology for immersive interaction in networked virtual environments (NVEs), which combines two concepts: *area of interest (AoI)* and *aural soundscape* to make proximal and spatialized audio interactions. Specifically, AoI limits the distribution area of audio streams as avatars are more likely to interact with others in proximity (the distance is measured by virtual coordinates), and aural soundscape allows distributive audio rendering from different sources to match the visual landscape.

9.3.2.2 Data Quality of UGC and Physical Input

Low-quality data input from physical sensors and the UGCs produced by avatars can deteriorate the quality-of-service (QoS) of Metaverse services and the quality-of-experience (QoE) of users. Effective quality control mechanisms are important to offer efficient Metaverse services and maintain sustainability of the creator economy.

In the literature, game theory and AI methods have been widely utilized to motivate users' high-quality data contribution or service offering, which can offer some lessons in the Metaverse design. For example Su et al. [52] propose a deep RL (DRL)-based incentive mechanism to encourage users' high-quality model contribution in distributed AI paradigms with consideration of both non-IID effects and collaboration between edge/cloud servers.

For accurate DT synchronization with its physical counterpart, Han et al. [19] propose a hierarchical game for dynamic DT synchronization in the Metaverse, where end devices collectively gather the status information of physical objects and VSPs decide proper synchronization intensities. In their work, every user selects the optimal VSP in the lower-level evolutionary game, and every VSP makes the optimal synchronization strategy in the upper-level differential game based on users' strategies and value of DT. For dynamic Metaverse applications, the information freshness (e.g. age of information) can be further considered in data/service offering.

9.3.2.3 Secure Data Sharing in XR Environment

Metaverse applications are usually multiuser such as multiplayer gaming and remote collaboration. Aimed for secure content sharing under multiuser AR applications, Ruth et al. [43] study an AR content sharing control mechanism and implement a prototype on HoloLens to allow AR content sharing among remote or colocated users with inbound and outbound control. By rigorously exploring user's design space on various AR apps, the authors also define various mapping manners of AR contents into the real world. In WebVR (a VR-based 3D virtual world on HTML canvases), Lee et al. [27] identify three new ad fraud threats (i.e. blind spot tracking, gaze and controller cursor-jacking, and abuse of an auxiliary display) in content sharing. User studies on 82 participants show the success rates range from 88.23% to 100%. Besides, a defense mechanism named AdCube is presented in [27] via visual confinement of 3D ad entities and sandboxing technique.

9.3.2.4 Provenance of UGC

Data provenance can realize the traceability of historical archives of a piece of UGC, which is essential to evaluate data quality, trace data source, reproduce data generation process, and conduct audit trail to quickly identify data responsible subjects. In the Metaverse, UGC provenance information such as the source, circulation, and intermediate processing information is often stored in disparate data silos (e.g. distinct blockchains), making it difficult to monitor and track in real time.

Existing works on IoT data provenance can offer some lessons for UGC provenance design in the Metaverse. Satchidanandan and Kumar [45] design a dynamic watermarking techniques which exploit indelible patterns imprinted in the medium to detect misbehaviors (e.g. signal tampering) of malicious sensors or actuators. Besides, advanced watermarking technique can be utilized for intellectual property protection and ownership authentication in the Metaverse. Liang et al. [33] present a blockchain-based cloud file provenance architecture named ProvChain with three stages, i.e. collection, storage, and verification of provenance information.

In the Metaverse, the life cycle of UGCs involves the ternary worlds and multiple sub-Metaverses, which can be more complex than that in traditional IoT. Moreover, smart contracts are anticipated to play an important role in enforcing UGC provenance across various Metaverse platforms, and more research efforts on its functionality, efficiency, and security are required. Besides, the scalability, trust, and efficiency (e.g. response delay) are still challenging issues in the provenance of massive UGCs in the large-scale Metaverse.

9.4 Privacy Threats and Countermeasures in Metaverse

9.4.1 Privacy Threats in Metaverse

When enjoying digital lives in the Metaverse, user privacy including location privacy, habit, living styles, and so on, may be offended during the life cycle of data services including data perception, transmission, processing, governance, and storage.

(1) Pervasive Data Collection: To immersively interact with an avatar, it requires pervasive user profiling activities at an unreasonably granular level [16] including facial expressions, eye/hand movements, speech and biometric features, and even brain wave patterns. Besides, via advanced XR and HCI technologies, it can facilitate the analysis of physical movements and user attributes and even enable user tracking [47]. For example, the motion sensors and four built-in cameras in the Oculus helmet help track the head direction and movement, draw our rooms, as well as monitor our positions and environment in real time with submillimeter accuracy, when we browse the Roblox and interact with other avatars. If this device is hacked by attackers, severe crimes can be committed on the basis of these large-volume of sensitive data.

Another example is the attractive virtual office (e.g. Horizon Workroom and Microsoft Mesh), which may arise significant security and privacy risks to employees. On the one hand, employee conversations, the e-mails they send, the URLs they visit, their behaviors, and even the tones of their voices may be monitored by the managers. On the other hand, the immersive workplace may be prone to other security and privacy issues such as intrusions, snooping, and impostors.

(2) Privacy Leakage in Data Transmission: In Metaverse systems, abundant personally identifiable information collected from wearables (e.g. HMDs) are transferred via wired and wireless communications, the confidentiality of which should be prohibited from unauthorized individuals/services [40]. Although communications are encrypted, and information is confidentially transmitted, adversaries may still access the raw data by eavesdropping on the specific channel and even track users' locations via advanced inference attacks [63].

(3) Privacy Leakage in Data Processing: In Metaverse services, the aggregation and processing of massive data collected from human bodies and their surrounding environments are essential for the creation and rendering of avatars and virtual environments, in which users' sensitive information may be leaked [29]. For example, the aggregation of private data (belonging to different users) to a central storage for training personalized avatar appearance models may offend user privacy and violate existing real-world regulations such as General Data Protection Regulation (GDPR) [59]. Besides, adversaries may infer users' privacy (e.g. preferences) by analyzing and linking the published processing results (e.g. synthetic avatars) in various virtual spaces such as Roblox and Fortnite.

(4) Privacy Leakage in Cloud/Edge Storage: The storage of massive users' private and sensitive information (e.g. user profiling) in cloud servers or edge devices may also raise privacy disclosure issues. For example, hackers may deduce users' privacy information by frequent queries via differential attacks [59] and even compromise the cloud/edge storage via distributed denial-of-service (DDoS) attacks. In 2006, a customer database of the Second Life (a Metaverse game) was hacked, and the user data were breached including unencrypted usernames and addresses, as well as encrypted payment details and passwords [3].

(5) Rogue or Compromised End Devices: In the Metaverse, more wearable sensors will be placed on human bodies and their surroundings to allow avatars to make natural eye contact, capture hand gesture, reflect facial expression, and so on, in real time. A significant risk is that these wearable devices can have a completely authentic sense of who you are, how you talk, behave, feel, and express yourself. The use of rogue or compromised wearable end devices (e.g. VR glasses) in the Metaverse is becoming an entryway for data breaches and malware invasions, and the problem may be more severe with the popularity of wearable devices for entering the Metaverse [47]. Under the manipulation of rogue or compromised end devices, the avatars in the Metaverse may turn into a source of data collection, thereby infringing user privacy.

(6) Threats to Digital Footprints: As the behavior pattern, preferences, habits, and activities of avatars in the Metaverse can reflect the real statuses of their physical counterparts, attackers can collect the digital footprints of avatars and exploit the similarity linked to real users to facilitate accurate user profiling and even illegal activities. Besides, Metaverse usually offers the third person view with a wider viewing angle of their avatar's surroundings than that in the real world [28], which may infringe on other players' behavior privacy without awareness. For example, an avatar may conduct the virtual stalking/spying attack in Roblox by following your avatar and recording all your digital footprints, e.g. purchasing behaviors, to facilitate social engineering attacks.

(7) Identity Linkability in Ternary Worlds: As the Metaverse assimilates the reality into itself, the human, physical, and virtual worlds are seamlessly integrated into the Metaverse, causing identity linkability concerns across the ternary worlds. For example, a malicious player A in Roblox can track another player B by the name appeared above the corresponding avatar of player B and infer his/her position in the real world. Another example is that hackers may track the position of users via compromised VR headsets or glasses [47].

(8) Threats to Accountability: XR and HCI devices intrinsically gather more sensitive data such as locations, behavior patterns, and surroundings of users than traditional smart devices. For example, in Pokémon Go (a location-based AR game), players can discover, capture, and battle Pokémon using mobile devices with GPS. The accountability in the Metaverse is important to ensure users' sensitive data are handled with privacy compliance. For Metaverse service providers, the audit process of the compliance of privacy regulations (e.g. GDPR) for accountability can be clumpy and time-consuming under the centralized service offering architecture. Besides, it is hard for VSPs to ensure the transparency of regulation compliance during the life cycle of data management, especially in the new digital ecology of Metaverse.

(9) Threats to Customized Privacy: Similar to existing Internet service platforms, distinct users generally exhibit customized privacy preferences for different services or interaction objects [50] under distinct sub-Metaverses. For example, a user in Roblox may be more sensitive to monetary trading activities than social activities. Besides, users/avatars may be more sensitive in interacting with strangers than acquaintances, friends, or relatives. However, challenges exist in developing customized privacy preservation policies for personal data management while considering avatars in the Metaverse as individual information subjects [64], as well as the characteristics of users and sub-Metaverses.

9.4.2 Privacy Countermeasures in Metaverse

9.4.2.1 Privacy in Metaverse Games

AR/VR games are the current most popular Metaverse application for users. AR/VR games usually contain three steps: the game platform (i) collects sensory data from users and their surroundings, (ii) identifies objects according to these contexts, and lastly (iii) performs rendering on game senses for immersiveness.

Existing works have demonstrated the security and safety concerns related to Metaverse games using case studies [12] and qualitative studies [47]. Bono et al. [12] offer two case studies (i.e. *Second Life* and *Anarchy Online*) and show that a hacker can exploit the features and vulnerabilities of massive multiplayer online (MMO) Metaverse games to fully compromise and take over players' devices (e.g. laptops). Shang et al. [47] identify a novel user location tracking attack in location-based AR games (e.g. Pokémon Go) by solely exploiting the network traffic of the player, and real-world experiments on 12 volunteers validate that the proposed attack model attains fine-grained geolocation of any player with high accuracy. Besides, three possible mitigation approaches are presented in [47] to alleviate attack effects.

9.4.2.2 Privacy-Preserving UGC Sharing and Processing

Existing privacy-preserving schemes for data sharing and processing mainly focus on four fields: differential privacy (DP), federated learning (FL), cryptographic approaches (e.g. secure multiparty computation (SMC), homomorphic encryption (HE), and zero-knowledge proof (ZKP)), and trusted computing.

Privacy-preserving UGC processing in the Metaverse has also attracted various attention. Based on Okamoto–Uchiyama HE, Li et al. [29] present a verifiable privacy-preserving method for data processing result prediction in edge-enabled cyber-physical social system (CPSS). Besides, batch verification is supported for multiple prediction results at one time to reduce communication burdens. Wang et al. [62] leverage the trusted computing technique to design a privacy-preserving off-chain data processing mechanism, where private UGC datasets are processed in an off-chain trusted enclave and the exchange of processed results and payment are securely executed via the designed fair exchange smart contract.

9.4.2.3 Confidentiality Protection of UGC and Physical Input

The confidentiality of UGCs (inside the Metaverse) along with physical inputs (to the Metaverse) should be ensured to prevent private data leakage and sensitive data exposure. The authentication and access control (in Section 9.2.3) and privacy computing technologies (in Section 9.4.2.2) are enablers to maintain UGC confidentiality in the Metaverse. For confidentiality of physical inputs, Raguram et al. [41] propose a novel threat named *compromising reflections*, which can automatically reconstruct user typing on virtual keyboards, thereby compromising data confidentiality and user privacy. Experiment results show that compromising reflections of a device's screen (e.g. sunglass reflections) are sufficient for automatic and accurate reconstruction with no limitation on the motion of handheld cameras even in challenging scenarios such as a bus and even at long distances (e.g. 12 m for sunglass reflections).

9.4.2.4 Digital Footprints Protection

In the Metaverse, privacy inside avatars' digital footprints can be classified into three types [16]: (i) personal information (e.g. avatar profiling), (ii) virtual behaviors, and (iii) interactions or communications between avatars or between avatar and nonplayer character (NPC). Avatars' digital footprints can be tracked via virtual stalking/spying attacks in the Metaverse to disclose user's real identity and other private information, e.g. shopping preferences, location, and even banking details. A potential solution is *avatar clone* [39], which creates multiple virtual clones of the avatar which appear identical to confuse the attackers. Nevertheless, it brings other challenging issues such as managing multiple representations of each user and managing millions of clones roaming around the Metaverse.

Another potential solution is to *disguise* by periodically changing avatar's appearance to confuse attackers, or *mannequin* by replacing the avatar with a single clone (e.g. bot) which imitates user's behavior and *teleport* user's true avatar to another location when being tracked. Other privacy preservation mechanisms [16] include invisibility, private enclave, lockout. *Invisibility* indicates the avatar is made to be temporarily invisible in case of suspected stalking. *Private enclaves* allow certain locations inside the Metaverse to be occupied by individuals, which are unobserved by others. In private enclaves, owners have control over who can enter into the enclave by teleporting, thereby offering a maximum level of privacy. *Lockout* means certain areas inside the Metaverse are temporarily locked out for private use. After the lock expires, the restriction is lifted and other users are allowed to enter the area.

9.4.2.5 Personalized Privacy-Preserving Metaverse

As users/avatars are featured with personalized privacy demands and service preferences, existing privacy computing technologies (in Section 9.4.2.2) should

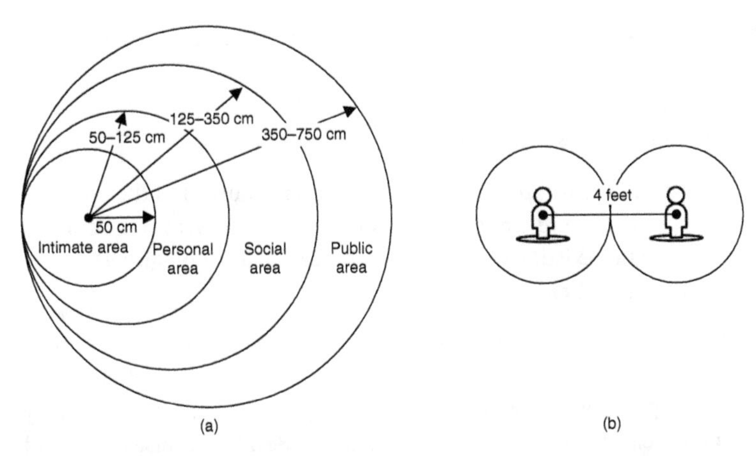

Figure 9.4 Illustration of personal space in real and virtual worlds. (a) Four types of personal spaces: public area (350–750 cm), social area (125–350 cm), personal area (50–125 cm), and intimate area (within 50 cm). (b) Meta's personal boundary function for avatars with default private border of 2 feet.

also take their customized privacy/service profiles into account in designing privacy-enhanced Metaverse. Existing works on personalized privacy computing are mainly based on similarity [64], randomized response [50], personalized FL [38], and so on. With the growth of Metaverse, more research on new personalized privacy preservation methods is required to serve new applications and the new ecology in the Metaverse.

9.4.2.6 Privacy-Enhancing Advances in Industry

In the Metaverse, there have been incidents such as VR groping and VR sexual harassments in Horizon Worlds. In the real world, people potentially keep an appropriate distance from others to maintain personal spaces when socializing. According to the interpersonal intimacy, psychologist Stanley Hall quantified and divided four types of personal spaces: public area (350–750 cm), social area (125–350 cm), personal area (50–125 cm), and intimate area (within 50 cm), as shown in Figure 9.4(a). It means that for less-familiar people, the more personal space we require. Similarly, each avatar also requires personal space even in the virtual world. Recently, Meta announced the *private boundary* function in its Metaverse platforms Horizon Venues and Horizon Worlds to avoid groping and harassments, where the default personal border for every avatar is a 2-foot circle. As shown in Figure 9.4(b), avatars need to keep at least 4 feet (about 1.2 m) away from others to maintain private space.

9.5 Network-Related Threats and Countermeasures in Metaverse

9.5.1 Threats to Metaverse Network

In the Metaverse, traditional threats (e.g. physical-layer security) to the communication networks can also be effective, as the Metaverse evolves from the current Internet and incorporates existing wireless communication technologies. Here, we list some typical threats as below.

(1) SPoF: In the construction of Metaverse systems, the centralized architecture (e.g. cloud-based system) brings convenience for user/avatar management and cost saving in operations. Nevertheless, it can be prone to the SPoF caused by the damage of physical root servers and DDoS attacks. Besides, it raises trust and transparency challenges in trust-free exchange of virtual goods, virtual currencies, and digital assets across various virtual worlds in the Metaverse.

(2) DDoS: As the Metaverse includes massive tiny wearable devices, adversaries may compromise these Metaverse end-devices and make them part of a botnet [10] (e.g. Mirai) to conduct DDoS attacks to make network outage and service unavailability by overwhelming the centralized server with giant traffic within short time periods, as depicted in the upper part of Figure 9.5. Besides, owing to the constrained communication pressure and storage space on the blockchain, part of NFT functions may be performed on off-chain systems in practical applications, where adversaries may launch DDoS attacks to cause service unavailability of the NFT system.

(3) Sybil Attacks: Sybil adversaries may manipulate multiple faked/stolen identities to gain disproportionately large influence on Metaverse services (e.g. reputation service, blockchain consensus, and voting-based service in digital governance) and even take over the Metaverse network, thereby compromising system effectiveness, as shown in the lower part of Figure 9.5. For example, adversaries may be able to out-vote genuine nodes by producing sufficient Sybil identities to refuse to deliver or receive some blocks, thereby effectively blocking other nodes from a blockchain network in the Metaverse.

9.5.2 Situational Awareness in Metaverse

Situational awareness is an effective tool for security monitoring and threat early warning in large-scale complex systems such as the Metaverse. In the Metaverse, local situational awareness is essential for monitoring a single security domain and global situational awareness can assist early warning of large-scale distributed threats targeted at multiple sub-Metaverses.

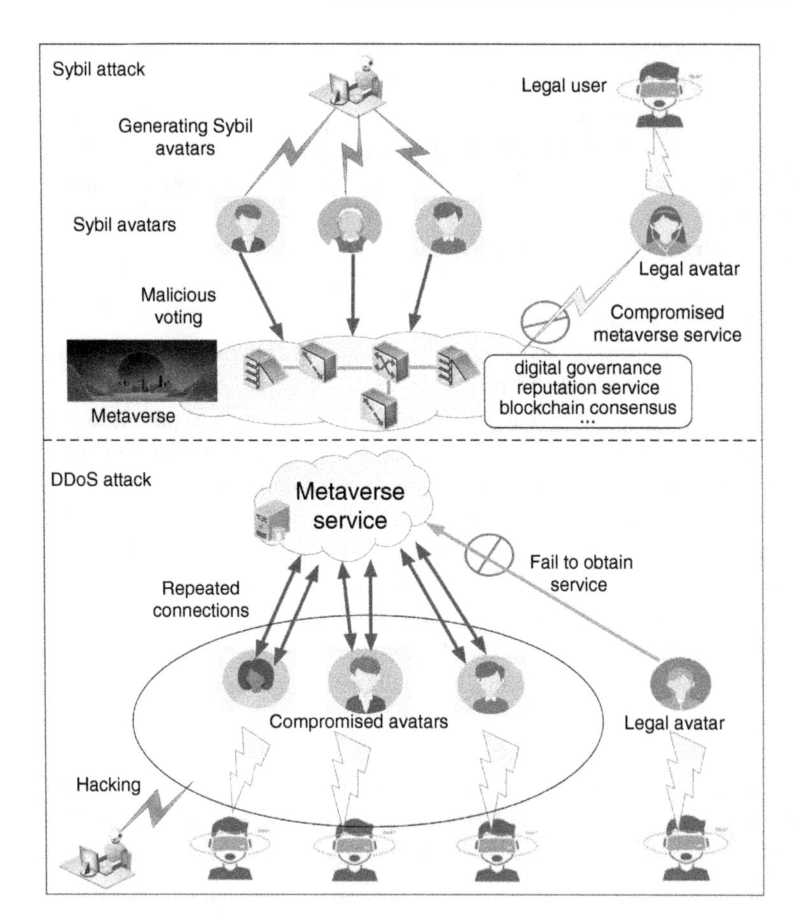

Figure 9.5 An illustrative example of Sybil attack and DDoS attack in Metaverse.

9.5.2.1 Local Situational Awareness

Situational awareness for devices and systems has received increasing attention in the Metaverse. Lv et al. [37] present a smart intrusion detection model to detect attack behaviors on 3D VR-based industrial control systems based on support vector machine (SVM). Experimental results on a simulated VR industrial scenario show that its average accuracy can keep above 90%. To be further adaptive to wearable devices with extreme size and energy constraints, Zhang et al. [69] propose a multilayered lightweight anomaly detection method by exploiting radio-frequency wireless communications to/from them to identify potentially malicious transactions.

To summarize, existing security measures can be categorized into two groups: *reactive* approaches (aim to counter past known attacks) and *proactive* approaches (aim to mitigate future unknown attacks). In general cases, reactive defenses built on timely attack trapping, frequent retraining, and decision verification can be more convenient and effective than pure proactive defenses. Besides, proactive defenses can be classified into two paradigms [11]: *security by design* defenses (against white-box attacks) and *security by obscurity* defenses (against black-box attacks). The above defense approaches can provide some lessons to resist unknown/new threats in the Metaverse.

9.5.2.2 Global Situational Awareness

Global situational awareness can facilitate understanding global security statuses in defending large-scale attacks in the Metaverse. Shahsavari et al. [46] propose a data-driven approach for global situational awareness based on multiclass SVM classifier to extract malicious events from collected raw metering data. Profiling of potential attack behaviors is another challenge in the Metaverse. Krishnan et al. [24] combine digital twin and software-defined network (SDN) to build a behavioral monitoring and profiling system where security strategies are evaluated on digital twins before being deployed in the real network.

Honeynets consisting of collaborative honeypots offer an alternative solution for building a secure Metaverse to defend against large-scale distributed attacks. Zarca et al. [68] propose SDN-enabled virtual honeynet services with higher degree of scalability and flexibility, and the efficiency of the proposed approach is validated using real implementations and tests. As shown in Figure 9.6, based on specific security policies, security virtual network functions (VNFs) (e.g. virtual honeynet, intrusion detection system (IDS), intrusion protection system (IPS), and firewall) can be configured and instanced on demand reactively or proactively,

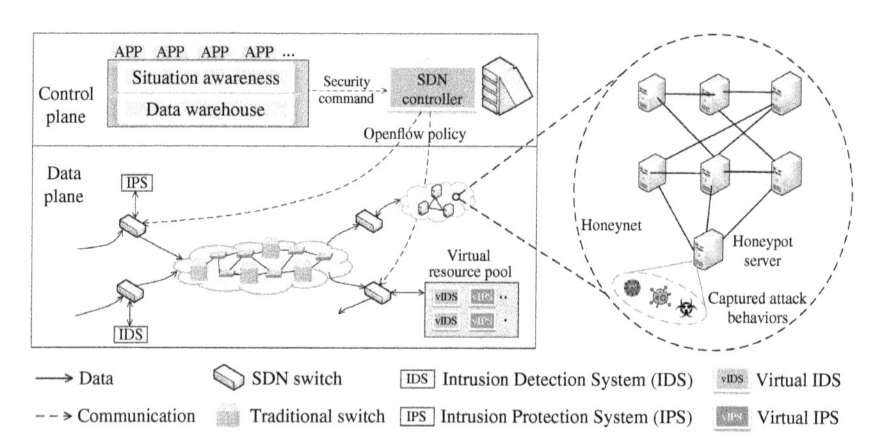

Figure 9.6 Illustration of software-defined network (SDN)-enabled virtual honeynet services for collaborative situational awareness [68].

coordinated by the SDN controller. Thereby, appropriate defense mechanisms (including situation monitoring, attack trapping, and security resource allocation) can be provisioned quickly and feasibly to enable self-protection, self-repair, and self-healing. However, the trust issues and resilience of compromised domain operators in aggregating local situational awareness into the global one require further investigation.

9.6 Economy-Related Threats and Countermeasures in Metaverse

9.6.1 Threats to Metaverse Economy

Various attacks may threaten the creator economy in the Metaverse from the service trust, digital asset ownership, and economic fairness aspects.

(1) Service Trust Issues in UGC and Virtual Object Trading: In the open Metaverse marketplace, avatars may be distrustful entities without historical interactions. There exist inherent fraud risks (e.g. repudiation and refusal-to-pay) during UGC and virtual object trading among different stakeholders in the Metaverse. Besides, in the construction of virtual objects via digital twin technologies, the Metaverse has to guarantee that the produced and deployed digital copies are authentic and trustworthy [35]. For example, malicious users/avatars may buy UGCs or virtual objects in Roblox and illegally sell the digital duplicates of them to others to earn profits. In addition, adversaries may exploit vulnerabilities in Metaverse systems to commit fraud and undermine service trust. An example is that the Metaverse project *Paraluni* based on Binance Smart Chain (BSC) lost over US$1.7 million in 2022 due to the reentrancy flaw in smart contracts [2].

(2) Threats to Digital Asset Ownership: Due to the lack of central authority and the complex circulation and ownership forms (e.g. collective ownership and shared ownership [42]) in the distributed Metaverse system, it poses huge challenges for the generation, pricing, trusted trading, and ownership traceability in the life cycle of digital assets in the creator economy. Empowered by blockchain technology, the indivisible, tamper-proof, and irreplaceable NFT offers a promising solution for asset identification and ownership provenance in the Metaverse [58]. However, NFTs also face threats such as ransomware, scams, and phishing attacks. For example, adversaries may mint the same NFT on multiple blockchains at the same time. Besides, evil actors may cash out their shares after inflating the value of NFTs, or they may sell NFTs to gain benefits before minting anything, where these decentralized finance (De-Fi) scams cause US$129 million lost in 2020 [1].

(3) Threats to Economic Fairness in Creator Economy: Well-designed incentives [65] are benign impetuses to promote user participation and open creativity in resource sharing and digital asset trading in the creator economy. The following three adversaries who threaten economic fairness are considered.

- **Strategic** users/avatars may manipulate the digital market in the Metaverse to make enormous profits by breaking the supply and demand status [65]. For example, in Metaverse auctions, strategic avatars may overclaim its bid, instead of its true valuation, to manipulate the auction market and win the auction.
- **Free-riding** users/avatars may unfairly gain revenues and enjoy Metaverse services without contributing to the Metaverse market [32], thereby compromising the sustainability of creator economy. For example, a free-riding avatar may submit meaningless local updates in collectively training an intelligent 3D navigation model under distributed AI and unfairly enjoy the benefits from the trained Metaverse model.
- **Collusive** users/avatars in the Metaverse may collude with each other or with the VSP to perform market manipulation and gain economic benefits [66]. For example, collusive avatars may collude to manipulate the results of Metaverse auctions and earn illegal revenues.

9.6.2 Open and Decentralized Creator Economy

Creator economy is an essential component of the Metaverse to maintain its sustainability and promote avatars' open creativity. Besides, it should be built on a decentralized architecture to prevent centralization risks, e.g. SPoF, non-transparency, and control by a few entities. Specifically, the Metaverse economy should simultaneously achieve three goals: (i) make data/assets from different sources mutually identifiable, trustworthy, and verifiable; (ii) design suitable incentive mechanisms for data/assets circulation to form a benign data sharing and coordination pattern; (iii) allow data subjects, data controller, data processor, and the user have the right to negotiate the rules and mechanisms of data protection and applications.

9.6.2.1 Trusted UGC/Asset/Resource Trading

Blockchain technologies (e.g. NFT and smart contract) provide a decentralized solution to construct the sustainable creator economy. NFT is the irreplaceable and indivisible token in the blockchain [58] and is regarded as the unique tradable digital asset associated with virtual objects (e.g. land parcel and digital drawing). For example, in the game Cryptokitties, players can buy virtual pet cats with unique genetic attributes identified by NFT and breed them. Besides, smart contracts enable the automatic transaction enforcement and financial settlement in trading virtual objects, items, and assets. For example, De Biase et al. [15] propose a swarm economy model for digital resource sharing which incorporates heterogeneous smart devices' spontaneous collaboration and dynamic organization. A blockchain-based transaction model is also developed in [15] for transparent and immutable currency audit, thereby ensuring trading trust among distrustful devices.

Apart from the trust-free blockchain approaches, trust or reputation management offers a quantifiable solution to evaluate the trustworthiness of participants and services with less computation/energy/storage consumption. Das and Islam [14] propose dynamic trust models and metrics based on user interactions including direct/indirect trust (derived from local/recommendation experience) and recent/historical trust (considering time decay effects). To achieve "trust without identify," Wang et al. [57] present an anonymous trust and reputation management system in mobile crowdsensing.

9.6.2.2 Economic Fairness for Manipulation Prevention

Collaboration is essential to the creator economy. Nevertheless, it is hard to promote collaboration among all individual users/avatars without sufficient incentives. Besides, the economic fairness in Metaverse markets may be violated by strategic, free-riding, and collusive users/avatars.

Strategy-proof incentive mechanisms, e.g. truthful auctions [65] and truthful contracts [61], can prevent strategic users/avatars from market manipulating. However, truthful participation also violates user's privacy, e.g. the true bid in auctions may reveal user's true valuation on the items. Existing strategy-proof and privacy-preserving auctions mainly depend on cryptographic mechanisms (e.g. ZKP and HE) and DP approaches, which may either bring large system burdens for energy-limited wearable devices or large data utility decrease in practical Metaverse applications. A trade-off mechanism between privacy and utility is needed for users/avatars with diverse preferences in the Metaverse.

Multiuser/avatar collusion prevention is also important for fair creator economy. Existing collusion-resistant mechanisms mainly focus on AI-based collusion behavior detection [48], cryptographic approaches [36], game theory [66], and optimization theory [31], which can be beneficial for collusion defense in Metaverse services. Future research efforts are required in designing fair mechanisms with the combination of strategy-proofness, collusion-resistance, and free-rider prevention, along with privacy preservation in the Metaverse.

9.6.2.3 Ownership Traceability of Digital Assets

In the Metaverse, blockchain provides a promising solution to manage the complex asset provenance and ownership tracing in the life cycle of digital assets by recording the evidence of content/asset originality and involved operations on the public ledgers. As the recorded historical activities on blockchain ledgers are maintained by the majority of entities in the Metaverse, it is ensured to be democratic, immutable, transparent, auditable, and nonrepudiable. Besides, smart contracts offer an intelligent traceability solution by coding the ownership management logic into scripts which are run atop the blockchain. In addition to private ownership, there can exist multiple types of ownership forms in the Metaverse such as collective ownership and shared ownership [42], which raise extra challenges in ownership management of virtual objects and Metaverse assets. In current Metaverse projects, there have been increasing interest in utilizing NFT for asset identification and ownership provenance [58]. Nevertheless, NFTs also face vulnerabilities such as cross-chain fraud, inflation attack, phishing, and ransomware. An example is that bad actors may concurrently mint the same NFT on multiple blockchains.

9.7 Threats to Physical World and Human Society and Countermeasures in Metaverse

The threats occurring in the Metaverse may also affect the physical world and threaten human society.

9.7.1 Threats to Physical World and Human Society

(1) Threats to Personal Safety: In the Metaverse, hackers can attack wearable devices, XR helmets, and other indoor sensors (e.g. cameras) to obtain the life routine and track the real-time position of users to facilitate burglary, which may threaten their safety. A report released by the XR Security Initiative (XRSI) shows that an adversary can manipulate a VR device to reset the hardware's physical boundaries [4]. Thereby, a user in Metaverse can be potentially pushed toward a flight of stairs or misdirected into dangerous physical situations (e.g. a street).

Besides, the Metaverse can open up new opportunities for misconducts and crimes. In the Metaverse, risks of physical trauma may be limited, but users could be mentally scarred. For example, due to the immersive realism of Metaverse, hackers can suddenly display harmful and scary content (e.g. ghost pictures) in the virtual environment in front of the avatar, which may lead to the death of fright of the corresponding user.

(2) Threats to Infrastructure Safety: By sniffing the software or system vulnerabilities in the highly integrated Metaverse, hackers may exploit the compromised devices as entry points [56] to invade critical national infrastructures (e.g. power grid systems and high-speed rail systems) via APT attacks [21].

(3) Social Effects: Although Metaverse offers an exciting digital society, severe side effects can also raise in human society such as user addiction, rumor prevention, child pornography, biased outcomes, extortion, cyberbullying, cyberstalkers, and even simulated terrorist camps. For example, the immersive Metaverse can provide future potentials for extremists and terrorists by making it easier to recruit and meet up, offering new ways for training and coordination, and lowering costs for finding new targets.

9.7.2 Physical Safety

9.7.2.1 Cyber Insurance-Based Solutions

Cyber insurance offers a financial instrument for risk mitigation of critical infrastructures in cyberthreats. To resolve the high premium stipulation in traditional insurance offered by insurance companies, Lau et al. [26] propose the coalitional insurance in power systems where the coalitional premium is computed by considering loss distributions, vulnerabilities, and budget compliance in an insurance coalition. Feng et al. [17] integrate cyber insurance into blockchain services to prevent potential damages under attacks, where a sequential game theoretical framework is developed to model the interactions among users, blockchain platform, and cyber-insurer. However, when applying to the Metaverse, the scalable and dynamic insurance coalition formation along with fair premium design under diverse cyber threats (e.g. anti-forensics) require further investigation.

9.7.2.2 CPSS-Based Solutions

Apart from the single cyber perspective, existing CPSS-based solutions afford lessons for cyberthreat defense and physical safety protection in the Metaverse from the perspective of interactions between cyber and physical worlds. Vellaithurai et al. [56] introduce cyber-physical security indices for security measurement of power grid infrastructures. The cyber probes (e.g. IDS) are deployed on host systems to profile system activities, where the generated logs along with the topology information are to build stochastic Bayesian models using belief propagation algorithms. Different from CPSS, Metaverse is an immersive and hyper-spatiotemporal virtual space with a sustainable economy ecosystem, which adds extra challenges in solution migration.

9.7.3 Society Management

9.7.3.1 Misinformation Spreading Mitigation

The extremely rapid information spreading (e.g. gossip) in the Metaverse makes the so-called "butterfly effect" more challenging in social governance and public safety in the real world. As an attempt to address this issue, Zhu et al. [72] propose to minimize the misinformation influence in online social networks (OSNs) by dynamically selecting a series of nodes to be blocked from the OSN. However, it only works in traditional static OSNs, and it is challenging to be applied in the fully interactive Metaverse with a huge and time-varying social graph structure.

9.7.3.2 Human Safety and Cyber Syndromes

The full immersiveness in Metaverse can also raise immersion concerns, e.g. occlusion and chaperone attack, as well as cybersickness [60]. Casey et al. [13] investigate a new attack named *human joystick attack* in immersive VR systems such as Oculus Rift and HTC Vive. In their work, adversaries can modify VR environmental factors to deceive, disorient, and control immersed human players and move them to other physical locations without consciousness. Valluripally et al. [55] present a novel cybersickness mitigation method and several design principles in social VR learning scenarios via threat quantification and attack-fault tree model construction. However, the ethical issues and adaptations to different attack-defense strategies are not considered in their work, which is an important factor for future Metaverse construction. Besides, more research efforts are required on the mitigation of other immersion risks to human body and human society.

9.8 Governance-Related Threats and Countermeasures in Metaverse

Driven by the above threats, it raises huge governance demands and poses huge regulation challenges to Metaverse lawmakers and regulators.

9.8.1 Threats to Metaverse Governance

In analogy to the social norms and regulations in the real world, the interactions among avatars (e.g. content creation, social activities, and virtual economy) in the Metaverse should align with the digital norms and regulations to ensure compliance. In the supervision and governance process of Metaverse, the following threats may deteriorate system efficiency and security.

(1) New Laws and Regulations for Virtual Crimes: Essentially, it is difficult to decide whether a virtual crime is the same as a real one. Thereby, it is hard to directly apply the laws and regulations in real life to enforce penalization for criminal actions [20] such as abusive language, virtual harassment, virtual stalking/spying, and so on. For example, if an avatar is verbally abusive in the Metaverse, it can be easily regarded as verbal abuse either in virtual or real worlds. However, if an avatar attempts to virtually stalk or harass another user's avatar in the Metaverse, the definitions of these crimes may be adapted from the real ones, as well as the appropriate punishments, which should be reconsidered for Metaverse lawmakers and regulators.

(2) Misbehaving Regulators: Regulators may misbehave and cause system paralysis, and their authorities also need supervision. Dynamic and effective punishment/reward mechanisms should be enforced for misbehaving/honest regulators, respectively. To ensure sustainability, punishment and reward rules should be maintained by the majority of avatars in a decentralized and democratic manner. Automatic regulations implemented by smart contracts without reliance on trusted intermediaries may be a promising solution. However, it also raises new issues such as information disclosure, mishandled exceptions, and susceptibility to short address attacks and reentrancy attacks [8].

(3) Threats to Collaborative Governance: To avoid the concentration of regulation rights, collaborative governance under hierarchical or flat mode is more suitable for large-scale Metaverse maintenance [22]. Nevertheless, collusive regulators may undermine the Metaverse system even under collaborative governance scenarios. For example, they can collude to make a certain regulator partitioned from the network via wormhole attacks.

9.8.2 Digital Governance in Metaverse

Apart from the laws or regulations (i.e. "hard law"), the "soft law" is also significant to adjust social relations and regulate user's behaviors in public Metaverse governance. The soft law refers to legal norms including autonomy and self-discipline norms and advocacy rules created by various organizations. Almeida et al. [6] highlight three principles in the digital governance of content moderation ecosystems: (i) open, transparent, and consensus-driven, (ii) respect human rights, and (iii) publicly accountable.

9.8.2.1 AI Governance

With the pervasive fusion of perception, computing, and actuation, AI will play a leading role to allow digital self-governance of individuals and society in the Metaverse in a fully automatic manner. AI approaches can be employed for detecting misbehaving entities and abnormal or Sybil accounts in the Metaverse. Besides, the outcomes of AI governance algorithms can be biased and unfair (e.g. race bias), thereby arising ethical concerns. Gasser and Almeida [18] propose a three-layer AI governance model from the sociological perspective, where the bottom technical layer allows the data governance and algorithm accountability; the middle ethical layer guides decision-making and data processing via ethical criteria and norms; and the top social and legal layer addresses the allocation of responsibilities in regulation. Nevertheless, the concrete governance protocols and algorithms with ethic-compliance (e.g. how to define a malicious behavior/avatar) require more research efforts. To summarize, both technological and sociological insights are required to build an AI-governed future Metaverse.

9.8.2.2 Decentralized Governance

For governance in the large-scale Metaverse maintenance, centralized regulatory can face multiple technical and standard obstacles and difficulty in the compatibility of transnational regulations. Collaborative governance can avoid concentration of regulation rights and promote democracy for avatars. Blockchain technologies offer potential decentralized solutions for collaborative governance in the Metaverse, where smart contracts offer a straightforward approach for decentralized governance in an automatic manner. Bai et al. [9] present a blockchain-based decentralized framework in digital city governance to encourage users' active engagement and witness in all administrative processes. In their approach, a verifier group is dynamically selected from digital citizens for transaction verification in the hybrid blockchain. A private-prior peer prediction mechanism is devised for collusion prevention among verifiers.

Based on SDN, Huang et al. [22] design a decentralized data lifecycle governance architecture, where UGC owners can implement customized governance rules for data usage to VSPs, aiming to promote an open environment to satisfy users' diverse requirements. The implementation of AI governance under decentralized architectures is a future trend for Metaverse governance. Besides, tailored blockchain solutions to Metaverse governance are required including Metaverse-specific consensus protocols, new on/off-chain data storage mechanisms, law-compliant regulated blockchain, etc.

9.9 Conclusions and Future Research Directions

In this chapter, we have investigated the security and privacy threats to the Metaverse, as well as the critical challenges in security defenses and privacy preservation. Furthermore, we have reviewed existing/potential solutions in designing tailored security and privacy countermeasures for the Metaverse. We expect that this chapter can shed light on the security and privacy provision in Metaverse applications and inspire more pioneering research in this emerging area. Future research directions in the Metaverse include the following aspects:

- With the continuity of ubiquitous cyber-physical attack surfaces in the Metaverse, current bring-in security defenses can be fragile and costly in practical use. How to design endogenous secure Metaverse for provisioning *secure by design* mechanisms with self-protection, self-evolution, and autoimmunity capabilities is an open problem.

- By analyzing the Metaverse system as a whole, the orchestration of cloud-edge-end computing is essential to facilitate seamless security provision and privacy protection while enhancing the QoE for Metaverse users/avatars, which still requires further investigation.
- Distinct sub-Metaverses may deploy services on heterogeneous blockchains to meet QoS requirements, resulting in severe interoperability concerns. The design of cross-chain interoperable and regulatory Metaverse remains to be investigated.

Bibliography

1 Don't get rugged: DeFi scams go from zero to $129 million in a year to become top financial hack. URL https://www.techrepublic.com/article/dont-get-rugged-defi-scams-go-from-zero-to-129-million-in-a-year-to-become-top-financial-hack/.

2 Hackers exploited reentrancy vulnerability to attack Paraluni, making more than $1.7 million. URL https://webscrypto.com/hackers-exploited-reentrancy-vulnerability-to-attack-paraluni-making-more-than-1-7-million-about-1-3-of-which-has-flowed-into-tornado/.

3 Metaverse breached: Second Life customer database hacked. URL https://techcrunch.com/2006/09/08/metaverse-breached-second-life-customer-database-hacked/.

4 Metaverse rollout brings new security risks, challenges. URL https://www.techtarget.com/searchsecurity/news/252513072/Metaverse-rollout-brings-new-security-risks-challenges.

5 NFT investors lose $1.7m in OpenSea phishing attack. URL https://threatpost.com/nft-investors-lose-1-7m-in-opensea-phishing-attack/178558/.

6 Virgílio Almeida, Fernando Filgueiras, and Danilo Doneda. The ecosystem of digital content governance. *IEEE Internet Computing*, 25(3):13–17, May-June 2021.

7 D. Antonioli, N. Tippenhauer, and K. Rasmussen. BIAS: Bluetooth impersonation attacks. In *IEEE Symposium on Security and Privacy (SP)*, pages 549–562, May 2020.

8 Nicola Atzei, Massimo Bartoletti, and Tiziana Cimoli. A survey of attacks on Ethereum smart contracts (SoK). In *International Conference on Principles of Security and Trust*, pages 164–186, 2017.

9 Yuhao Bai, Qin Hu, Seung-Hyun Seo, Kyubyung Kang, and John J. Lee. Public participation consortium blockchain for smart city governance. *IEEE Internet of Things Journal*, 9(3):2094–2108, February 2022.

10 E. Bertino and N. Islam. Botnets and Internet of Things security. *Computer*, 50(2):76–79, February 2017.

11 B. Biggio and F. Roli. Wild patterns: Ten years after the rise of adversarial machine learning. *Pattern Recognition*, 84:317–331, December 2018.

12 Stephen Bono, Dan Caselden, Gabriel Landau, and Charlie Miller. Reducing the attack surface in massively multiplayer online role-playing games. *IEEE Security & Privacy*, 7(3):13–19, May-June 2009.

13 Peter Casey, Ibrahim Baggili, and Ananya Yarramreddy. Immersive virtual reality attacks and the human joystick. *IEEE Transactions on Dependable and Secure Computing*, 18(2):550–562, March-April 2021.

14 Anupam Das and Mohammad Mahfuzul Islam. SecuredTrust: A dynamic trust computation model for secured communication in multiagent systems. *IEEE Transactions on Dependable and Secure Computing*, 9(2):261–274, March-April 2012.

15 Laisa Caroline Costa De Biase, Pablo C. Calcina-Ccori, Geovane Fedrecheski, Gabriel M. Duarte, Phillipe Soares Santos Rangel, and Marcelo Knörich Zuffo. Swarm economy: A model for transactions in a distributed and organic IoT platform. *IEEE Internet of Things Journal*, 6(3):4561–4572, June 2019.

16 Ben Falchuk, Shoshana Loeb, and Ralph Neff. The social metaverse: Battle for privacy. *IEEE Technology and Society Magazine*, 37(2):52–61, June 2018.

17 Shaohan Feng, Wenbo Wang, Zehui Xiong, Dusit Niyato, Ping Wang, and Shaun Shuxun Wang. On cyber risk management of blockchain networks: A game theoretic approach. *IEEE Transactions on Services Computing*, 14(5):1492–1504, September-October 2021.

18 Urs Gasser and Virgilio A. F. Almeida. A layered model for AI governance. *IEEE Internet Computing*, 21(6):58–62, November/December 2017.

19 Yue Han, Dusit Niyato, Cyril Leung, Dong In Kim, Kun Zhu, Shaohan Feng, Sherman Xuemin Shen, and Chunyan Miao. A dynamic hierarchical framework for IoT-assisted digital twin synchronization in the metaverse. *IEEE Internet of Things Journal*, 2022. doi: 10.1109/JIOT.2022.3201082.

20 Adel Hendaoui, Moez Limayem, and Craig W. Thompson. 3D social virtual worlds: Research issues and challenges. *IEEE Internet Computing*, 12(1):88–92, January-February 2008.

21 Pengfei Hu, Hongxing Li, Hao Fu, Derya Cansever, and Prasant Mohapatra. Dynamic defense strategy against advanced persistent threat with insiders. In *IEEE Conference on Computer Communications (INFOCOM)*, pages 747–755, 2015.

22 Gang Huang, Chaoran Luo, Kaidong Wu, Yun Ma, Ying Zhang, and Xuanze Liu. Software-defined infrastructure for decentralized data lifecycle governance: Principled design and open challenges. In *IEEE International Conference on Distributed Computing Systems (ICDCS)*, pages 1674–1683, July 2019.

23 Jostein Jensen and Martin Gilje Jaatun. Federated identity management - we built it; why won't they come? *IEEE Security & Privacy*, 11(2):34–41, March-April 2013.

24 Prabhakar Krishnan, Kurunandan Jain, Rajkumar Buyya, Pandi Vijayakumar, Anand Nayyar, Muhammad Bilal, and Houbing Song. MUD-based behavioral profiling security framework for software-defined IoT networks. *IEEE Internet of Things Journal*, 9(9):6611–6622, 2022.

25 Sanjeev Kumar, Jatin Chhugani, Changkyu Kim, Daehyun Kim, Anthony Nguyen, Pradeep Dubey, Christian Bienia, and Youngmin Kim. Second life and the new generation of virtual worlds. *Computer*, 41(9):46–53, September 2008.

26 Pikkin Lau, Lingfeng Wang, Zhaoxi Liu, Wei Wei, and Chee-Wooi Ten. A coalitional cyber-insurance design considering power system reliability and cyber vulnerability. *IEEE Transactions on Power Systems*, 36(6):5512–5524, November 2021.

27 Hyunjoo Lee, Jiyeon Lee, Daejun Kim, Suman Jana, Insik Shin, and Sooel Son. AdCube: WebVR Ad fraud and practical confinement of Third-Party Ads. In *30th USENIX Security Symposium (USENIX Security 21)*, pages 2543–2560, August 2021.

28 Ronald Leenes. Privacy in the metaverse: Regulating a complex social construct in a virtual world. In *The Future of Identity in the Information Society*, pages 95–112. Springer, July 2008.

29 Xiong Li, Jiabei He, P. Vijayakumar, Xiaosong Zhang, and Victor Chang. A verifiable privacy-preserving machine learning prediction scheme for edge-enhanced HCPSs. *IEEE Transactions on Industrial Informatics*, 18(8):5494–5503, August 2022.

30 Zi Li, Qingqi Pei, Ian Markwood, Yao Liu, and Haojin Zhu. Secret key establishment via RSS trajectory matching between wearable devices. *IEEE Transactions on Information Forensics and Security*, 13(3):802–817, March 2018.

31 Kun Li, Shengling Wang, Xiuzhen Cheng, and Qin Hu. A misreport- and collusion-proof crowdsourcing mechanism without quality verification. *IEEE Transactions on Mobile Computing*, 21(9):3084–3095, 2022.

32 Minglu Li, Jiadi Yu, and Jie Wu. Free-riding on BitTorrent-like peer-to-peer file sharing systems: Modeling analysis and improvement. *IEEE Transactions on Parallel and Distributed Systems*, 19(7):954–966, July 2008.

33 Xueping Liang, Sachin Shetty, Deepak Tosh, Charles Kamhoua, Kevin Kwiat, and Laurent Njilla. ProvChain: A blockchain-based data provenance architecture in cloud environment with enhanced privacy and availability. In *IEEE/ACM International Symposium on Cluster, Cloud and Grid Computing (CCGRID)*, pages 468–477, May 2017.

34 G. Liang, S. R. Weller, J. Zhao, F. Luo, and Z. Y. Dong. The 2015 Ukraine blackout: Implications for false data injection attacks. *IEEE Transactions on Power Systems*, 32(4):3317–3318, July 2017.

35 Siyi Liao, Jun Wu, Ali Kashif Bashir, Wu Yang, Jianhua Li, and Usman Tariq. Digital twin consensus for blockchain-enabled intelligent transportation systems in smart cities. *IEEE Transactions on Intelligent Transportation Systems*, 2021. doi: 10.1109/TITS.2021.3134002.

36 Jing Liu and Bo Yang. Collusion-resistant multicast key distribution based on homomorphic one-way function trees. *IEEE Transactions on Information Forensics and Security*, 6(3):980–991, September 2011.

37 Zhihan Lv, Dongliang Chen, Ranran Lou, and Houbing Song. Industrial security solution for virtual reality. *IEEE Internet of Things Journal*, 8(8):6273–6281, April 2021.

38 Jed Mills, Jia Hu, and Geyong Min. Multi-task federated learning for personalised deep neural networks in edge computing. *IEEE Transactions on Parallel and Distributed Systems*, 33(3):630–641, March 2022.

39 Huansheng Ning, Hang Wang, Yujia Lin, Wenxi Wang, Sahraoui Dhelim, Fadi Farha, Jianguo Ding, and Mahmoud Daneshmand. A survey on metaverse: The state-of-the-art, technologies, applications, and challenges. *arXiv preprint arXiv:2111.09673*, 2021.

40 Aleksandr Ometov, Sergey V. Bezzateev, Joona Kannisto, Jarmo Harju, Sergey Andreev, and Yevgeni Koucheryavy. Facilitating the delegation of use for private devices in the era of the internet of wearable things. *IEEE Internet of Things Journal*, 4(4):843–854, August 2017.

41 Rahul Raguram, Andrew M. White, Yi Xu, Jan-Michael Frahm, Pierre Georgel, and Fabian Monrose. On the privacy risks of virtual keyboards: Automatic reconstruction of typed input from compromising reflections. *IEEE Transactions on Dependable and Secure Computing*, 10(3):154–167, May-June 2013.

42 Hubert Ritzdorf, Claudio Soriente, Ghassan O. Karame, Srdjan Marinovic, Damian Gruber, and Srdjan Capkun. Toward shared ownership in the cloud. *IEEE Transactions on Information Forensics and Security*, 13(12):3019–3034, December 2018.

43 Kimberly Ruth, Tadayoshi Kohno, and Franziska Roesner. Secure multi-user content sharing for augmented reality applications. In *28th USENIX Security Symposium (USENIX Security 19)*, pages 141–158, August 2019.

44 Efat Samir, Hongyi Wu, Mohamed Azab, Chunsheng Xin, and Qiao Zhang. DT-SSIM: A decentralized trustworthy self-sovereign identity management framework. *IEEE Internet of Things Journal*, 9(11):7972–7988, 2022.

45 Bharadwaj Satchidanandan and P. R. Kumar. Dynamic watermarking: Active defense of networked cyberphysical systems. *Proceedings of the IEEE*, 105(2):219–240, February 2017.

46 Alireza Shahsavari, Mohammad Farajollahi, Emma M. Stewart, Ed Cortez, and Hamed Mohsenian-Rad. Situational awareness in distribution grid using micro-PMU data: A machine learning approach. *IEEE Transactions on Smart Grid*, 10(6):6167–6177, 2019.

47 Jiacheng Shang, Si Chen, Jie Wu, and Shu Yin. ARSpy: Breaking location-based multi-player augmented reality application for user location tracking. *IEEE Transactions on Mobile Computing*, 21(2):433–447, February 2022.

48 Haiying Shen, Yuhua Lin, Karan Sapra, and Ze Li. Enhancing collusion resilience in reputation systems. *IEEE Transactions on Parallel and Distributed Systems*, 27(8):2274–2287, August 2016.

49 Meng Shen, Huisen Liu, Liehuang Zhu, Ke Xu, Hongbo Yu, Xiaojiang Du, and Mohsen Guizani. Blockchain-assisted secure device authentication for cross-domain industrial IoT. *IEEE Journal on Selected Areas in Communications*, 38(5):942–954, May 2020.

50 Haina Song, Tao Luo, Xun Wang, and Jianfeng Li. Multiple sensitive values-oriented personalized privacy preservation based on randomized response. *IEEE Transactions on Information Forensics and Security*, 15:2209–2224, December 2020.

51 Jangirala Srinivas, Ashok Kumar Das, Neeraj Kumar, and Joel J. P. C. Rodrigues. Cloud centric authentication for wearable healthcare monitoring system. *IEEE Transactions on Dependable and Secure Computing*, 17(5):942–956, September-October 2018.

52 Zhou Su, Yuntao Wang, Tom H. Luan, Ning Zhang, Feng Li, Tao Chen, and Hui Cao. Secure and efficient federated learning for smart grid with edge-cloud collaboration. *IEEE Transactions on Industrial Informatics*, 18(2):1333–1344, February 2022.

53 Zhou Su, Yuntao Wang, Qichao Xu, and Ning Zhang. LVBS: Lightweight vehicular blockchain for secure data sharing in disaster rescue. *IEEE Transactions on Dependable and Secure Computing*, 19(1):19–32, January-February 2022.

54 Fangmin Sun, Weilin Zang, Haohua Huang, Ildar Farkhatdinov, and Ye Li. Accelerometer-based key generation and distribution method for wearable IoT devices. *IEEE Internet of Things Journal*, 8(3):1636–1650, February 2020.

55 Samaikya Valluripally, Aniket Gulhane, Khaza Anuarul Hoque, and Prasad Calyam. Modeling and defense of social virtual reality attacks inducing cybersickness. *IEEE Transactions on Dependable and Secure Computing*, 19(6):4127–4144, October 2021.

56 Ceeman Vellaithurai, Anurag Srivastava, Saman Zonouz, and Robin Berthier. CPIndex: Cyber-physical vulnerability assessment for power-grid infrastructures. *IEEE Transactions on Smart Grid*, 6(2):566–575, March 2015.

57 Xinlei Wang, Wei Cheng, Prasant Mohapatra, and Tarek Abdelzaher. Enabling reputation and trust in privacy-preserving mobile sensing. *IEEE Transactions on Mobile Computing*, 13(12):2777–2790, December 2013.

58 Qin Wang, Rujia Li, Qi Wang, and Shiping Chen. Non-fungible token (NFT): Overview, evaluation, opportunities and challenges. *arXiv preprint arXiv:2105.07447*, 2021.

59 Yuntao Wang, Haixia Peng, Zhou Su, Tom H. Luan, Abderrahim Benslimane, and Yuan Wu. A platform-free proof of federated learning consensus mechanism for sustainable blockchains. *arXiv preprint arXiv:2208.12046*, 2022.

60 Fei-Yue Wang, Rui Qin, Xiao Wang, and Bin Hu. MetaSocieties in Metaverse: MetaEconomics and MetaManagement for MetaEnterprises and MetaCities. *IEEE Transactions on Computational Social Systems*, 9(1):2–7, February 2022.

61 Yuntao Wang, Zhou Su, Tom H. Luan, Ruidong Li, and Kuan Zhang. Federated learning with fair incentives and robust aggregation for UAV-aided crowdsensing. *IEEE Transactions on Network Science and Engineering*, 9(5):3179–3196, 2022.

62 Yuntao Wang, Zhou Su, Ning Zhang, Jianfei Chen, Xin Sun, Zhiyuan Ye, and Zhenyu Zhou. SPDS: A secure and auditable private data sharing scheme for smart grid based on blockchain. *IEEE Transactions on Industrial Informatics*, 17(11):7688–7699, November 2021.

63 Segev Wasserkrug, Avigdor Gal, and Opher Etzion. Inference of security hazards from event composition based on incomplete or uncertain information. *IEEE Transactions on Knowledge and Data Engineering*, 20(8):1111–1114, August 2008.

64 Zongda Wu, Guiling Li, Qi Liu, Guandong Xu, and Enhong Chen. Covering the sensitive subjects to protect personal privacy in personalized recommendation. *IEEE Transactions on Services Computing*, 11(3):493–506, May-June 2018.

65 Fan Wu, Tianrong Zhang, Chunming Qiao, and Guihai Chen. A strategy-proof auction mechanism for adaptive-width channel allocation in wireless networks. *IEEE Journal on Selected Areas in Communications*, 34(10):2678–2689, October 2016.

66 Zichuan Xu and Weifa Liang. Collusion-resistant repeated double auctions for relay assignment in cooperative networks. *IEEE Transactions on Wireless Communications*, 13(3):1196–1207, March 2014.

67 Jun Yu, Zhenzhong Kuang, Baopeng Zhang, Wei Zhang, Dan Lin, and Jianping Fan. Leveraging content sensitiveness and user trustworthiness to recommend fine-grained privacy settings for social image sharing. *IEEE Transactions on Information Forensics and Security*, 13(5):1317–1332, May 2018.

68 Alejandro Molina Zarca, Jorge Bernal Bernabe, Antonio Skarmeta, and Jose M. Alcaraz Calero. Virtual IoT honeynets to mitigate cyberattacks in SDN/NFV-enabled IoT networks. *IEEE Journal on Selected Areas in Communications*, 38(6):1262–1277, June 2020.

69 Meng Zhang, Anand Raghunathan, and Niraj K. Jha. MedMon: Securing medical devices through wireless monitoring and anomaly detection. *IEEE Transactions on Biomedical Circuits and Systems*, 7(6):871–881, December 2013.

70 Leo Yu Zhang, Yifeng Zheng, Jian Weng, Cong Wang, Zihao Shan, and Kui Ren. You can access but you cannot leak: Defending against illegal content redistribution in encrypted cloud media center. *IEEE Transactions on Dependable and Secure Computing*, 17(6):1218–1231, November-December 2020.

71 Tianming Zhao, Yan Wang, Jian Liu, Yingying Chen, Jerry Cheng, and Jiadi Yu. TrueHeart: Continuous authentication on wrist-worn wearables using PPG-based biometrics. In *IEEE Conference on Computer Communications (INFOCOM)*, pages 30–39, July 2020.

72 Jianming Zhu, Peikun Ni, and Guoqing Wang. Activity minimization of misinformation influence in online social networks. *IEEE Transactions on Computational Social Systems*, 7(4):897–906, August 2020.

73 Roger Zimmermann and Ke Liang. Spatialized audio streaming for networked virtual environments. In *ACM International Conference on Multimedia (MM)*, pages 299–308, October 2008.

10

IoT-Assisted Metaverse Services

Yue Han[1], Cyril Leung[2], and Dong In Kim[3]

[1] *Alibaba-NTU Singapore Joint Research Institute, Nanyang Technological University, Singapore*
[2] *Department of Electrical and Computer Engineering, University of British Columbia, Vancouver, BC, Canada*
[3] *Department of Electrical and Computer Engineering, Sungkyunkwan University, Suwon, Gyeonggi-do, Korea*

After reading this chapter you should be able to:

- Understand why we need Internet-of-Things (IoT) for the Metaverse.
- Understand how to use IoT for the Metaverse and possible challenges.
- Apply dynamic games in addressing the resource allocation problem in an IoT-assisted Metaverse synchronization.
- Future challenges and research issues for deploying IoTs in Metaverse services.

10.1 Why Need IoT for Metaverse Services

10.1.1 Metaverse and Virtual Services

The Metaverse can be viewed as a quasisuccessor to the Internet [4], i.e. a massively scalable interoperable network of real-time rendered 3D virtual worlds (VWs). It allows a large number of users to synchronously and persistently share high quality, immersive, seamless, and personalized experiences even as they travel across different VWs. Users can extend their everyday lives in the *parallel* VWs in the Metaverse with each VW providing a virtual service (e.g. virtual education) that augments a Metaverse user's life. The virtual services can span all walks of life. For convenience, we categorize them into two groups: (i) augmenting an individual's experience and (ii) augmenting an industry service or operation.

Metaverse Communication and Computing Networks: Applications, Technologies, and Approaches, First Edition.
Edited by Dinh Thai Hoang, Diep N. Nguyen, Cong T. Nguyen, Ekram Hossain, and Dusit Niyato.
© 2024 The Institute of Electrical and Electronics Engineers, Inc. Published 2024 by John Wiley & Sons, Inc.

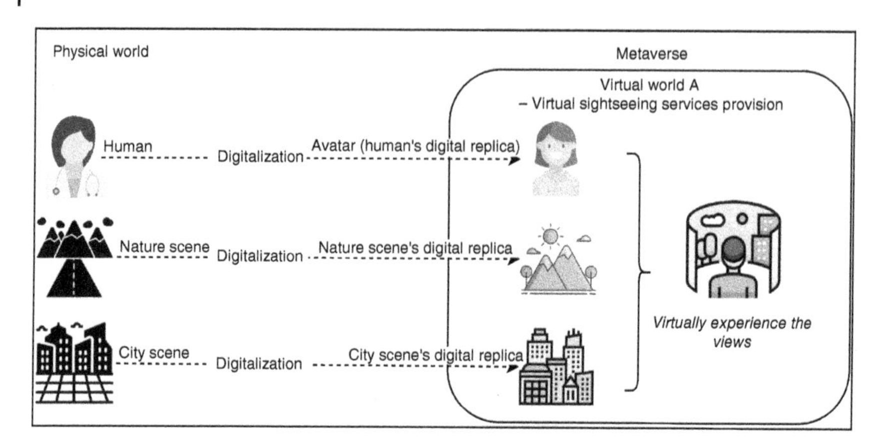

Figure 10.1 Virtual services in the Metaverse augment an individual's experience.

10.1.1.1 Augmenting an Individual's Experience

There is one type of virtual service in the Metaverse that can enrich a user's life by providing fantastic virtual experience, e.g. space traveling or virtual sightseeing. For example, as shown in Figure 10.1, a virtual service provider (VSP) *A* provides virtual sightseeing services [20, 21, 29] in a VW. For simplicity, we denote the VW that VSP *A* operates by the same notation, i.e. VW *A*. Here, a person in the physical world is represented in VW *A* by his *digital replica*, or *avatar*. Virtual scenes can be digital replicas of natural or metropolitan scenes or events, including, e.g. live street show. This type of virtual service would be of great interest to tourists during the COVID-19 pandemic period. Other experiential virtual services include virtual theme park [3], virtual concert [16], and virtual safari sightseeing [25], which digitize and replicate parts of the physical world in the parallel Metaverse to provide users vivid experiences anytime and anywhere.

10.1.1.2 Augmenting Industry Services or Operations

Existing industry product or service providers can extend/expand their business models in the Metaverse for higher efficiency and better process optimization, thereby enabling industrial transformation with additional profits and growth. For example, as shown in Figure 10.2, a real-world shoe manufacturer may provide a new customization service in VW *B* in the Metaverse, in which the production line is digitalized and replicated so that a user can virtually inspect and personalize the product. In VW *C*, a driver training service provider can replicate realistic road conditions so that a user's driving skills can be realistically assessed without any real danger.

For all the abovementioned examples, a realistic representation of real-world entities (e.g. products, cities, nature) is essential. This is because of their

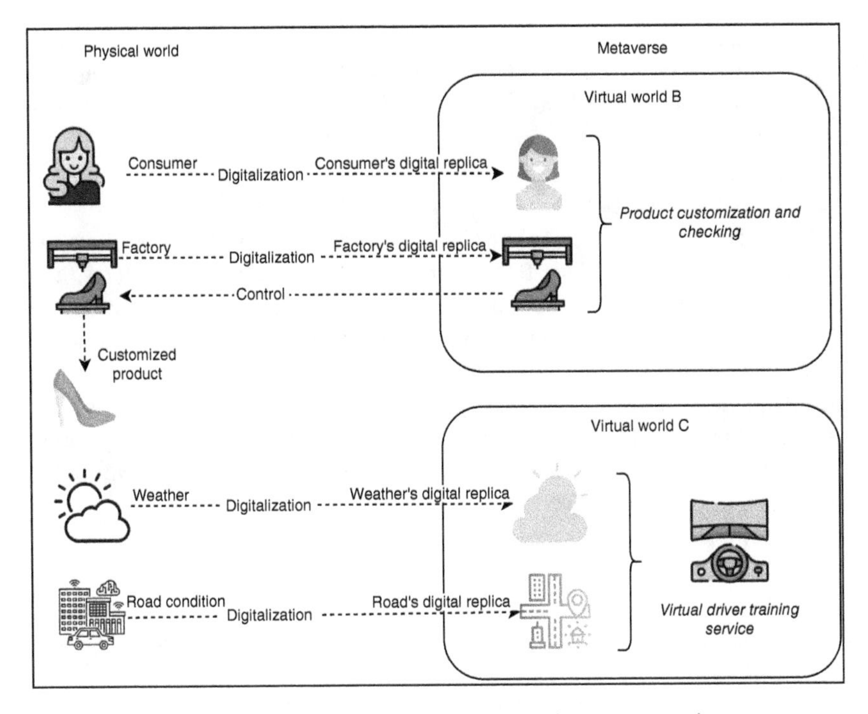

Figure 10.2 Virtual services in the Metaverse augment industry operations.

time-varying states, e.g. the city scenes in Figure 10.1, and the road conditions in Figure 10.2 have states which are constantly changing. Using digital replicas with lagging states in virtual services can cause disutility for users and VSPs, e.g. poor user experience due to outdated city views. Thus, sufficiently up-to-date digital replicas are essential.

To address this issue, digital twins (digital replicas of real-world entities) can be a promising solution, as it attempts to address (i) *digitalization*, i.e. how to digitize and represent the real world in the Metaverse and (ii) *synchronization*, i.e. how to maintain a sufficiently high degree of synchronization between physical entities and their digital replicas so as to ensure the usefulness of the digital replicas in Metaverse's virtual services. These two questions are discussed in the remaining of this chapter.

10.1.2 Digital Twins

Unlike entities in the real world that are composed of atoms, the primary existence in the Metaverse is information measured in bits. As such, a virtual replica, also known as digital twins (DTs) of real-world entities is needed. DT is not a new concept and has been used in various industries, e.g. digital farming [22], in which a digital farm, a virtual representation of a real farm is created to study how to reduce energy costs and improve productivity, in addition to impact on climate protection, food security, and resource management.

10.1.2.1 Definition of DTs

DTs were proposed in 2003 [12] by Grieves in his course on "product life cycle management," which defines DTs by physical products, virtual products, and their connections. In 2012, the National Aeronautics and Space Administration (NASA) defined DTs as "integrated multiphysical, multiscale, probabilistic simulations of an as-built vehicle or system using the best available physical models, sensor updates, and historical data" [11]. Regardless of the specific definition, it is clear that a key feature of a DT is its *mirroring* of a physical object, which means that *continuous updates* from physical space to virtual space are needed (physical → virtual), as the physical object's state changes over time.

There is no consensus regarding updates from DTs to physical objects (virtual → physical). On the one hand, the authors in [9, 15, 28] believe that there are *bidirectional* updates across cyberspace and physical space for DTs (virtual ↔ physical). To emphasize the bidirectional updates, the authors in [9] use *digital models* to refer to the case where there is no update across the two spaces and use *digital shadows* to refer to the case where there is a *one-directional* update from physical space to the cyberspace, i.e. a change in the physical objects' states causes changes in the digital objects but not vice versa. On the other hand, several other works [10, 12, 33] use DTs as simulation-based only. That is, they treat DTs as digital shadows (i.e. physical → virtual). Meanwhile, the control of physical assets via their DTs (physical ← virtual) are studied around another concept called cyber–physical systems (CPS) [2]. CPS can be leveraged to support large distributed control, e.g. automated traffic control and ubiquitous healthcare monitoring and delivery. Given the ambiguity in the definitions of DTs, a number of papers have called for more clarity on the difference between DTs, digital shadows, and CPS [9, 15]. In this chapter, we mainly use the term "DT" to refer to the *digital existence in the Metaverse*, i.e. a digital replication of a physical object (physical → virtual world).

10.1.2.2 Difference Between Metaverse and DTs

The short answer is that the Metaverse is more than DTs: the Metaverse is a platform, like the Internet, while DTs are only one key component, e.g. websites. The Metaverse is able to leverage DTs as one of its fundamental tools to represent the real world to study realistic problems in virtual space. For example, the DTs of the real world can provide advanced optimization and further decision-making support. This allows the VSPs in the Metaverse to understand interrelations within its industry production system and the consequent effects on the performance, without interrupting real-world production, taking into account social and environment sustainability and human health and well-being. In addition, the Metaverse allows more futuristic application scenarios to be studied, e.g. fusion of DTs across firms or industries due to the Metaverse's interoperability as well as considering the DTs with nonfungible tokens, etc.

10.1.2.3 Position of IoT in Metaverse DT Construction

In order to construct a DT, there are usually several steps involved:

- *Collecting* information, such as geometric information (e.g. size, shape), physical properties of a real-world entity, and state information (e.g. temperature, humidity) by wireless sensors and IoT devices.
- *Processing* data, extracting useful information to facilitate intelligent decision-making. This may involve advanced analytics methods such as artificial intelligence (AI) and machine learning (ML). To reduce processing time, advanced computing paradigms (e.g. parallel computing) are needed as well.
- *Storing* the processed data in a distributed manner, e.g. through blockchain technology and the edge-cloud architecture.
- *Fusion* of processed information from various sources to develop a global DT system, as some DT systems may contain several sub-DT systems, e.g. a smart city twin may involve several building DTs.
- *Analytics and learning* based on e.g. AI and ML are important. For example, they can be applied to the simulation process to reconstruct a DT's trajectory in a VW or to interpolate data points in the 3D model. This step relies heavily on the algorithm adopted by the VSP and the level of computational capability deployed, e.g. at the edge cloud or centralized data center.
- *State synchronization* with real-time data transfers between the virtual and real worlds to keep the states of the DTs in the Metaverse up-to-date is needed to make sure that the DT in the Metaverse is relevant, useful, and reliable.

We see that IoT devices and wireless sensors play an important role in the development of Metaverse DT based on the abovementioned steps. The reasons are as follows:

- **Flexible and on-demand deployment**: Mobile IoT devices are known for their flexible and on-demand deployment and ubiquitous connectivity. They can provide a wide sensing area coverage together with high flexibility in deployment which reduces the cost of deployment of sensors for VSPs.
- **On-device computation and learning**: IoT devices also provide on-device computation capability to process the raw data. Recent advances in federated learning [23] further support the on-device learning paradigm based on AI and ML. This can help some VSPs to develop DTs to address data privacy or data regulation concerns.

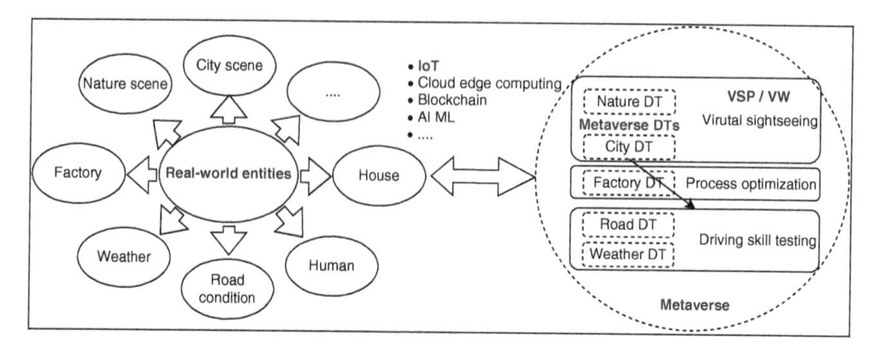

Figure 10.3 Relationship among IoT, DTs, VSPs, and Metaverse.

- **Proximity to a data source for real-time services**: The IoT devices' proximity to the data source and computation capability relieve the heavy burden on the backhaul network, thereby supporting VSPs' real-time services.

Therefore, we can see that IoT devices play a critical role in enabling the Metaverse's DTs, which bring about the convergence between the physical and virtual worlds. We illustrate the relationships among IoT, DTs, and the Metaverse in Figure 10.3. In Section 10.2, we will discuss how to use IoT for constructing the DTs and the potential challenges.

10.2 How to Use IoT for Metaverse DTs

To leverage IoT for constructing DTs in the Metaverse, we may take advantage of the recent emerging distributed data acquisition paradigm, called mobile crowd sensing (MCS).

10.2.1 Mobile Crowdsensing

MCS is a paradigm that can support the monitoring, control, and surveillance of physical assets without the deployment of *fixed* sensing and computing resources. MCS development depends to a large extent on wide-spread IoT and advanced communication technologies, which allow for a large number of connected devices available at the network edge. According to Cisco [1], by 2023, the number of hand-held or personal mobile-ready devices is estimated to reach 8.7 billion, and machine-to-machine (M2M) connections (e.g. asset tracking systems and GPS systems in cars) will reach 4.4 billion. The global rise in mobile devices and connectivity suggests an unprecedented upsurge in mobile

IoT, which links individuals, businesses, data, and things to make networked connections more valuable than ever. As a result, it is critical to supplement existing sensing infrastructures with cost-effective, sustainable, and nondedicated sensing technologies [6], such as participatory or opportunistic sensing using smartphone sensors [5] and UAVs [8, 18, 27, 30, 31].

10.2.2 Scenarios

To facilitate Metaverse services to develop their DTs, IoT-enabled MCS can be helpful in the following use cases:

- **External assistance when pushing interoperability process**: Unlike DTs in Industry 4.0, in which the owner of a physical asset (e.g. a motor) creates a DT to serve its own purpose, the Metaverse is known for its interoperability, which applies to all its sub-VWs and DTs contained in VWs. As such, the DTs are served not only by one VSP but also by other relevant VSPs. Normally, DTs are created by the owner of a physical asset. However, it is expected that not all physical asset owners will fully integrate with the required interoperability standard during the current budding phase of the Metaverse. In this case, IoT devices can provide external help to facilitate the collection of state information about physical objects, thereby allowing VSPs to develop DTs with interoperable standards.
- **Improving shared DT quality by having additional state information**: As noted above, DTs can be traded and shared across various VSPs for interoperability. However, these VSPs may have different DTs requirements (e.g. for freshness of information [14] or reliability [15]). Unsatisfactory quality of the DTs can affect the quality of experience (QoE) or service quality (QoS) for virtual businesses. As a result, VSPs may require additional data (or data with additional dimensions) for those shared DTs. IoT-enabled MCS can help such VSPs to improve the quality of the shared DTs, by collecting additional information on their physical counterparts on an on-demand basis.
- **Meeting ad hoc services requests by the VSPs**: As an example, virtual sightseeing services may wish to include some seasonal festival celebration activities. It can be expensive and challenging to deploy traditional fixed sensors to collect the status data of physical objects. The seasonal ad hoc services enabled by IoT crowdsensing can be helpful for those VSPs.

10.2.3 Challenges of Using IoT-Enabled MCS for Metaverse DTs

While MCS is a promising approach, there are a number of factors that need to be addressed.

10.2.3.1 Incentives

This is the first vital factor in MCS, as devices that perform the sensing tasks will incur costs (e.g. time, energy, and potential privacy leakage). Therefore, an effective incentive mechanism is necessary to encourage sensor-mounted devices to participate.

10.2.3.2 Data Quality

Heterogeneous devices may have different sensing capabilities based on the type of sensors mounted on the devices. The diverse quality of data provided by MCS needs to be considered in the design of the incentive mechanism.

10.2.3.3 Resource Management

IoT devices are known for their limited resources such as energy and computation capability. Therefore, how to allocate scarce resources to different MCS tasks is a key concern for IoT devices to maximize their utilities.

In addition to the abovementioned challenges for MCSs, there are some additional concerns that need to be well addressed for Metaverse DTs with IoT-enabled MCSs.

10.2.3.4 DT Value for the Virtual Business

We use the term "DT" value to refer to the usefulness of the DT for a VSP. For example, there can be some *temporal features* associated with a DT, for which *reliability* decreases with time [15]. The DT value may drop with time if there is no state synchronization. For example, if there is no synchronization, the temperature or rainfall of a weather DT gradually becomes less reliable. Other examples include traffic volumes for a traffic DT and locations for a car DT. Therefore, state synchronization between physical entities and their Metaverse DTs is critically required to maintain the value of the DT for VSPs.

10.2.3.5 VSPs' Tolerance to Nonupdated DTs

VSPs have different levels of tolerance to nonupdated DTs, due to (i) their business types, (ii) the extents to which they use AI, as well as (iii) computation capabilities. For example, an irregularly updated weather twin may have little impact on a virtual travel service provider, but a significant impact on virtual driver training by producing an incorrect assessment result. A VSP with a robust AI algorithm and a higher computation capacity may accurately predict the patterns of DT states in the near future, and thus may not require as many updates thereby allowing VSPs to optimize their synchronization intensities. How to represent, model, and take into account the VSP's tolerance levels in determining the optimal synchronization intensity is a crucial and largely open research problem.

10.2.3.6 Decision Sequence Among VSPs in the Metaverse Ecosystem

Multiple VSPs exist in the Metaverse ecosystem, and they are in fact similar to the real-world economic market players due to the transactions and sharing of DTs among VSPs. There may be some dominant players in the Metaverse market, such as VSPs with large business volumes. As such, the decision sequence among VSPs and how this affects the optimal synchronization strategy that VSPs apply for the DTs remain unknown.

In the following, we present a resource management solution for IoT-enabled MCS for Metaverse DT synchronization based on a game theoretical approach. The solution provides a general framework to model the crowd of mobile IoT devices' strategic behaviors in participating in the MCS campaign for assisting VSPs to collect state information about the DTs. The incentive mechanism design is proposed in the framework. In addition, the framework considers three issues regarding the Metaverse DTs listed above, namely (i) DT value dynamics, (ii) VSP's tolerance to poorly updated DTs, and (iii) decision sequence among VSPs in the Metaverse ecosystem. The detailed system model, solution, and experimental results are given in Section 10.3.

10.3 A Dynamical Hierarchical Game-Theoretical Approach for IoT-Assisted Metaverse Synchronization

The system model is shown in Figure 10.4. It includes (i) N edge devices (e.g. UAVs), denoted by the set $\mathcal{N} = \{1, \dots, n, \dots, N\}$ and (ii) M VSPs, denoted by the set $\mathcal{M} = \{1, \dots, m, \dots, M\}$. Hereafter, we consider UAVs as an example of mobile IoT devices.

Each VSP can use a set of DTs, which are critical to its virtual business profit. As time elapses, the difference between a DT's state and that of its physical counterpart may increase, thereby lowering the usefulness (value) of a DT to a VSP. To model this decay of DT values, we use $\theta_m > 0$ to represent the value decay rate of the DTs of VSP m. Furthermore, different VSPs can have different DT value decay rates, due to, e.g. the types of their virtual services, computational capabilities, and the extent to which they use AI.

Let $z_m(t) \geq 0$ denote the DT's instant valuation, i.e. the utility that a DT brings to VSP m at time instant t. For simplicity, we model the rate of change of the DT values, which is the first-order time derivative $\dot{z}_m(t) = z_m(t)/t$ as follows:

$$\dot{z}_m(t) = \eta_m(t) - \theta_m z_m(t), \quad m \in \mathcal{M}, \tag{10.1}$$

where $\eta_m(t)$ denotes the intensity, or rate, at which synchronization activities are carried at time t. One may interpret the DT value dynamics in (10.1) as follows: if VSP m determines not to synchronize DTs at all, i.e. $\eta_m(t) = 0$, then the value

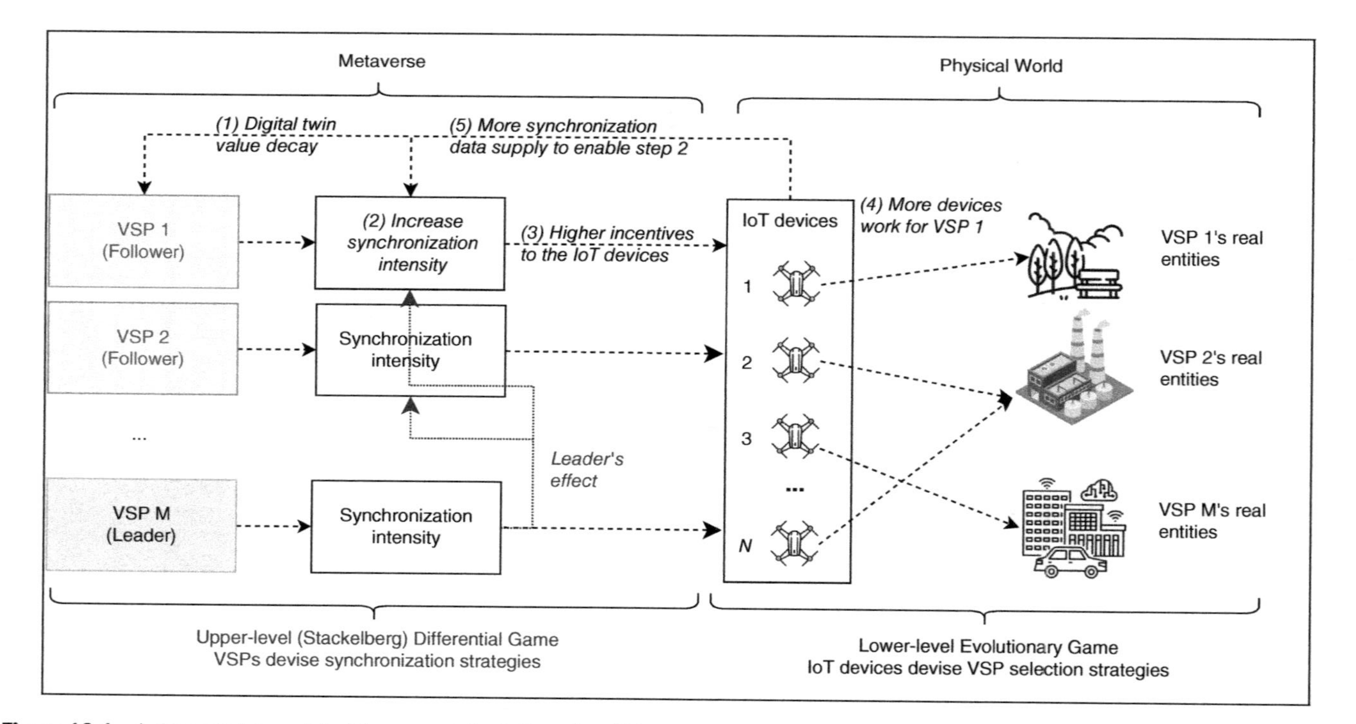

Figure 10.4 A dynamic hierarchical framework for IoT-assisted Metaverse synchronization.

of DTs deteriorates at the (time-independent) rate θ_m. By using a positive rate of synchronization, i.e. $\eta_m(t) > 0$, the VSP can slow down, or even reverse, the process of deterioration of its DTs. For convenience, we use $\mathbf{z}(t) = [z_m(t)]_{m\in\mathcal{M}}$ to denote the vector of DT values of all VSPs and let the vector $\boldsymbol{\eta}(t) = [\eta_m(t)]_{m\in\mathcal{M}}$ denote the synchronization strategies of all VSPs at time instant t.

To synchronize DTs with their real counterparts, the states data of the real counterparts are needed. Here, we consider that a set of N UAVs which can be motivated to assist VSPs synchronization tasks by sensing the corresponding real entities states for the benefit of the VSPs. For simplicity, we consider a group of UAVs[1] of the same type, e.g. characterized by the same sensing quality (i.e. resulting in similar data quality) and unit energy cost [24]. Based on synchronization requests from the M VSPs, each of the N UAVs can select a VSP to work for. UAVs that select the same VSP are expected to collectively sense the physical entities of interest to the VSP and share the incentives provided by the VSP. It is expected that when VSP m's synchronization intensity increases, VSP m will allocate more incentives to motivate UAVs to work for it. Therefore, the total incentive pool from VSP m should be positively correlated with the synchronization intensity chosen by VSP m.

Since the synchronization intensity is controlled by the VSP, we refer to it as *a control variable, a control, or a strategy* for a VSP. It is a function of the time t. By changing its control strategy for the synchronization intensity, a VSP can affect the states of the system, including DT value states in addition to the UAV's VSP selection strategy. Each VSP chooses a control which optimizes its utility. To determine the optimal control strategy of the synchronization intensity and its associated DT values and the UAV's VSP selections, a dynamic hierarchical framework is proposed as follows:

- **Lower-level evolutionary game**: At the lower-level, we investigate VSP selection strategies for UAVs. Each UAV is considered to have bounded rationality, that is, to select a strategy that is satisfactory rather than optimal [32]. This is to address the case in which UAV decisions are suboptimal with potentially incomplete information about the game (e.g. the payoffs received by other UAVs), especially when the number of UAVs is large. In this regard, we formulate the problem as an evolutionary game in Section 10.3.1, which models the strategy adaptation process of the UAVs.
- **Upper-level (Stackelberg) differential game**: At the upper level, we investigate the optimal synchronization strategies of the VSPs. As the synchronization intensity jointly affects the synchronization data supply and the value status of the DTs, the VSPs need to devise the optimal control strategy for

1 The extension to the scenario with heterogeneous types of UAVs is straightforward, as the set of UAVs can always be partitioned into multiple subpopulations so that the UAVs within a subpopulation are of the same type.

synchronization intensities so that the accumulated utilities, discounted at the present time, are maximized. We adopt a differential game approach to solve the problem in Section 10.3.2. Additionally, when there is a dominant VSP that has the privilege of choosing its strategy first, we formulate the problem as a Stackelberg differential game and solve it based on control theory.

10.3.1 Lower-Level Evolutionary Game

An evolutionary game is formulated based on a set of populations with evolutionary dynamics. The various aspects of the evolutionary game are as follows:

- **Players and populations**: Each UAV $n \in \mathcal{N}$ is a player in the evolutionary game. In addition, the set $\mathcal{N}$ is referred to as the population of players.
- **Strategy**: The VSP selection is a strategy that can be implemented by the player.
- **Population states**: Population states are the strategy distribution for the population, denoted by a vector $\boldsymbol{x}(t) = [x_m(t)]_{m \in \mathcal{M}}$. The component $x_m(t)$ denotes the fraction of UAVs in the population that choose to work for VSP m at time instant t. As the population states are subject to $\sum_{m \in \mathcal{M}} x_m(t) = 1$, the state space, i.e. the set of all the possible population states, is a unit simplex $\Delta \in \mathbb{R}^{M-1}$.
- **Utility functions**: Utility function $u_m(\boldsymbol{x}(t), \eta_m)$ describes the utility that a UAV can receive given the population states $\boldsymbol{x}(t)$ and the synchronization strategy η_m.

Each VSP m has d_m DTs and chooses a synchronization intensity $\eta_m(t)$, where $\eta_m(t) \geq 0$, at time t.[2] Let the time horizon for the analysis be defined as $\mathcal{T} = [0, T]$. We refer to $\boldsymbol{\eta}(t)$, $t \in \mathcal{T}$, as the control path over the time horizon $\mathcal{T}$. For notational simplicity, we omit the time variable t where there is no ambiguity. Note that determination of the optimal synchronization strategy path over $\mathcal{T}$ can be explored as an optimal control problem, which is to maximize the present value of the accumulated utility that the VSP can obtain, as described in Section 10.3.2. In this section, we assume that the optimal synchronization intensity path has been determined by the VSPs.

For VSP m, the incentive pool that it allocates to its associated UAVs is given as follows:

$$R_m(t) = \eta_m(t) d_m g(\theta_m), \tag{10.2}$$

where $g(\cdot)$ is a monotonically increasing function representing the weight affected by the value decay rate θ_m. One can interpret (10.2) as both synchronization rate η_m and the number of DTs d_m have a positive correlation with the incentive pool. That is, the higher the synchronization intensity or the number of DTs that a VSP has,

2 Note that when the control is a function of time only, it is referred to as an open-loop solution, which is explained later in Section 10.3.2.2.

the more incentives the VSP should offer UAVs. Similarly, the higher the decay rate θ_m is, the more incentives a VSP should offer. We consider an affine mapping for $g(\cdot)$, e.g. $g(\theta_m) = g_0 + g_1\theta_m$, where g_1 is a positive number to represent the positive correlation between the decay rate and the incentive.

Given the population states $\boldsymbol{x}$, there are Nx_m UAVs that select VSP m to sense the data and assist in its synchronization tasks. With a uniform incentive sharing scheme [13], each UAV receives an incentive in the amount of $\frac{R_m(t)}{Nx_m(t)}$. Let c_m represent the energy cost incurred by the sensing task for VSP m, e.g. UAV's energy cost flying from the base to the target region and the communication cost [17]. The utility received by a UAV selecting VSP m is $u_m(\boldsymbol{x}, \eta_m) = \frac{R_m(t)}{Nx_m(t)} - c_m = \frac{\eta_m(t)d_m g(\theta_m)}{Nx_m(t)} - c_m$.

The utility information for selecting each of the VSPs at the current time t can be exchanged among UAVs, e.g. at their base or device-to-device (D2D) communication in the air. A UAV may then adjust its VSP selection strategy at time $t+1$. The evolutionary process of the VSP selection strategy can be modeled by *replicator dynamics* [32], which is a set of ordinary differential equations, given as follows:

$$\dot{x}_m = \delta x_m(u_m - \bar{u}), \quad m \in \mathcal{M}, \tag{10.3}$$

where δ is the learning rate of the UAVs, $\dot{x}_m := dx_m(t)/dt$, and $\bar{u} := \sum_{m\in\mathcal{M}} x_m u_m$ denotes the average utility that a UAV population can have. Again, we omit arguments in $u_m(\boldsymbol{x}(t), \eta_m)$, and $x_m(t)$ for simplicity.

It can be seen from (10.3) that the population state $\boldsymbol{x}_m(t)$ evolves when the payoff received by a UAV is different from the population average utility. If the reward received by a device that selects VSP m is higher than the average utility, i.e. $u_m > \bar{u}$, then the population state $x_m(t)$ increases since more UAVs will select VSP m, i.e. $\dot{x}_m > 0$. The evolutionary process stops when $\dot{x}_m = 0$ for all $m \in \mathcal{M}$. This is called the *stationary state* or *evolutionary equilibrium* (EE). The stationary states can be achieved by either $x_m = 0$ for all VSP except one or $u_m = \bar{u}, \forall u \in \mathcal{M}$.

10.3.2 Upper-Level Differential Game for VSPs

For the upper-level game, we consider two types of decision-making sequences among VSPs: (i) simultaneous decision and (ii) sequential decision-making. Both settings are solved by the open-loop Nash solution.

10.3.2.1 Simultaneous Decision-Making Setting

Suppose that the M VSPs choose their synchronization strategies at the same time, and each player competes to maximize the objective functional $\mathfrak{J}_m$, i.e. the present value of utility derived over a finite or infinite time horizon, by designing a synchronization strategy η_m that is under the VSP's control. The choice of synchronization intensity by a player, say VSP m, influences (i) the evolution of the UAV population states $\boldsymbol{x}(t)$, (ii) the value states of the DT $z_m(t)$, and (iii) the objective functional of the other VSPs in the set $\mathcal{M}$, i.e. $\mathfrak{J}_{m'}$, $m' = 1, 2, \ldots, m-1, m+1, \ldots, M$. The influence to (i) and (ii) is captured via a set of differential equations (the system dynamics). The derivation of $\mathfrak{J}_m$ is given as follows.

With $x_m(t)N$ UAVs assisting VSP m in sensing the current states of its real world twins and $z_m(t)$ being the current values of its DTs, the current utility rate J_m at time t for VSP m can be described as follows:

$$J_m(\boldsymbol{x}, \boldsymbol{z}, \boldsymbol{\eta}, t) = \omega_m^1 J_m^1 + \omega_m^2 J_m^2 - \omega_m^3 J_m^3 - \omega_m^4 J_m^4, \tag{10.4}$$

$$J_m^1 = x_m Nb\alpha_m,\ J_m^2 = \beta_m z_m d_m,\ J_m^3 = (z_m - v_m)^2,\ J_m^4 = (x_m Nb - \eta_m d_m k_m)^2. \tag{10.5}$$

That is, J_m can be interpreted as the weighted sum of four utility components J_m^1, J_m^2, J_m^3, and J_m^4, in which J_m^1 and J_m^2 are the positive utilities, and J_m^3 and J_m^4 are the disutilities. $\omega_m^i \geq 0, i = \{1, 2, 3, 4\}$ are the weight parameters to form the objective function J_m. We explain the utility components as follows:

J_m^1: represents the gains generated by acquisition of new data of the size $x_m Nb$, where b represents the average amount of data that a UAV transmits to a VSP. Here, with synchronization data from the UAVs, VSP as a data supplier to the Metaverse platform, can benefit by selling the data to the Metaverse platform. The platform, as an intermediary to provide the data interoperability, can benefit the other VSPs to construct the DTs for their own use. Therefore, there is a portion of revenue inflow for VSP m that is linked to the data supply, or the data contribution from the UAVs. Let α_m denote the unit data price for the VSP m, then we have $J_m^1 = \alpha_m x_m Nb$ in (10.5).

J_m^2: represents the gains (e.g. virtual business profit) generated by the DTs with value of z_m. Here, we consider that the business is positively correlated with the DTs values. Therefore, the gains can be evaluated as $J_m^2 = \beta_m z_m d_m$, where β_m denotes the unit preference value that VSP m has toward a unit increase in the value of the DTs. In addition, β_m is considered to concave-upward w.r.t. θ_m, e.g., $\beta_m = e^{10\theta_m}$. This is to indicate DTs are valued more when the VSP is more sensitive to the nonupdated DTs, i.e., a higher valued decay rate.

J_m^3: represents a penalty term, reflecting disutility when DT values are far away from the preferred values, e.g. the twins data are not fresh enough (under-synchronized) or too fresh than needed (over-synchronized, leading to excessive synchronization cost). Let v_m denote the VSP's desired values of its DTs, and J_m^3 can be defined as $(z_m - v_m)^2$ accordingly [35].

J_m^4: represents the disutility caused by UAVs' insufficient data supply. Let d_m be the number of DTs of VSP m, and k_m be the data rate requested by each DT on average. Then, with the synchronization intensity η_m, overall, the total amount of data that VSP m requires from the UAVs is $d_m\eta_m k_m$. However, since there are $x_m N$ devices that choose VSP m, the total data contribution to VSP m is $x_m Nb$ as stated earlier. The gap $(d_m\eta_m k_m - x_m Nb)$ results in disutility to the VSP m. For example, when an insufficient number of UAVs select VSP m, UAVs can complete the synchronization task at lower sampling rates [19], resulting in lower quality synchronization data and affecting the utility of the VSP m. We adopt the square term to counter the data over-contribution.

The objective functional $\mathfrak{J}_m(\boldsymbol{\eta})$ to be maximized for VSP m is defined by the discounted cumulative payoff over the time horizon $\mathcal{T}$, expressed as follows:

$$\begin{aligned} \mathfrak{J}_m(\boldsymbol{\eta}) &= \int_0^T e^{-\rho t} J_m(\boldsymbol{x}(t), \boldsymbol{z}(t), \boldsymbol{\eta}(t), t)\mathrm{d}t \\ &= \int_0^T e^{-\rho t}\{\omega_m^1 x_m(t)Nb\alpha_m + \omega_m^2 z_m(t)\beta_m \\ &\quad - \omega_m^3(z_m(t) - v_m)^2 - \omega_m^4[x_m(t)Nb - \eta_m(t)d_m k_m]^2\}\mathrm{d}t, \end{aligned} \tag{10.6}$$

where $\rho \geq 0$ denotes the constant time preference rate (or discount rate) for VSPs. $J_m(\boldsymbol{x}(t), \boldsymbol{z}(t), \boldsymbol{\eta}(t), t)$ is the instantaneous utility derived by choosing the synchronization intensity value $\boldsymbol{\eta}(t)$ at time t when the current states of the game is $\boldsymbol{x}(t)$ and $\boldsymbol{z}(t)$, as explained earlier in (10.4).

Therefore, the optimal synchronization intensity control problem for VSP m can be formulated as

$$\max_{\eta_m} \quad \mathfrak{J}_m(\boldsymbol{\eta}) \tag{10.7}$$

$$\text{subject to} \quad \dot{x}_m(t) = \delta x_m(t)(u_m(t) - \bar{u}(t)), \quad \forall m \in \mathcal{M} \tag{10.8}$$

$$\dot{z}_m(t) = \eta_m(t) - \theta_m z_m(t), \quad \forall m \in \mathcal{M} \tag{10.9}$$

$$\boldsymbol{x}(0) = \boldsymbol{x}_0, \quad \boldsymbol{z}(0) = \boldsymbol{z}_0 \tag{10.10}$$

$$\boldsymbol{x}(t) \in \Delta, \boldsymbol{z}_m(t) \geq 0, \quad \eta_m(t) \geq 0, \tag{10.11}$$

for $m = 1, 2, \ldots, M$, where the column vectors $\boldsymbol{x}(0)$ and $\boldsymbol{z}(0)$ are the initial states for the population states of UAVs and DT value states.

10.3.2.2 Open-Loop Nash Solutions

A Nash solution or Nash equilibrium is an M-tuple of synchronization strategies $\boldsymbol{\eta} = [\eta_1, \eta_2, \ldots, \eta_M]$ such that, given the opponents' equilibrium synchronization strategies, no VSP has an incentive to change its own strategy. Denote the synchronization strategies of VSPs other than m as $\boldsymbol{\eta}_{-m} := [\eta_1, \eta_2, \ldots, \eta_{m-1}, \eta_{m+1}, \ldots, \eta_M]$.

In the differential game, the Nash solution is defined by a set of M admissible trajectories $\boldsymbol{\eta}^* := [\eta_1^*, \eta_2^*, \dots, \eta_M^*]$, which have the property that

$$\mathfrak{J}_m(\boldsymbol{\eta}^*) = \max_{\eta_m} \mathfrak{J}_m(\eta_1^*, \dots, \eta_{m-1}^*, \eta_m, \eta_{m+1}^*, \dots, \eta_M^*), \tag{10.12}$$

for $m = 1, 2, \dots, M$.

Next, we adopt the **open-loop solutions** for the above Nash differential game. The open-loop Nash solution to the optimal control problem refers to the case where the control paths are functions of time t only, satisfying (10.12). For simplicity, hereafter, we use a column vector $\boldsymbol{y}$ to represent the system states $\boldsymbol{x}$ and $\boldsymbol{z}$, i.e. $\boldsymbol{y} = [x_1, x_2, \dots, x_M, z_1, z_2, \dots, z_M]^T$. Then, the constraints defined in (10.8)–(10.11) can be replaced by the following conditions:

$$\dot{\boldsymbol{y}}(t) = [\dot{x}_1, \dots, \dot{x}_M, \dot{z}_1, \dots, \dot{z}_M]^T, \tag{10.13}$$

$$\boldsymbol{y}(0) = [\boldsymbol{x}(0)^T, \boldsymbol{z}(0)^T]^T, \tag{10.14}$$

$$\boldsymbol{y}(t) \in \mathcal{Y} := \Delta \times \mathbb{R}_+^M, \quad \eta_m(t) \in \mathbb{R}_+. \tag{10.15}$$

This means that the process that is to solve the open-loop Nash solution is to solve the optimal control problem, $\mathscr{P}1$, defined by

$$\begin{aligned} \max_{\eta_m} \quad & \mathfrak{J}_m(\eta_m, \boldsymbol{\eta}_{-m}^*) \\ \text{subject to} \quad & (10.13), (10.14), (10.15), \end{aligned} \tag{10.16}$$

for $m = 1, 2, \dots, M$. To solve $\mathscr{P}1$, we first define a (current-value) Hamiltonian function H as follows:

$$H_m(\boldsymbol{y}, \eta_m, \lambda_m, t) = J_m(\boldsymbol{y}, \eta_m, \boldsymbol{\eta}_{-m}^*, t) + \lambda_m \dot{\boldsymbol{y}}, \tag{10.17}$$

for $m = 1, 2, \dots, M$. The domain of H_m is the set $\{(\boldsymbol{y}, \eta_m, \lambda_m, t) | \boldsymbol{y} \in \mathcal{Y}, \eta_m \in \mathbb{R}_+, \lambda_m \in \mathbb{R}^{2M}, t \in \mathcal{T}\}$. Here, the row vector $\lambda_m = [\lambda_{m1}, \lambda_{m2}, \dots, \lambda_{m2M}]$ is called the (current-value) adjoint variables or costate variables. The maximized Hamiltonian function $H^* : \mathcal{Y} \times \mathbb{R}^{2M} \times \mathcal{T} \to \mathbb{R}$ is

$$H_m^*(\boldsymbol{y}, \lambda_m, t) = \max \{H_m(\boldsymbol{y}, \eta_m, \lambda_m, t) | \eta_m \geq 0\}. \tag{10.18}$$

A necessary and sufficient condition for the optimal control is given by the augmented maximum principle, stated as in Theorem 10.1. See [7] for a proof.

Theorem 10.1 *Consider an optimal control problem $\mathscr{P}1$ and define the Hamiltonian function H_m and the maximized Hamiltonian function H_m^* as above. The state space Θ is a convex set and the scrap value function S is continuously differentiable and concave (note that $S \equiv 0$ in $\mathscr{P}1$). If there exists an absolutely* *continuous function* $\lambda_m : [0, T] \to \mathbb{R}^{2M}$ *for all* $m \in \mathcal{M}$, *such that the maximum condition*

$$H_m(\boldsymbol{y}, \eta_m^*, t) = H_m^*(\boldsymbol{y}, \lambda_m, t), \tag{10.19}$$

the adjoint (costate) equation

$$\dot{\lambda}_m = \rho \lambda_m - \frac{\partial H_m^*(\boldsymbol{y}, \lambda_m, t)}{\partial \boldsymbol{y}}, \tag{10.20}$$

and the transversality condition

$$\lambda_m(T) = S'(\boldsymbol{y}(T)) = 0 \tag{10.21}$$

are satisfied, and such that the function H_m^* *is concave and continuously differentiable w.r.t.* x *for all* $t \in \mathcal{T}$, *then* $\eta_m(\cdot)$ *is an optimal control path. If further the set of feasible controls does not depend on* $\boldsymbol{y}$ *(which is true for* $\mathscr{P}1$ *as* $\eta_m \in \mathbb{R}_+$*), (10.20) can be replaced by*

$$\dot{\lambda}_m = \rho \lambda_m - \frac{\partial H_m(\boldsymbol{y}, \eta_m^*, t)}{\partial \boldsymbol{y}}. \tag{10.22}$$

Furthermore, to solve (10.19), we note that the function H_m in $\mathscr{P}1$ is strictly concave w.r.t. η_m. Therefore, we can instead solve for η_m^* by the first-order optimality conditions, given as follows:

$$\left.\frac{\partial H_m(\boldsymbol{y}, \eta_m, t)}{\partial \eta_m}\right|_{\eta_m = \eta_m^*} = 0. \tag{10.23}$$

After η_m^* is obtained using (10.23), a boundary value problem of a system of ordinary differential equations can be defined by $\dot{\boldsymbol{y}}$ in (10.13), $\dot{\lambda}$ in (10.22), together with their boundary values defined in (10.14) and (10.21). The states for this new dynamic systems are $\boldsymbol{y}$ and λ, which can be numerically solved using *bvp4c* in Matlab, or *scipy.integrate.solve_bvp* in python.

10.3.2.3 Hierarchical Decision-Making Setting

We now consider a more complicated realistic case, in which some VSPs are allowed by the Metaverse to choose their synchronization strategies earlier than the other VSPs. We refer to such VSPs as *leaders*. The VSPs that observe leaders' strategies and then make their decisions are called *followers*. To model this sequential (hierarchical) strategic interactions among the VSPs, or a hierarchical game, we adopt the Stackelberg differential game.

Let $\mathcal{L}$ denote the set of leaders and $\mathcal{F}$ the set of followers, such that $\mathcal{L} \cap \mathcal{F} = \emptyset$ and $\mathcal{L} \cup \mathcal{F} = \mathcal{M}$. Let $\boldsymbol{\eta}^L = [\eta_i^L]_{i \in \mathcal{L}}$ denote the synchronization strategies of leaders and $\boldsymbol{\eta}^F = [\eta_m^F]_{m \in \mathcal{F}}$ the synchronization strategies of followers, with $\boldsymbol{\eta}^{L*}$ and $\boldsymbol{\eta}^{F*}$

denoting the corresponding optimal strategies. At time 0, the leaders announce the synchronization strategy path $\boldsymbol{\eta}^L(t)$. The followers, taking these synchronization strategy paths as given, choose their synchronization strategies $\boldsymbol{\eta}^F(t)$ so as to maximize their objective functional.

Followers' Problem $\mathscr{P}_F$

Given the leaders' optimal synchronization strategy paths $\boldsymbol{\eta}^L$, the followers' problem $\mathscr{P}_F$, is the same as $\mathscr{P}_1$ defined in Section 10.3.2. In particular, for a follower $m \in \mathcal{F}$, the optimal control problem is

$$\begin{aligned} \max_{\eta_m^F} \quad & \mathfrak{J}_m(\eta_m^F, \boldsymbol{\eta}^L, \eta_{\mathcal{F}\backslash\{m\}}^{F*}) \\ \text{subject to} \quad & (10.13), (10.14), (10.15), \end{aligned} \tag{10.24}$$

where $\mathcal{F}\backslash m := \{i \in \mathcal{F}, i \neq m\}$ is the set of followers other than VSP m.

For the follower $m \in \mathcal{F}$, its Hamiltonian function, denoted by H_m^L, is the same as (10.17), that is $H_m^L = H_m$. Then, the optimal synchronization strategy η_m^{F*} for the follower m is equivalent to η_m^*, which is the solution to (10.23), and adjoint equations $\dot{\lambda}_m$ satisfy (10.22). Due to the strict concavity of H_m^F, η_m^{F*} can be uniquely determined by (10.23), as a function of $\boldsymbol{y}$, $\boldsymbol{\eta}^L$, $[\eta_j^{F*}]_{j\in\mathcal{F}\backslash m}$ and t, for all $m \in \mathcal{F}$. That is, we can write

$$\eta_m^{F*} = \tilde{\mathfrak{g}}_m(\boldsymbol{y}, \lambda_m, \boldsymbol{\eta}^L, \boldsymbol{\eta}_{\mathcal{F}\backslash\{m\}}^{F*}, t), \quad \forall m \in \mathcal{F}, \tag{10.25}$$

which can be further simplified by substituting $[\eta_j^{F*}]_{j\in\mathcal{F}\backslash m}$ based on (10.25) into $\tilde{\mathfrak{g}}_m(\cdot)$. Therefore, we can express η_m^{F*} as a function of $\boldsymbol{y}$, λ_m and $\boldsymbol{\eta}^L$ as follows:

$$\eta_m^{F*} = \mathfrak{g}_m(\boldsymbol{y}, \lambda_m, \boldsymbol{\eta}^L, t), \quad \forall m \in \mathcal{M}, \tag{10.26}$$

or the vector function $\boldsymbol{\eta}^{F*} = \mathfrak{g}(\boldsymbol{y}, \Lambda, \boldsymbol{\eta}^L, t)$, where $\Lambda := [\lambda_m]_{m\in\mathcal{F}}$ represents all the adjoint (costate) variables in $\mathscr{P}_F$.

Substituting (10.26) into (10.22), we obtain

$$\dot{\lambda}_m = \rho\lambda_m - \frac{\partial H_m(\boldsymbol{y}, \mathfrak{g}_m(\boldsymbol{y}, \lambda_m, \boldsymbol{\eta}^L, t), t)}{\partial \boldsymbol{y}}, \quad m \in \mathcal{F}. \tag{10.27}$$

Equations (10.13), (10.14), (10.15), (10.21), (10.27), and (10.26) characterize the follower's best response to the leaders control path $\boldsymbol{\eta}^{L*}$.

Leaders' Problem $\mathscr{P}_L$

As for the leaders' problem $\mathscr{P}_L$, for any leader $i \in \mathcal{L}$, it knows the followers' best responses. Therefore, in contrast to the simultaneous differential game, the system dynamics additionally include the adjoint equations of the followers' problem. Again, similar to the follower's game, among leaders, we obtain the Nash equilibrium. As such, given the best responses of all the followers, and the other opponent leaders play their strategies $\boldsymbol{\eta}^L_{\mathcal{L}\setminus i}$, the optimal control problem for the leader $i \in \mathcal{L}$ is formulated as follows:

$$\begin{aligned} \max_{\eta_i^L} \quad & \mathfrak{J}_i(\eta_i^L, \boldsymbol{\eta}^{F*}, \boldsymbol{\eta}^L_{\mathcal{L}\setminus\{i\}}) \\ \text{subject to} \quad & (10.13), (10.14), (10.15), (10.21), (10.27), (10.26), \end{aligned} \tag{10.28}$$

where $\boldsymbol{\eta}^{F*} = \mathsf{g}(\boldsymbol{y}, \Lambda, \boldsymbol{\eta}^L, t)$ is stated before.

Then, we define the Hamiltonian function for leader i as

$$H_i^L(\boldsymbol{y}, \Lambda, \boldsymbol{\eta}^L, \boldsymbol{\psi}_i, \boldsymbol{\phi}_i, t) = J_i(\eta_i^L, \mathsf{g}(\boldsymbol{y}, \Lambda, \boldsymbol{\eta}^L, t), t), \boldsymbol{\eta}^L_{\mathcal{L}\setminus i}) + \boldsymbol{\psi}_i \dot{\boldsymbol{y}} + \sum_{m \in \mathcal{F}} \boldsymbol{\phi}_{mi} \dot{\boldsymbol{\lambda}}_m^T, \tag{10.29}$$

for $i \in \mathcal{L}$, where the row vector $\boldsymbol{\psi}_i = [\psi_{ij}]_{j=1}^{2M}$ is the adjoint variables for the states $\boldsymbol{y}$, row vector $\boldsymbol{\phi}_{mi} = [\phi_{mij}]_{j=1}^{2M}$ is the adjoint variables for the adjoint variables λ_m, and $\boldsymbol{\phi}_i = [\boldsymbol{\phi}_{mi}]_{m \in \mathcal{F}}$. Note that the last term in (10.29), $\boldsymbol{\phi}_{mi} \dot{\boldsymbol{\lambda}}_m^T$, is an inner product as $\dot{\lambda}_m$ is a row vector, and $(\cdot)^T$ is the transpose operation.

We then have the optimality conditions (again applying Theorem 10.1)

$$\frac{\partial H_i^L(\boldsymbol{y}, \Lambda, \boldsymbol{\eta}^L, \boldsymbol{\psi}_i, \boldsymbol{\phi}_i, t)}{\partial \eta_i^L} = 0, \tag{10.30}$$

$$\dot{\boldsymbol{\psi}}_i = \rho \boldsymbol{\psi}_i - \frac{\partial H_i^L(\boldsymbol{y}, \Lambda, \boldsymbol{\eta}^L, \boldsymbol{\psi}_i, \boldsymbol{\phi}_i, t)}{\partial \boldsymbol{y}}, \tag{10.31}$$

$$\dot{\boldsymbol{\phi}}_{mi} = \rho \boldsymbol{\phi}_{mi} - \frac{\partial H_i^L(\boldsymbol{y}, \Lambda, \boldsymbol{\eta}^L, \boldsymbol{\psi}_i, \boldsymbol{\phi}_i, t)}{\partial \boldsymbol{\phi}_{mi}}, \tag{10.32}$$

for $i \in \mathcal{L}$ and $m \in \mathcal{F}$.

Similarly, the concavity of the leader's Hamiltonian function H_i^L in (10.29) ensures that η_i^{L*} can be uniquely expressed as a function of $\boldsymbol{y}, \Lambda, \boldsymbol{\eta}_{\mathcal{L}\setminus\{i\}}, \boldsymbol{\psi}_i, \boldsymbol{\phi}_i, t$ for all $i \in \mathcal{L}$. As such, after simplification, we have $\eta_i^{L*} = \mathsf{h}_i(\boldsymbol{y}, \Lambda, \boldsymbol{\psi}_i, \boldsymbol{\phi}_i, t)$, or the vector function $\boldsymbol{\eta}^{L*} = \mathsf{h}(\boldsymbol{y}, \Lambda, \Psi, \Phi, t)$, where $\Psi = [\boldsymbol{\psi}]_{i \in \mathcal{L}}$ and $\Phi = [\boldsymbol{\phi}_i]_{i \in \mathcal{L}}$. By backward induction, namely, substituting $\boldsymbol{\eta}^{L*} = \mathsf{h}(\boldsymbol{y}, \Lambda, \Psi, \Phi, t)$ into (10.26), (10.32), and (10.31), we can obtain the dynamics of the system states in (10.13), (10.27), (10.32), and (10.31) with only the states and time variable, i.e. $\boldsymbol{y}, \Lambda, \Psi, \Phi$ and t. Together with the boundary conditions for the system states, a two-point boundary value problem is defined, which can be solved numerically as stated earlier in Section 10.3.2.

10.3.3 Simulation Results

We conducted simulations to compare three cases, (i) simultaneous moves VSPs, i.e. a simultaneous differential game, (ii) hierarchical moves VSPs, and (iii) a static Stackelberg game with an evolutionary game, as a benchmark. In the static

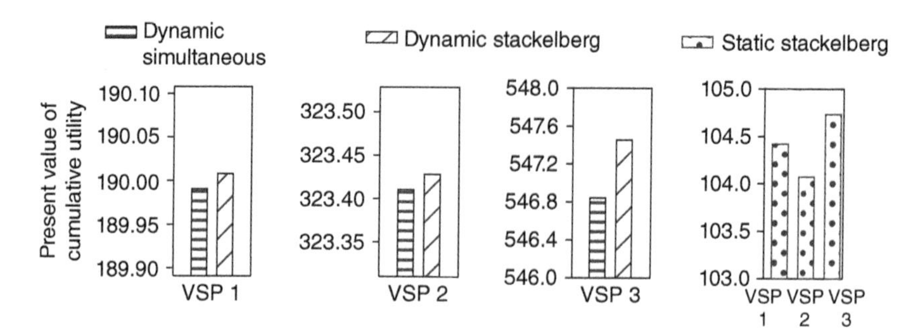

Figure 10.5 Comparison of the cumulative payoffs discounted at the present time for Stackelberg differential game, simultaneous differential game, and static Stackelberg game (baseline).

Stackelberg game, there is no dynamic interaction between VSPs and UAVs: VSPs perform a one-step optimization at the very beginning, and then UAVs populations evolve with the static synchronization intensity. In other words, in the static Stackelberg game, $\eta_m(t) = C_m, \forall t$, where C_m is some constant that optimizes VSP m's strategy. The steps used to obtain a solution for a static Stackelberg game can be found in [26]. For illustration, we consider the case of one leader and two followers in the Stackelberg differential game and three simultaneous VSPs in the simultaneous differential game. As for the experiment setting, we consider VSPs 1–3 from the previous experiment and allow VSP 3 to be the leader. Additionally, we set $\delta = 0.02$ and $\rho = 1$.

As shown in Figure 10.5, both the Stackelberg differential game and simultaneous differential game provide higher discounted cumulative payoffs than the static Stackelberg game. The reason is that both the Stackelberg differential game and simultaneous differential game capture the dynamic interactions between VSPs and UAVs and thus optimize the synchronization intensity control over time. It can also be observed that the Stackelberg differential game yields slightly higher discounted cumulative payoffs than the simultaneous differential game, especially for the leader. The reason is that the leader's decision can affect the followers' decisions, and thus the synchronization intensity control obtained by VSP 3 (leader) under the Stackelberg differential game can help improve the overall discounted utilities.

10.4 Conclusions and Future Research Directions

In this chapter, we discussed (i) why IoT is needed and (ii) how to use IoT for Metaverse services, followed by (iii) a proposed hierarchical game-theoretic approach. The first aspect covered two types of Metaverse services, Metaverse digital twins, and IoT in Metaverse DT construction, while the second aspect covered IoT-enabled MCS for constructing Metaverse DTs and its challenges and potential concerns.

As Metaverse develops and its internal ecosystem evolves alongside its integration with all types of existing industries, it is expected that more application scenarios will emerge where DTs will play an important role. With its mobile sensing capabilities, connectivity to everything, and growing computing capabilities, IoT can be a key tool in gathering data from DTs' real-world counterparts. This raises many interesting research questions on the next step of realizing the Metaverse DT so as to synchronize the two worlds:

- **Privacy concerns of the data**: IoT data may be highly sensitive, e.g., human health data collected by smart watches. How to develop DT when IoT devices are constrained in how they can gather or transmit data needs to be carefully addressed for the next phase of Metaverse DT development.
- **Fusion of the data**: It is possible to have a DT system that contains sub-DTs, as such heterogeneous data can be collected by different types of IoTs, e.g. images, text, and so on. In this case, how to combine data to develop a Metaverse DT is another important question to be addressed. In addition, data from other DTs can often be helpful in developing new DTs by providing contextual information. How to leverage such data is an important question.
- **AI on temporal data**: IoT data have some temporal features that may exhibit a certain degree of similarity to time-series data, such as speech data and natural language. As such, AI algorithms based on sequential learning, e.g. recurrent neural networks, transformers, and gated recurrent units, could be helpful in extracting additional information from the time-series patterns. The additional information may facilitate the prediction of DTs state in the near future thereby enabling less frequent data transfers between the two worlds. This can substantially relieve the burden on the wireless network.
- **Semantic-aware data transmission**: The frequent update of VWs is the prerequisite for maintaining Metaverse DTs reliably, especially for time-sensitive information applications, which incur undue transmission overhead given the constrained resources (spectrum and energy) for IoT devices. Consequently, it is most important to reduce the overhead substantially via semantic-aware data transmission that can be realized in distributed learning [34]. The latter is also beneficial to protecting the privacy of sensitive IoT data. Moreover, the lifetime of IoT devices can prolong to meet the sustainability of low-power IoT networks.

Overall, the development of Metaverse DT is crucial in bringing about convergence between the virtual and real worlds. Leveraging IoT for sensing, computing, and synchronization for Metaverse DTs provides flexibility for VSPs. With the increased adoption of AI and ML, smarter information transfer across the two worlds can be achieved, thereby reducing the burden on the communication infrastructure.

Bibliography

1 Cisco Annual Internet Report - Cisco Annual Internet Report (2018–2023) White Paper. URL https://www.cisco.com/c/en/us/solutions/collateral/executive-perspectives/annual-internet-report/white-paper-c11-741490.html. (accessed 2022-05-13).

2 Cyber-Physical Systems (CPS) US National Science Foundation. URL https://www.nsf.gov/pubs/2010/nsf10515/nsf10515.htm. (accessed 2022-03-03).

3 Virtual theme park rides you can experience from home - Tips - The Jakarta Post. URL https://www.thejakartapost.com/travel/2020/03/31/virtual-theme-park-rides-you-can-experience-from-home.html. (accessed 2022-03-16).

4 Matthew Ball. *The Metaverse: And How It Will Revolutionize Everything*. Liveright, New York, NY, July 2022. ISBN 978-1-324-09203-2.

5 Andrea Capponi, Claudio Fiandrino, Burak Kantarci, Luca Foschini, Dzmitry Kliazovich, and Pascal Bouvry. A survey on mobile crowdsensing systems: Challenges, solutions, and opportunities. *IEEE Communications Surveys & Tutorials*, 21(3):2419–2465, 2019.

6 Venkat Surya Dasari, Burak Kantarci, Maryam Pouryazdan, Luca Foschini, and Michele Girolami. Game theory in mobile crowdsensing: A comprehensive survey. *Sensors*, 20(7):2055, January 2020. doi: 10.3390/s20072055.

7 Engelbert J. Dockner, Steffen Jorgensen, Ngo Van Long, and Gerhard Sorger. *Differential Games in Economics and Management Science*. Cambridge University Press, Cambridge, 2000. ISBN 978-0-521-63732-9. doi: 10.1017/CBO9780511805127.

8 Alessio Fascista. Toward integrated large-scale environmental monitoring using WSN/UAV/crowdsensing: A review of applications, signal processing, and future perspectives. *Sensors*, 22(5):1824, 2022.

9 Aidan Fuller, Zhong Fan, Charles Day, and Chris Barlow. Digital twin: Enabling technologies, challenges and open research. *IEEE Access*, 8:108952–108971, 2020. ISSN 2169-3536. doi: 10.1109/ACCESS.2020.2998358.

10 Thomas Gabor, Lenz Belzner, Marie Kiermeier, Michael Till Beck, and Alexander Neitz. A simulation-based architecture for smart cyber-physical systems. In *Proceedings of the IEEE International Conference on Automation and Computing*, pages 374–379, 2016. doi: 10.1109/ICAC.2016.29.

11 Edward H. Glaessgen and D. S. Stargel. The digital twin paradigm for future NASA and U.S. Air force vehicles. In *53rd AIAA/ASME/ASCE/AHS/ASC Structures, Structural Dynamics and Materials Conference - Special Session on the Digital Twin*, 2012.

12 Michael Grieves. Digital twin: Manufacturing excellence through virtual factory replication. *White paper*, 1:1–7, 2014.

13 Yue Han, Dusit Niyato, Cyril Leung, Chunyan Miao, and Dong In Kim. A dynamic resource allocation framework for synchronizing metaverse with IoT service and data. In *ICC 2022 - IEEE International Conference on Communications*, pages 1196–1201, 2022. doi: 10.1109/ICC45855.2022.9838422.

14 Clement Kam, Sastry Kompella, Gam D. Nguyen, Jeffrey E. Wieselthier, and Anthony Ephremides. Information freshness and popularity in mobile caching. In *Proceedings of the ISIT*, pages 136–140, 2017. doi: 10.1109/ISIT.2017.8006505.

15 Latif U. Khan, Walid Saad, Dusit Niyato, Zhu Han, and Choong Seon Hong. Digital-twin-enabled 6G: Vision, architectural trends, and future directions. *arXiv:2102.12169 [cs]*, November 2021.

16 Putu Sandra Devindriati Kusuma. Children virtual concert in the COVID-19 pandemic. *Journal of Music Science Technology and Industry*, 3(2):247–260, October 2020. ISSN 2622-8211. doi: 10.31091/jomsti.v3i2.1160.

17 Wei Yang Bryan Lim, Jianqiang Huang, Zehui Xiong, Jiawen Kang, Dusit Niyato, Xian-Sheng Hua, Cyril Leung, and Chunyan Miao. Towards federated learning in UAV-enabled internet of vehicles: A multi-dimensional contract-matching approach. *IEEE Transactions on Intelligent Transportation Systems*, 22(8):5140–5154, August 2021. ISSN 1558-0016. doi: 10.1109/TITS.2021.3056341.

18 Chi Harold Liu, Chengzhe Piao, and Jian Tang. Energy-efficient UAV crowdsensing with multiple charging stations by deep learning. In *IEEE INFOCOM 2020-IEEE Conference on Computer Communications*, pages 199–208, 2020.

19 Xiao Liu, Houbing Song, and Anfeng Liu. Intelligent UAVs trajectory optimization from space-time for data collection in social networks. *IEEE Transactions on Network Science and Engineering*, 8(2):853–864, April 2021. ISSN 2327-4697, 2334-329X. doi: 10.1109/TNSE.2020.3017556.

20 Junyu Lu, Xiao Xiao, Zixuan Xu, Chenqi Wang, Meixuan Zhang, and Yang Zhou. The potential of virtual tourism in the recovery of tourism industry during the COVID-19 pandemic. *Current Issues in Tourism*, 25(3):441–457, 2022.

21 David Mirk and Helmut Hlavacs. Using drones for virtual tourism. In *International Conference on Intelligent Technologies for Interactive Entertainment*, pages 144–147, 2014.

22 Abozar Nasirahmadi and Oliver Hensel. Toward the next generation of digitalization in agriculture based on digital twin paradigm. *Sensors*, 22(2):498, January 2022. ISSN 1424-8220. doi: 10.3390/s22020498.

23 Swarna Priya Ramu, Parimala Boopalan, Quoc-Viet Pham, Praveen Kumar Reddy Maddikunta, Thien Huynh-The, Mamoun Alazab, Thanh Thi Nguyen, and Thippa Reddy Gadekallu. Federated learning enabled digital twins for smart cities: Concepts, recent advances, and future directions. *Sustainable Cities and Society*, 79:103663, April 2022. ISSN 22106707. doi: 10.1016/j.scs.2021.103663.

24 Reza Shakeri, Mohammed Ali Al-Garadi, Ahmed Badawy, Amr Mohamed, Tamer Khattab, Abdulla Khalid Al-Ali, Khaled A. Harras, and Mohsen Guizani. Design challenges of multi-UAV systems in cyber-physical applications: A comprehensive survey and future directions. *IEEE Communications Surveys & Tutorials*, 21(4):3340–3385, 2019. ISSN 1553-877X, 2373-745X. doi: 10.1109/COMST.2019.2924143.

25 Joanne Shurvell. Take A Virtual African Safari In 2020 In Real-Time From Your Sofa. URL https://www.forbes.com/sites/joanneshurvell/2020/05/15/take-a-virtual-african-safari-in-2020-in-real-time-from-your-sofa/. (accessed 2022-03-16).

26 M. Simaan and J. B. Cruz. On the Stackelberg strategy in nonzero-sum games. *Journal of Optimization Theory and Applications*, 11(5):533–555, 1973. ISSN 1573-2878. doi: 10.1007/BF00935665.

27 Xi Tao and Abdelhakim Senhaji Hafid. Trajectory design in UAV-aided mobile crowdsensing: A deep reinforcement learning approach. In *ICC 2021-IEEE International Conference on Communications*, pages 1–6, 2021.

28 Fei Tao, He Zhang, Ang Liu, and A. Y. C. Nee. Digital twin in industry: State-of-the-art. *IEEE Transactions on Industrial Informatics*, 15(4):2405–2415, April 2019. ISSN 1941-0050. doi: 10.1109/TII.2018.2873186.

29 L. P. Voronkova. Virtual tourism: On the way to the digital economy. In *IOP Conference Series: Materials Science and Engineering*, volume 463, page 042096, 2018.

30 Bowen Wang, Yanjing Sun, Dianxiong Liu, Hien M. Nguyen, and Trung Q. Duong. Social-aware UAV-assisted mobile crowd sensing in stochastic and dynamic environments for disaster relief networks. *IEEE Transactions on Vehicular Technology*, 69(1):1070–1074, 2019.

31 Yuntao Wang, Zhou Su, Ning Zhang, and Abderrahim Benslimane. Learning in the air: Secure federated learning for UAV-assisted crowdsensing. *IEEE Transactions on Network Science and Engineering*, 8(2):1055–1069, 2020.

32 Jörgen W. Weibull. *Evolutionary Game Theory*. MIT press, 1997.

33 Stephan Weyer, Torben Meyer, Moritz Ohmer, Dominic Gorecky, and Detlef Zühlke. Future modeling and simulation of CPS-based factories: An example from the automotive industry. *IFAC-PapersOnLine*, 49(31):97–102, January 2016. ISSN 2405-8963. doi: 10.1016/j.ifacol.2016.12.168.

34 Huiqiang Xie and Zhijin Qin. A lite distributed semantic communication system for Internet of Things. *IEEE Journal on Selected Areas in Communications*, 39(1):142–153, 2021. ISSN 1558-0008. doi: 10.1109/JSAC.2020.3036968.

35 Kun Zhu, Ekram Hossain, and Dusit Niyato. Pricing, spectrum sharing, and service selection in two-tier small cell networks: A hierarchical dynamic game approach. *IEEE Transactions on Mobile Computing*, 13(8):1843–1856, August 2014. ISSN 1536-1233. doi: 10.1109/TMC.2013.96.

11

Quantum Technologies for the Metaverse: Opportunities and Challenges

Mahdi Chehimi and Walid Saad

Wireless@VT, Bradley Department of Electrical and Computer Engineering, Virginia Tech, Arlington, VA, USA

After reading this chapter you should be able to:

- Explain the basic operating principles of the different quantum technologies.
- Identify how quantum algorithms can enable low-latency in the Metaverse.
- Explain how quantum machine learning helps extracting the contextual meaning of big data in the Metaverse.
- Describe the role of quantum communications for the Metaverse and its security.

11.1 Introduction

To date, the rapid evolution of communication networks has been predominantly driven by a chase for more bandwidth. Today, the emergence of new Internet of Everything (IoE) services such as holographic teleportation, digital twins (DTs), and extended reality (XR) enables the creation of an immersive Metaverse. With all these technologies, intelligently utilizing the available computing resources in the communication systems has become a key necessity to deliver such fundamentally complex and smart applications [13, 56]. Furthermore, those applications require extreme enhancements in the capacity of today's cellular communication systems, which can be achieved by the adoption of higher frequency bands (specifically terahertz (THz)) [14]. In addition, enabling the various services of the immersive Metaverse requires the coexistence of various wireless functions such as communications, sensing, and intelligent control [15].

The various technologies of the sixth-generation communication systems (6G) and the different IoE applications will play a fundamental role in developing the

Metaverse Communication and Computing Networks: Applications, Technologies, and Approaches, First Edition.
Edited by Dinh Thai Hoang, Diep N. Nguyen, Cong T. Nguyen, Ekram Hossain, and Dusit Niyato.
© 2024 The Institute of Electrical and Electronics Engineers, Inc. Published 2024 by John Wiley & Sons, Inc.

envisioned fully immersive Metaverse, which will have a transformative impact on our daily life activities, the financial sector, education process, and healthcare system [49]. In particular, the Metaverse represents an immersive experience that integrates the different XR technologies, i.e. virtual reality (VR), mixed reality (MR), and augmented reality (AR), to develop a unique user experience that mixes both the physical and the digital realities [52]. In addition, the Metaverse includes developing an interactive digital mirror of physical systems using DTs, which will enable intelligent and precise control of the physical world through the lens of machine learning (ML) and artificial intelligence (AI) applied in the digital world [38].

Indeed, the Metaverse necessitates an overarching *AI-native communication infrastructure* to fulfill its diversified, stringent requirements (e.g. high-resolution sensing, real-time control, high rates and high reliability low-latency) [15, 57]. In fact, there exist multiple challenges that render the development of the fully immersive Metaverse and reaping its full potential, such as [30]: (i) the achievable data throughput, (ii) the computational and communication latency, (iii) the contextual analysis and understanding of the Metaverse's big data, and (iv) the increased security and privacy risks. These limitations must be carefully addressed to build the fully immersive Metaverse, which requires advancing today's electronics to achieve more powerful and faster computational capabilities for future technologies. However, recent indicators show that Moore's law started being violated and that current electronics reached the limits, beyond which, the quantum effects can no longer be ignored in their design [2]. This implies that a major leap in the computing and communication infrastructure for the future Metaverse will require *quantum technologies*.

The field of quantum computing has recently emerged as the most promising futuristic technology that will drastically transform the way computations are performed. In the last few years, governments, academia, and the industry have shown great interest in this field and dedicated significant funds to support the development of real quantum computers [53]. Particularly, technology giants such as IBM [47], Google [3], and Microsoft are leading the developments in this research area, in addition to many startups and big companies like D-Wave [40]. The recent advancements in the area of quantum computing and quantum information science (QIS) indicate a great potential for applying such technologies in the Metaverse. In particular, today's *Noisy Intermediate-Scale Quantum* (NISQ) devices were proven to achieve a superior performance compared to their classical counterparts in different areas, which promise to enable a powerful quantum advantage [24]. The application of quantum technologies in the Metaverse stems from the great benefits they can bring about to overcome the pressing challenges facing the Metaverse. Particularly, the advantages of quantum computers over the classical ones span a wide range of research problems, which include but are not limited to

- **Computational speedups**: Quantum computers can achieve up to exponential speedups in performing certain computations compared to their classical counterparts. Moreover, several quantum optimization algorithms can achieve a superior performance in terms of computational complexity and execution time compared to the counterpart classical algorithms [50].
- **Information extraction and contextual ML**: The great ability of quantum computers to handle exponentially large data dimensions is utilized to enhance the performance of ML models. Moreover, novel designs including various quantum circuits and gates that resemble the way classical ML models operate resulted in the niche field of *quantum machine learning* (QML). Particularly, various QML models and approaches were recently proven to achieve a superior performance compared to their classical counterparts in certain problems [7]. Moreover, when operating on high-dimensional data, QML models are capable of extracting and understanding hidden data patterns that classical ML models are unable to discover [19].
- **Unbreakable security**: The unique nature of quantum physics and the probabilistic nature of quantum measurements provide quantum cryptography and quantum communication protocols with unbreakable security [6]. For instance, if an eavesdropper tries to intercept a quantum communication signal, it is incapable of extracting the embedded information since measuring a quantum state destroys it [50].
- **Enhanced communication throughput**: The principles of *quantum superposition* and higher-order quantum states enable the embedding of large volumes of classical data in few quantum communication resources (quantum states). Henceforth, integrating quantum communications with the classical infrastructure has the potential to enhance the communication throughput by sending more data using less communication resources [22].

The main contribution of this chapter is to investigate the potential opportunities for applying quantum technologies to serve the immersive Metaverse and overcome its challenges. Toward this overarching goal, the chapter begins by giving a brief background review about quantum information, which is necessary to understand the operation of the different quantum technologies and how they can be deployed to enhance the Metaverse. Then, three main areas of quantum technologies are analyzed as potential candidates to serve the Metaverse, namely quantum computing and quantum optimization, QML, and quantum communications. For each technology, its basics are explained, the Metaverse challenges it addresses are identified, and a vision of the role it will play in the different applications of the Metaverse is given. Finally, the chapter ends with some concluding

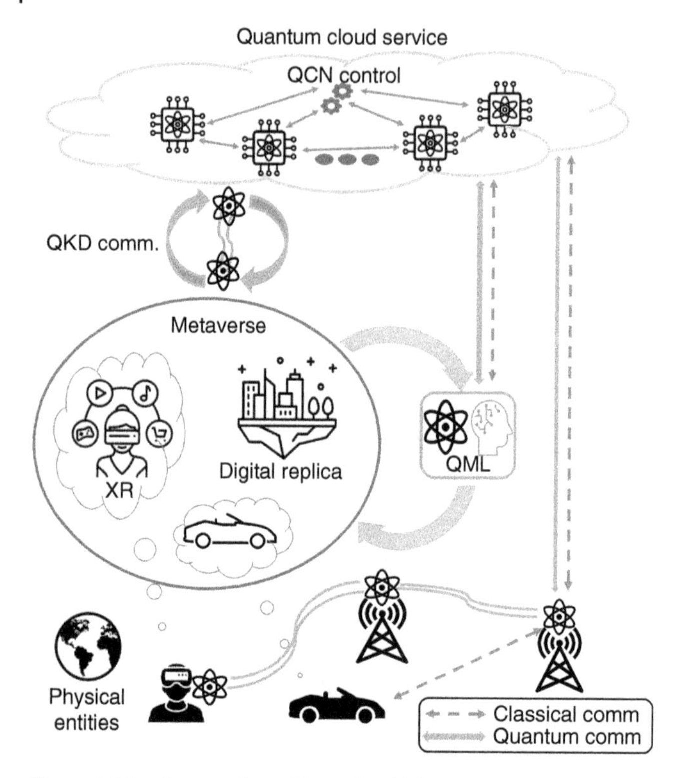

Figure 11.1 An overview of how the different quantum technologies can serve the Metaverse and its applications like DTs and the XR. The figure shows the quantum cloud service, which includes a QCN controller that controls multiple quantum-computing units that perform heavy computations for the Metaverse, in addition to the QML models that are applied on the big data of the Metaverse. Moreover, quantum communication links for quantum-secured data transmissions are presented in the networks, in addition to the QKD quantum cryptographic protocol.

remarks and future promising research directions that are necessary to reap the full potential of quantum technologies in the Metaverse. A general overview of the chapter and how quantum technologies can serve the Metaverse is given in Figure 11.1.

11.2 Preliminaries

Quantum technologies mainly include quantum computers, quantum sensors, and quantum communication protocols, which are used to build networks of quantum devices communicating with each others by sending quantum states along with classical data, i.e. quantum communication networks (QCNs).

Those networks are currently being deployed on a small-scale for specific applications such as quantum key distribution (QKD) cryptographic protocol. When QCNs mature and become capable of sending quantum data between various quantum devices separated by long distances all over the world, they will converge to what is known as the *quantum Internet (QI)* [18].

All those quantum technologies rely on unique principles of *QIS*, which are inherently different from the classical principles since they use concepts from quantum mechanics that have no classical counterparts. This requirement of some quantum-specific principles makes the process of designing and building QCNs uniquely challenging. In this section, we will introduce the required preliminaries in order to understand the way quantum communications is performed and the general underlying principles behind the operation of the different quantum technologies.

The basic element for all quantum computations and communication protocols is known as the *qubit*, which carries purely quantum data, or embeddings of classical data. Unlike classical bits which can only take a binary value, qubits can be in any superposition of both "0" and "1" bits. A general quantum state (qubit) is defined, in Dirac's *bra-kit* notation [28], as

$$|\psi\rangle = c_0 |0\rangle + c_1 |1\rangle, \tag{11.1}$$

where $|c_0|^2 + |c_1|^2 = 1$. This superposition nature of quantum states will potentially enable quantum communications to enhance the communication throughput, especially when considered with higher-order quantum states [19]. Although a qubit can be in any superposition of "0" and "1," when a qubit is measured, its quantum state collapses to either the "0" or "1" classical bits, and is no more in a superposition. However, the quantum measurement operation is nontrivial and has a probabilistic nature. In particular, the probability of obtaining the state $|0\rangle$ is $|c_0|^2$, and the probability of obtaining the state $|1\rangle$ is $|c_1|^2$.

One key resource in QCNs is *quantum entanglement*, or the spooky action at a distance [31]. When an entangled pair of qubits is generated, they become connected no matter how far they are separated. Measuring one qubit causes the state of the other qubit to collapse immediately to a specific quantum state. One of the most general and basic entangled pairs are known as Bell states which are generated using some dedicated hardware. There are four Bell pairs, and they are maximally entangled [25] pairs of qubits that represent a superposition of the $|0\rangle$ and $|1\rangle$ quantum states. They take one of the following forms:

$$|\phi_{\pm}\rangle = \frac{1}{\sqrt{2}}(|00\rangle \pm |11\rangle), \tag{11.2}$$

$$|\psi_{\pm}\rangle = \frac{1}{\sqrt{2}}(|01\rangle \pm |10\rangle). \tag{11.3}$$

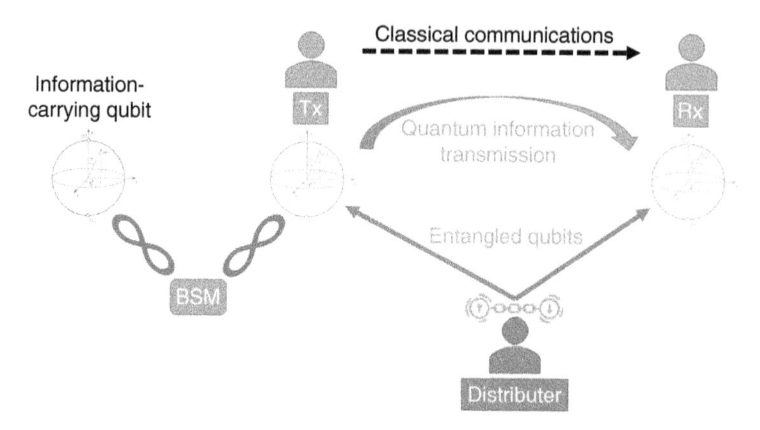

Figure 11.2 The quantum teleportation protocol.

Instead of directly sending a qubit to the desired receiver in a QCN, efficient QCN designs rely on sharing entangled pairs of qubits between different nodes. In such networks, information is transmitted using the *quantum teleportation* protocol [50] (explained in Figure 11.2) which allows the creation of a connection between distant parties in a quantum network that can be used at any time to send quantum information. This is extremely important because it does not require the availability of a good quantum channel between the two communicating parties at all times. Before applying the teleportation protocol, an entangled pair of qubits is generated, and the qubits are split between the transmitter and the receiver such that each side has one qubit. The receiver does not perform measurement, and both parties store their qubits in their corresponding quantum memories. When information is to be transmitted, the transmitter apply some unitary quantum gates (for more details on quantum gates and circuits, refer to [50]) and performs Bell state measurements (BSMs) on the information-carrying qubit and the stored entangled qubit. This measurement results in a classical outcome that corresponds to one of the four Bell states. At this point, the state of the receiver's entangled qubit collapses to the transmitted quantum state. The measurement outcome is communicated classically with the receiver, which applies some quantum unitaries and performs measurement to decode the information that was sent, which completes the quantum teleportation protocol.

In general, there are two main types of qubits in any quantum information system: (i) matter qubits, and (ii) flying qubits. The matter qubits are the ones responsible for storage and performing the quantum computations on each quantum device. Furthermore, the flying qubits are the ones that carry the quantum information between distant points and are generally sent between quantum devices over quantum channels. Quantum channels are realized by either fiber optics or free space optical (FSO) channels.

There are mainly two approaches for representing quantum states. The first one is the discrete-variable representation, like the qubits presented earlier, which can be implemented by mapping information into a two-level quantum system such as the spin of an electron or the polarization of a single photon, which is mainly used in quantum communications. In this case, the photon, or qubit, becomes a superposition of the vertical and horizontal polarizations [50], which requires single-photon sources to create the qubits, and single-photon detectors to detect them. Another approach for representing the quantum states is the continuous-variable representation, which relies on laser light and encoding the information in its quadrature variables [9]. In order to detect these continuous variables, homodyne/heterodyne laser detectors are used. Both approaches are still under development and none has demonstrated a superior performance. However, the majority of the recent developments in quantum computing happened with quantum computers relying on discrete-variable qubits. Regarding the physical realization of qubits, there are various technologies that are adopted in the industry, and, today, there are functioning NISQ devices based on each approach. Those technologies include, but are not limited to superconducting qubits, ion traps, cold atoms, silicon spins, and photonic qubits. However, today's NISQ devices are lossy and do not have error-correction capabilities, and we are still far away from building fault-tolerant quantum computers [35].

Due to a principle known as the *No-cloning theorem*, an arbitrary quantum state cannot be copied, nor broadcast to multiple destinations simultaneously [50]. This feature highly affects performing quantum computations and designing QCNs, especially ones that include quantum repeaters [10]. Another challenge that faces quantum technologies is the phenomena of *quantum decoherence*, which corrupts the evolution of quantum systems. Due to decoherence, quantum states deteriorate over time, and they get lost and become useless after a certain "coherence" time period. This effect happens because of the unavoidable interactions between a qubit and its surrounding environment [50]. Apparently, quantum technologies suffer from various unique losses that render their advancements more challenging than their classical counterparts. Henceforth, careful attention is needed while designing quantum systems in order to comply with their underlying quantum physics principles that govern their operation [18].

This was a brief background on QIS and the basics of quantum computing and communications, which are necessary to understand the potential role that quantum technologies can play for solving the challenges facing the Metaverse. Readers who are interested in more thorough analysis of the mentioned concepts are referred to some fundamental books in the field [39, 50].

In order to identify where the different quantum technologies fit in the Metaverse, we need to distinguish between two main aspects present in the Metaverse. Particularly, the Metaverse has both technical and social characteristics since it integrates multiple cutting-edge technologies like XR and DT, and it represents a new social form that gives it a sociality aspect [52]. Here, the different quantum technologies can play a vital role to mainly serve the technical aspect of the Metaverse, which will be discussed in the remaining of this chapter. In addition, the Metaverse's challenges that quantum technologies will help resolving will be investigated, and the applications they will enhance will be identified.

11.3 Quantum Computing for a Faster Metaverse

Quantum computers can perform complex computations in a fast manner that reduces the computational complexity and the computing time compared to their classical counterparts. This computational advantage brought by quantum computers can be leveraged in order to minimize the delays and latency in the different applications of the Metaverse.

11.3.1 Quantum Computing Speedups

In general, the main advantage of quantum computing is its inherent parallelism due to the unique underlying quantum physics principles such as quantum superposition. This parallelism allows performing heavy computations in an extremely rapid manner compared to the strongest classical supercomputer. For instance, Shor's factoring algorithm and Grover's search algorithm can achieve exponential and polynomial computational speedups, respectively [50]. Such computational speedups can be directly seen in enabling applications like unfolding proteins, which include extremely large amounts of potential conformations. For instance, around the order of 10^{47} conformations exist in a chain that includes 100 amino acids, which can, in theory, be folded in any combination of trillions of potential ways, that no working classical computer can handle. Here comes quantum computers, with their high-dimensional spaces, to propose an efficient approach to solve the protein folding problem, which was verified in [55].

Moreover, some quantum optimization algorithms were shown [48] to achieve a superior performance to their classical counterparts, particularly in combinatorial problems and portfolio optimization problems. A key example is the famous *quantum approximate optimization algorithm* (QAOA) [63], which is a hybrid classical-quantum algorithm that has nontrivial guarantees of its superior performance [34]. However, QAOA is not the only successful quantum algorithm. Various other algorithms were proven to be successful at reducing the computational complexity of difficult tasks, such as the *Harrow–Hassidim–Lloyd* (HHL) algorithm among others [8].

11.3.2 Quantum Computing for Low Latency

In general, the computational speedups brought by quantum computers and the different quantum algorithms could be leveraged to enable ultralow-latency applications and fast (near instantaneous) response in the Metaverse. The Metaverse itself relies on collecting gigantic data from IoE sensors, actuators, and different devices. As a digital reality, the Metaverse can be represented as a mesh of a huge number of nodes that incorporate the various sensors' data. Hence, each of these node represents an N-dimensional vector, where each dimension corresponds to a feature collected by the sensors (e.g. temperature and color). Since the Metaverse is a direct representation of the physical reality, it must be updated periodically in order to ensure an accurate synchronization between the physical and the digital realities [26]. Henceforth, each node in the Metaverse, i.e. an N-dimensional vector, must be periodically updated and analyzed, which requires significant speedups in performing the updates, the processing computations, and the communication. Here, quantum computers can be greatly utilized in order to speed-up such computations and processes for the Metaverse. This can be achieved by deploying quantum computing devices on the central nodes that are responsible for updating the Metaverse.

11.3.3 Quantum Computing for Synchronized DTs

Moreover, a central technology in the Metaverse is DTs, which are responsible for mapping the physical reality into a digital image through collecting data from the various IoE sensors. In order to be used in the Metaverse, the digital reality that is generated by DTs must be teleported using holographic teleportation into the Metaverse. This teleportation step requires huge computational resources and has significantly stringent low-latency requirements [32]. Herein, utilizing quantum computers and quantum algorithms to speed up the holographic teleportation performance of DTs is a critical application of quantum technologies in the Metaverse.

In fact, DTs mainly operate on the edge of communication networks; hence, edge computing is a pillar technology for the Metaverse. Particularly, if quantum-computing services can be available at the edge of communication networks, tremendous speedups for DTs will be achieved and low-latency synchronization will be guaranteed. Moreover, the variational quantum optimization algorithms might be deployed to optimally utilize the available resources on the edge, and to process the collected sensor data for the DTs [38].

11.3.4 Quantum Computing for Responsive XR

Another essential Metaverse technology that will crucially need the computational speedups of quantum computers is XR. In general, ultimate XR services, such as VR, MR, and AR [1], require high data rates and reliable low-latency simultaneously (end-to-end latency [14]). Here, there are two main sources of delay in the XR system. First, there is the communication delay needed to transfer the large volumes of XR data between the users and the processing unit. The other delay is the computing delay stemming from the processing and analysis of the data. Recently, it was shown that the end-to-end communication delay can be substantially reduced by operating at higher frequency bands, particularly the terahertz (THz) frequency band [14]. Nonetheless, the issue of computational delay remains unresolved since we are reaching a particular threshold that we cannot go below with the regular classical computing capabilities. This delay represents a hurdle to the operation of the XR services in the Metaverse.

In this regard, if the classical computing devices responsible for the XR services were replaced or assisted by quantum computing devices, we will be able to achieve the utmost timely performance of the XR. Moreover, combining quantum computing with high-frequency communications might ultimately enable achieving an almost negligible motion-to-photon latency of XR, which is an interesting area that needs further investigation in order to have a unique immersive experience in the Metaverse [13]. Furthermore, the quantum computers and quantum optimization algorithms can play a significant role in speeding up the back-and-forth interactions between actions/reactions taken by the XR users in real life and the corresponding actions/reactions taken by their extended digital version in the Metaverse. This task will require substantial computational speedups and low latency in order to have a responsive XR user experience, which can be achieved through quantum computing.

11.3.5 Challenges for Quantum Computing in the Metaverse

A major challenge facing the utilization of quantum computing speedups and the different quantum algorithms for the Metaverse resides in the fact that accessing quantum computing resources is nontrivial and requires building networks of quantum computers that operate over the cloud, where scheduling algorithms are utilized to control the quantum resources. Furthermore, quantum states deteriorate over time when they interact with their surrounding environments. Hence, when quantum information is transferred between quantum computers, this operation must be reliably performed in a timely manner, which is critical to guarantee accurate computations and reliable results for the operation of the Metaverse.

Furthermore, an open area of research that requires careful investigation is the one concerned with developing a hybrid computational framework for the Metaverse. Developing such a framework is a challenging task since it has to incorporate both quantum and classical computers. Moreover, this framework needs to include novel control strategies that monitor the interplay between quantum and classical computers to perform fast computations with efficient utilization of the available resources. An open question in this area is regarding the decisions of which computing tasks in the Metaverse need to be prioritized, which ones must be run on quantum (or classical) computers, and which ones will need to run hybrid computing algorithms. This distribution task of the different Metaverse computations is necessary to reap the full potential of both quantum and classical computers, particularly given the fact that the available computing resources for each Metaverse are limited and must be managed carefully.

11.4 Quantum Machine Learning for Contextual Metaverse

A contextual Metaverse is an immersive Metaverse where the gigantic amounts of data generated by the Metaverse applications are processed and analyzed, and their contextual meaning is extracted. Such intelligently extracted contextual information, that resembles the hidden meanings, or semantics, in the data, can then be communicated between the different Metaverse devices to minimize the required communication resources. Hence, the Metaverse applications would rely on performing more computations and less communications, which lead to a smarter Metaverse that can make use of the limited available communication resources.

The applications of the Metaverse, such as XR and DTs, generate big data of high dimensionality, which requires a significantly large amount of communication resources to transmit and update. Here, extracting the contextual meanings of such big data is of great importance to enhance the performance of the Metaverse. As explained previously, quantum technologies have an unprecedented ability to handle and process big data of exponentially high dimensions. Hence, quantum technologies are a perfect candidate to operate on the big data of the Metaverse, and they can play a vital role in developing the contextual Metaverse. This can be particularly achieved through the field of *QML*.

11.4.1 The Power of QML Models

QML is an emerging field that is gaining significant interest from the research community [7]. This field is grounded on the ability of quantum computers to handle high-dimensional data and linear algebra in a unique, inherently paralleled approach [11]. QML includes various hybrid algorithms that integrate both classical and quantum computations. Those algorithms cover different areas of ML from the quantum version of the support vector machine algorithm [54], to unsupervised quantum learning algorithms such as the

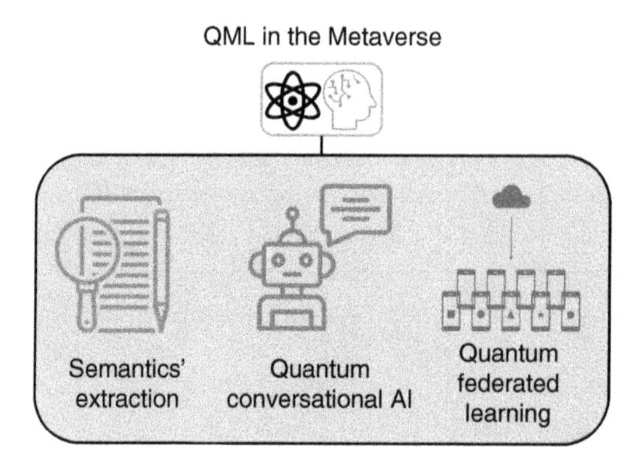

Figure 11.3 QML applications in the Metaverse.

quantum dimensionality reduction algorithm [46], and quantum reinforcement learning frameworks [29]. Those algorithms, in general, can achieve a superior performance compared to their classical counterparts in terms of their computational complexity, running time, or achieved accuracy.

Another important type of QML models are variational quantum circuits (VQCs) which are the quantum version of the classical neural networks [5]. Such models are sometimes referred to as *quantum neural networks (QNNs)*, which are hybrid models that include parametrized quantum circuits with learnable parameters that are trained and optimized using classical optimizers. Such QNN models include various structures, like the *quantum convolutional neural networks* (QCNNs) for classification tasks [23], and the *quantum long short-term memory* (QLSTM) [20] models, which operate on temporal time-series data. Such models can achieve a superior performance compared to the classical counterpart models under certain constraints.

There are various Metaverse applications where QML models can play a transformative role. Some key areas are discussed next and summarized in Figure 11.3.

11.4.2 QML for Semantics' Extraction

Of particular interest regarding the contextual Metaverse are unsupervised quantum clustering [62] and the supervised quantum kernel classification algorithms [54]. Such algorithms, when operating on high-dimensional big data, were shown to discover hidden data patterns and to extract statistical information that could not be discovered by their classical counterparts [7]. By deploying such QML models, large volumes of high-dimensional data generated in the Metaverse (through XR and DTs) can be understood, and their contextual meanings, or semantics, can be extracted and represented [60] such that the Metaverse is equipped with intelligent capabilities and semantic information that reduces the required quantum communication resources for its different applications.

For instance, a DT needs to be teleported to the Metaverse using holographic teleportation. Then, the teleported DT must be periodically updated and synchronized with the physical reality. All these steps incorporate the transmission and reception of large volumes of data with the Metaverse. Such DT data will be processed and analyzed with AI and ML tools in order to efficiently operate the physical reality. Here, QML models can be utilized on the edge of the communication network to extract hidden patterns from the DT data and to understand its contextual meanings so as to minimize the required communication resources for holographic teleportation.

11.4.3 QML for Quantum Conversational AI

In general, conversational AI systems are models that can handle text and speech data, and, hence, are the systems responsible for listening to and speaking with the individual users [45]. In the Metaverse, the immersive experience requires conversational AI systems to enable the generation of sophisticated avatars, or virtual characters and assistants, that operate in the Metaverse and are capable of interacting, listening to, and speaking with the different Metaverse users [41]. Hence, those models need to perform instantaneous analysis of the speech of each user, in addition to text-to-speech synthesis that can give assistance to the users in the Metaverse. In general, conversational AI models include the classical transformer networks [42], and recurrent neural networks [58], among many other models in natural language processing (NLP) [21].

In this regard, QML models can play a great role in advancing such conversational AI models in the Metaverse to have a superior performance. In particular, promising results were recently achieved in the area of quantum natural language processing (QNLP), which opens the door for great advances in this field [27]. Furthermore, *quantum recurrent neural networks* (QRNNs) were shown to achieve a superior performance compared to their classical counterparts [4], in addition to the previously mentioned QLSTM models [20]. By deploying such QML models in the Metaverse, the quantum-powered conversational AI models and avatars will have more powerful abilities to operate with less-computational complexity, to have a faster response, and to overcome the shortage of data in some cases. In particular, it was recently shown that QML models can guarantee a good generalization performance with few training data samples [12], which promises the Metaverse with great potential powers.

11.4.4 Quantum Federated Learning in the Metaverse

Last but not least, the different users and applications in the Metaverse may have private data that they do not wish to share over the communication network.

However, in order to develop a fully immersive experience in the Metaverse, the different users and applications need to cooperate and interact in the Metaverse. Moreover, due to shortage of data or the lack of generalizability in some applications and ML/QML models, the different users might need to collaboratively train their ML or QML models to benefit from each other's data. In such cases, the different Metaverse users will need to deploy collaborative distributed learning frameworks such as *federated learning* (FL) [43].

In the case of collaboratively training QML models (e.g. for contextual meaning extraction or quantum conversational AI applications), a *quantum federated learning* (QFL) framework is needed. In this regard, a comprehensive QFL framework was proposed in [17] and was proven to achieve a good performance while handling quantum data, which can embed large volumes of classical data or be collected from quantum IoE sensors. Such distributed QML frameworks will come in handy for the different Metaverse applications. For instance, QFL can enable secure and powerful collaborative quantum learning between multiple DTs that are simultaneously operated to control different physical objects in a synchronized manner.

11.4.5 Challenges Facing QML in the Metaverse

The deployment of QML models in the Metaverse is a nontrivial task that faces multiple challenges. For instance, embedding classical data samples into quantum states is a fundamental step that is mandatory to apply various QML models. This embedding process is currently a noisy process that might take a considerable time duration before being completed, which represents an issue for some Metaverse applications like the DT systems of critical nature that require low-noise, real-time feedback between the digital and the physical worlds. In such applications, the resulting noise would not be negligible, and the users have tight time constraints to perform the embeddings, which impose a serious challenge. Here, it becomes necessary to develop novel approaches for reliably embedding classical data into quantum states. Henceforth, advancing the research in QML will lead to enhancements in its applicability for the future Metaverse. Moreover, today's QML models have limited numbers of qubits due to the present noise in the current NISQ devices. Hence, such QML models suffer from scalability issues that make them suffer when operating on extremely large datasets that require a large number of qubits. This issue becomes severe in the case of NLP applications, which require training datasets of extremely large volumes. For instance, today's QLSTM models would require a long training time if it where given a very large dataset, since it can only reliably operate using a small number of qubits using today's quantum technologies. Thus, the great potential of the QML models in the Metaverse requires great efforts in developing scalable QML models that can operate using large numbers of qubits.

11.5 Quantum Communications for Secure Metaverse

Quantum communications represents novel means for secure communications that rely on unique quantum physics principles that have no classical counterpart to guarantee unbreakable security. Since today's wireless technologies are incapable of achieving such unbreakable security [61], quantum technologies are a must in the Metaverse as they can break the strongest existing classical encryption techniques, and because quantum cryptography cannot, in general, be broken [51].

11.5.1 Quantum-Enhanced Security

In general, quantum communications can secure data transmissions in two main ways (explained in Figure 11.4). First, quantum cryptographic protocols are used to share quantum-based secure keys that are generated and transferred using quantum states that are transmitted over quantum channels. Photonic quantum communication hardware are the mainly used technology to implement such protocols by developing quantum communication systems [6]. A key example in this regard is the QKD cryptographic protocol, which generates secret keys that secure the classical communications using quantum communications [59]. The quantum-based keys are immune against eavesdropping, and they can potentially achieve unbreakable security.

The other approach is based on the principle of embedding classical information into quantum states that are then transmitted (whether using direct quantum communications or using the quantum teleportation protocol) between distant quantum users, which can be integrated with security protocols like QKD. This approach makes use of the probabilistic nature of quantum states and the fact that quantum measurements collapse the measured quantum states and result in the

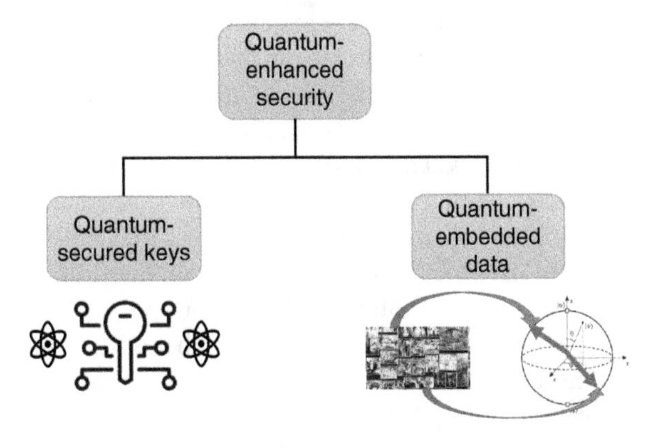

Figure 11.4 Quantum communication approaches to enhance security.

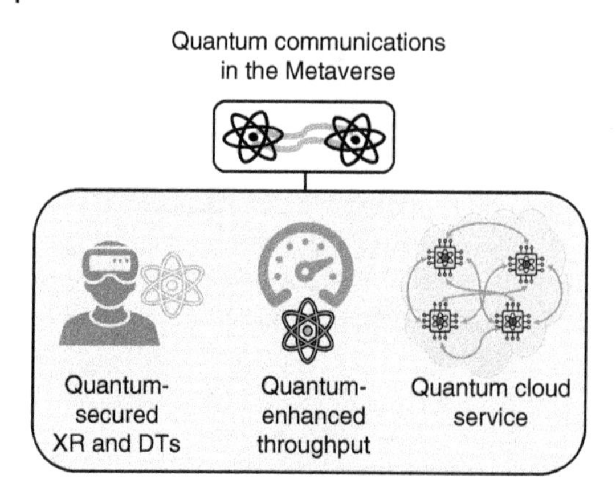

Figure 11.5 Quantum communication applications in the Metaverse.

loss of the embedded information, particularly if integrated with quantum-secured keys. Hence, any interception of the quantum states or attempts to measure them during their transmission process can be detected due to the unique underlying quantum physics principles. Henceforth, quantum communications represents a unique opportunity to send private data with unbreakable security so as to increase user security and privacy, which is fundamental for the different applications of the Metaverse [33]. Next, we delve into the details of some applications of quantum communications in the Metaverse, which are summarized in Figure 11.5

11.5.2 Quantum-Secured XR and DTs

In the Metaverse, there are various aspects that require increased security due to the confidentiality of the shared data. First of all, part of the IoE sensor data that are periodically collected to update the Metaverse itself might incorporate certain confidential features that must be secured. Hence, those specific features might be encrypted with quantum keys, or the IoE sensor data can be communicated using quantum states. Next, the Metaverse faces major concerns regarding the privacy of its users and their data. These concerns will significantly rise as today's XR technologies mature and become capable of accurately and instantaneously capturing every motion action of the users. Such data might be used to extract various information about users' habits, nature, and psychological behaviors, which put the Metaverse users at a high risk of losing their privacy. Here, if this data are embedded using quantum circuits into quantum states that are transferred over quantum networks and secured with quantum keys, then the private data will be transferred securely. In particular, if the transmitted information-carrying quantum states and quantum keys were intercepted and measured, e.g. by an eavesdropper, they would collapse, and the embedded raw data will be lost. Henceforth, quantum communications can play a fundamental role in guaranteeing the privacy of the Metaverse users by securing their private data transmissions.

Another Metaverse application that may heavily require quantum-based security is DTs applied in certain areas. For instance, the private DT data that are collected from a financial or a medical physical device, then transferred to the Metaverse, need extreme levels of security in order to ensure a private operation of such critical services in the Metaverse [44]. Moreover, some big companies use DTs to advance the models and designs of the devices they manufacture in their factories. Such DTs operating on cutting-edge technologies incorporate the detailed descriptions of confidential designs which must be carefully secured. Here, quantum communications represent a unique tool to achieve the required security for the different DT applications. By incorporating the large volumes of DT data into quantum states and securing them with quantum keys, one can ensure a secure transmission of the data against eavesdropping attacks, in addition to enhanced throughput, which is explained next.

11.5.3 Quantum-Enhanced Throughput

Beyond security, the second most distinguishing feature of quantum communications is its capability of achieving an enhanced data throughput and a higher information capacity [37]. This is based on the concepts of quantum superposition and high-dimensional quantum states which enable an exponential increase in the achievable data throughput that can be handled by *QCNs* compared to classical networks, which are central for the Metaverse data-hungry applications such as the XR [22]. In particular, the high-dimensional quantum states are composed of a superposition of l basis states. In general, such quantum states are represented as the vector $|\psi\rangle$ that lives in an l-dimensional Hilbert space $\mathcal{H}_l$, which is represented as

$$|\psi\rangle = c_0\,|0\rangle + c_1\,|1\rangle + c_2\,|2\rangle + \ldots + c_{l-1}\,|l-1\rangle\,, \tag{11.4}$$

where $|c_0|^2 + |c_1|^2 + |c_2|^2 + |c_3|^2 + \ldots + |c_{l-1}|^2 = 1$. This way, high-dimensional quantum states can hold up to the information held in 2^l classical information bits, and hence having a great potential of increasing the communication throughput which comes in handy when considering the Metaverse applications. As mentioned previously, the large volumes of high-dimensional data representing the Metaverse, and the data resulting from DTs are perfect candidates to benefit from the advantages brought by quantum communications, since they would result in savings in the number of communication resources required for the Metaverse.

11.5.4 Quantum Cloud Service

The last, but not least, benefit that quantum communications can bring to the Metaverse is based on the concept of the *quantum cloud service*. In order to reap all the benefits of the different quantum technologies, the different applications of the Metaverse need to have access to the powerful quantum computers. However, such quantum computers are extremely expensive, and some of them require very special conditions to operate with low errors, e.g. almost zero Kelvin temperature [50]. Accordingly, today's technology is extremely far away from developing affordable quantum computers for personal use. In the foreseeable future, it is generally agreed that quantum computing services will mainly be available in specialized centers of big companies, and people will be given online access to those quantum computing resources as a paid service in the cloud. This principle was recently established with companies such as IBM, Amazon, and D-Wave offering cloud access to their quantum computing resources in order to enable researchers to perform computations and advance research efforts in the field [36].

In general, in order for the quantum cloud to operate, the different quantum devices must be interconnected using QCNs so that they can share quantum states, establish secure connections, perform distributed quantum computations, and achieve collaborative quantum learning. Henceforth, quantum communications and QCNs will play a fundamental role in delivering the different quantum benefits to the Metaverse applications in the short term [18]. For instance, the availability of the quantum cloud would enable the DTs to benefit from the fast quantum computing algorithms to speedup the analysis of the data in the digital world, and to synchronize both the digital and the physical realities [38].

11.5.5 Challenges Facing Quantum Communications in the Metaverse

However, deploying QCNs and the quantum cloud to enhance the Metaverse requires careful development of control and management protocols of the quantum cloud and QCNs [16]. In particular, in order to handle the large volumes of data generated by the XR and DT services, the available quantum resources in QCNs must be carefully scheduled and allocated so as to satisfy the different users' requirements in a timely manner. This is crucially important because of the limited lifetime of quantum states and the principle of decoherence where quantum information gets lost over time [18].

Finally, the split between computing and communication resources in the Metaverse is not obvious in itself. Here, with the introduction of quantum computing and quantum communication services, this split in the Metaverse becomes more challenging. This area requires careful investigation from the research community in order to compromise between the available quantum communication resources, which have limited lifetime, and the quantum computing resources which are limited in capacity. This challenge adds to the challenge of developing a comprehensive hybrid quantum-classical computing framework that was described in Section 11.3. It is well known that the goal of the future communication networks is to become more computing-intensive, i.e. perform more computations on the edge, rather than being communication-intensive [13]. The main motivation behind this research direction is to save the limited communication resources. When quantum technologies are considered in the Metaverse, complex smart control algorithms are required to answer similar questions on which Metaverse applications must be secured using quantum communications, which need to be sent with few quantum communication resources, which must be analyzed on quantum computers before being transferred, and which need to run on hybrid quantum-classical models. This area is of fundamental importance for the future of the Metaverse and the future of quantum technologies and their integration with today's state-of-the-art technologies.

11.6 Conclusions and Future Research Directions

The future fully immersive, low-latency deployment of the Metaverse faces multiple challenges. This is due to reaching the physical limit of today's electronics below which quantum effects must be taken into consideration. This chapter describes the basics of quantum technologies and analyzes the role each quantum technology can play to address the challenges facing the Metaverse. In particular, quantum computers and quantum algorithms can minimize the computational delays, which enable ultralow-latency Metaverse applications. Moreover, QML is capable of extracting hidden patterns in the big data of the Metaverse, which would lead to a contextual Metaverse where the semantics of the data are extracted and communicated. Finally, quantum communications and QCNs guarantee unbreakable security for the Metaverse, in addition to ensuring reliable access to the quantum computing devices over the cloud. Henceforth, quantum technologies must be incorporated in the future Metaverse in order to overcome its different pressing challenges.

Toward this overarching goal, some promising future research directions include but are not limited to

- Incorporating quantum optimization techniques to enhance the performance of Metaverse-enabled VR devices in terms of responsiveness and delay minimization. In particular, certain quantum computing architectures, such as quantum annealers, can be a perfect fit for solving various combinatorial and portfolio optimization problems encountered during the interaction between VR users and virtual elements in the Metaverse. Moreover, other relevant applications can fall under the scope of holographic presence protocols in the Metaverse.

- Deploying novel quantum cryptographic techniques to enhance security levels of financial transactions in the Metaverse. This includes securing cryptocurrencies, nonfungible tokens (NFTs), etc., with quantum key distribution and quantum conference key agreement. These methods propel a top-notch security enhancement as they provide unbreakable security due to the unique principles of quantum mechanics.
- Working toward practical deployments of distributed quantum computing and learning on the edge. Incorporating such technologies over the network edge will radically enhance the computing performance necessary for the distributed Metaverse architectures. Among the challenges that face the Metaverse decentralization is the need to preserve the synchronization of the distributed architecture, with that of the real-world elements. This necessitates enhancing computational capabilities over the edge by acquiring distributed quantum edge computing and quantum devices.
- Advancing the quantum error correction (QEC) techniques to ensure reliable quantum communication performance. This is fundamental to ensure responsiveness and synchronized performance between the physical devices and their corresponding DTs in the Metaverse, which rely on QCNs to ensure security. QEC allows QCNs to overcome the various losses and noise encountered during the transmission of quantum states over optical fibers or FSO quantum channels, which is a main bottleneck that renders the practical deployments of QCNs.
- Incorporating quantum sensors for accurate modeling and seamless synchronization of the Metaverse. Quantum sensors are significantly much more sensitive to electromagnetic fields, whereby they have promising applications in clock synchronization and localization. By incorporating such sensors through adopting QCNs into the Metaverse, we are capable of achieving accurate modeling of the Metaverse. This can play a pivotal role in enhancing the physical-digital interactions between the real world and the Metaverse.

Bibliography

1 Ian F. Akyildiz and Hongzhi Guo. Wireless extended reality (XR): Challenges and new research directions. *ITU Journal on Future and Evolving Technologies*, 3:1–15, 2022.

2 Malik Amir, Christian Bauckhage, Alina Chircu, Christian Czarnecki, Nico Piatkowski, and Eldar Sultanow. What can we expect from Quantum (Digital) Twins? 2022.

3 Frank Arute, Kunal Arya, Ryan Babbush, Dave Bacon, Joseph C. Bardin, Rami Barends, Rupak Biswas, Sergio Boixo, Fernando G. S. L. Brandao,

David A. Buell et al. Quantum supremacy using a programmable superconducting processor. *Nature*, 574(7779):505–510, October 2019. ISSN 1476-4687. doi: 10.1038/s41586-019-1666-5.

4 Johannes Bausch. Recurrent quantum neural networks. Advances in Neural Information Processing Systems 33 (NIPS 2020), 1368–1379, 2020.

5 Marcello Benedetti, Erika Lloyd, Stefan Sack, and Mattia Fiorentini. Parameterized quantum circuits as machine learning models. *Quantum Science and Technology*, 4(4):043001, November 2019. ISSN 2058-9565. doi: 10.1088/2058-9565/ab4eb5.

6 Charles H. Bennett and Gilles Brassard. Quantum cryptography: Public key distribution and coin tossing. *arXiv preprint arXiv:2003.06557*, 2020.

7 Jacob Biamonte, Peter Wittek, Nicola Pancotti, Patrick Rebentrost, Nathan Wiebe, and Seth Lloyd. Quantum machine learning. *Nature*, 549(7671):195–202, 2017.

8 Panagiotis Botsinis, Dimitrios Alanis, Zunaira Babar, Hung Viet Nguyen, Daryus Chandra, Soon Xin Ng, and Lajos Hanzo. Quantum search algorithms for wireless communications. *IEEE Communications Surveys & Tutorials*, 21(2):1209–1242, 2018.

9 Samuel L. Braunstein and Peter Van Loock. Quantum information with continuous variables. *Reviews of Modern Physics*, 77(2):513, 2005.

10 Briegel H.-J., Dür W., Juan I. Cirac, and Peter Zoller. Quantum repeaters for communication. *arXiv preprint quant-ph/9803056*, 1998.

11 Michael Broughton, Guillaume Verdon, Trevor McCourt, Antonio J. Martinez, Jae Hyeon Yoo, Sergei V. Isakov, Philip Massey, Murphy Yuezhen Niu, Ramin Halavati, Evan Peters, Martin Leib, Andrea Skolik, Michael Streif, David Von Dollen, Jarrod R. McClean, Sergio Boixo, Dave Bacon, Alan K. Ho, Hartmut Neven, and Masoud Mohseni. TensorFlow quantum: A software framework for quantum machine learning, 2020.

12 Matthias C. Caro, Hsin-Yuan Huang, Marco Cerezo, Kunal Sharma, Andrew Sornborger, Lukasz Cincio, and Patrick J. Coles. Generalization in quantum machine learning from few training data. *Nature Communications*, 13(1):1–11, 2022.

13 Christina Chaccour and Walid Saad. Edge intelligence in 6G systems. In *6G Mobile Wireless Networks*, pages 233–249. Springer, 2021.

14 Christina Chaccour, Mehdi Naderi Soorki, Walid Saad, Mehdi Bennis, and Petar Popovski. Can terahertz provide high-rate reliable low latency communications for wireless VR? *IEEE Internet of Things Journal*, 9(12):9712–9729, 2022.

15 Christina Chaccour, Mehdi Naderi Soorki, Walid Saad, Mehdi Bennis, Petar Popovski, and Mérouane Debbah. Seven defining features of terahertz

(THz) wireless systems: A fellowship of communication and sensing. *IEEE Communications Surveys & Tutorials*, 24(2):967–993, 2022.

16 Mahdi Chehimi and Walid Saad. Entanglement rate optimization in heterogeneous quantum communication networks. In *2021 17th International Symposium on Wireless Communication Systems (ISWCS)*, pages 1–6. IEEE, 2021.

17 Mahdi Chehimi and Walid Saad. Quantum federated learning with quantum data. In *ICASSP 2022-2022 IEEE International Conference on Acoustics, Speech and Signal Processing (ICASSP)*, pages 8617–8621. IEEE, 2022.

18 Mahdi Chehimi and Walid Saad. Physics-informed quantum communication networks: A vision towards the quantum internet. *IEEE Network*, 36(5):32–38, September 2022.

19 Mahdi Chehimi, Christina Chaccour, and Walid Saad. Quantum semantic communications for resource-efficient quantum networking. *arXiv preprint arXiv:2205.02422*, 2022.

20 Samuel Yen-Chi Chen, Shinjae Yoo, and Yao-Lung L. Fang. Quantum long short-term memory. In *ICASSP 2022-2022 IEEE International Conference on Acoustics, Speech and Signal Processing (ICASSP)*, pages 8622–8626. IEEE, 2022.

21 Chowdhary K. R.. Natural language processing. In *Fundamentals of Artificial Intelligence*, pages 603–649. Springer, 2020.

22 Mostafa Zaman Chowdhury, Md Shahjalal, Shakil Ahmed, and Yeong Min Jang. 6G wireless communication systems: Applications, requirements, technologies, challenges, and research directions. *IEEE Open Journal of the Communications Society*, 1:957–975, July 2020.

23 Iris Cong, Soonwon Choi, and Mikhail D. Lukin. Quantum convolutional neural networks. *Nature Physics*, 15(12):1273–1278, 2019.

24 Samudra Dasgupta and Travis S. Humble. Characterizing the stability of NISQ devices. In *2020 IEEE International Conference on Quantum Computing and Engineering (QCE)*, pages 419–429. IEEE, 2020.

25 Julio I. de Vicente, Cornelia Spee, and Barbara Kraus. Maximally entangled set of multipartite quantum states. *Physical Review Letters*, 111(11):110502, 2013.

26 Sahraoui Dhelim, Tahar Kechadi, Liming Chen, Nyothiri Aung, Huansheng Ning, and Luigi Atzori. Edge-enabled metaverse: The convergence of metaverse and mobile edge computing. *arXiv preprint arXiv:2205.02764*, 2022.

27 Riccardo Di Sipio, Jia-Hong Huang, Samuel Yen-Chi Chen, Stefano Mangini, and Marcel Worring. The dawn of quantum natural language processing. In *ICASSP 2022–2022 IEEE International Conference on Acoustics, Speech and Signal Processing (ICASSP)*, pages 8612–8616. IEEE, 2022.

28 Rainer Dick. Notions from linear algebra and bra-ket notation. In *Advanced Quantum Mechanics*, pages 63–85. Springer, 2020.

29 Vedran Dunjko, Jacob M. Taylor, and Hans J. Briegel. Advances in quantum reinforcement learning. In *2017 IEEE International Conference on Systems, Man, and Cybernetics (SMC)*, pages 282–287. IEEE, 2017.

30 Yogesh K. Dwivedi, Laurie Hughes, Abdullah M. Baabdullah, Samuel Ribeiro-Navarrete, Mihalis Giannakis, Mutaz M. Al-Debei, Denis Dennehy, Bhimaraya Metri, Dimitrios Buhalis, Christy M. K. Cheung, et al. Metaverse beyond the hype: Multidisciplinary perspectives on emerging challenges, opportunities, and agenda for research, practice and policy. *International Journal of Information Management*, 66:102542, 2022.

31 Einstein A., Podolsky B., and Rosen N.. Can quantum-mechanical description of physical reality be considered complete? *Physical Review*, 47:777–780, May 1935. doi: 10.1103/PhysRev.47.777.

32 Abdulmotaleb El Saddik. Digital twins: The convergence of multimedia technologies. *IEEE Multimedia*, 25(2):87–92, 2018.

33 Adel S. Elmaghraby and Michael M. Losavio. Cyber security challenges in smart cities: Safety, security and privacy. *Journal of Advanced Research*, 5(4):491–497, 2014.

34 Edward Farhi, Jeffrey Goldstone, and Sam Gutmann. A quantum approximate optimization algorithm. *arXiv preprint arXiv:1411.4028*, 2014.

35 Amir Fruchtman and Iris Choi. Technical roadmap for fault-tolerant quantum computing. *NQIT Technical Roadmap*, 2016.

36 Constantin Gonzalez. Cloud based QC with Amazon Braket. *Digitale Welt*, 5(2):14–17, 2021.

37 Lajos Hanzo, Harald Haas, Sándor Imre, Dominic O'Brien, Markus Rupp, and Laszlo Gyongyosi. Wireless myths, realities, and futures: From 3G/4G to optical and quantum wireless. *Proceedings of the IEEE*, 100(Special Centennial Issue):1853–1888, 2012. doi: 10.1109/JPROC.2012.2189788.

38 Omar Hashash, Christina Chaccour, and Walid Saad. Edge continual learning for dynamic digital twins over wireless networks. In *2022 IEEE 23rd International Workshop on Signal Processing Advances in Wireless Communication (SPAWC)*, pages 1–5, 2022. doi: 10.1109/SPAWC51304.2022.9833928.

39 Dieter Heiss. *Fundamentals of Quantum Information: Quantum Computation, Communication, Decoherence and All That*, volume 587. Springer, 2008.

40 Feng Hu, Ban-Nan Wang, Ning Wang, and Chao Wang. Quantum machine learning with d-wave quantum computer. *Quantum Engineering*, 1(2):e12, 2019.

41 Thien Huynh-The, Quoc-Viet Pham, Xuan-Qui Pham, Thanh Thi Nguyen, Zhu Han, and Dong-Seong Kim. Artificial intelligence for the metaverse: A survey. *arXiv preprint arXiv:2202.10336*, 2022.

42 Max Jaderberg, Karen Simonyan, Andrew Zisserman, and Koray Kavukcuoglu. Spatial transformer networks. *Advances in Neural Information Processing Systems 28 (NIPS 2015)*, 2015.

43 Peter Kairouz, H. Brendan McMahan, Brendan Avent, Aurélien Bellet, Mehdi Bennis, Arjun Nitin Bhagoji, Kallista Bonawitz, Zachary Charles, Graham Cormode, Rachel Cummings, et al. Advances and open problems in federated learning. *Foundations and Trends® in Machine Learning*, 14(1–2):1–210, 2021.

44 Latif U. Khan, Walid Saad, Dusit Niyato, Zhu Han, and Choong Seon Hong. Digital-twin-enabled 6G: Vision, architectural trends, and future directions. *arXiv preprint arXiv:2102.12169*, 2021.

45 Pradnya Kulkarni, Ameya Mahabaleshwarkar, Mrunalini Kulkarni, Nachiket Sirsikar, and Kunal Gadgil. Conversational AI: An overview of methodologies, applications & future scope. In *2019 5th International Conference On Computing, Communication, Control And Automation (ICCUBEA)*, pages 1–7. IEEE, 2019.

46 Seth Lloyd, Masoud Mohseni, and Patrick Rebentrost. Quantum principal component analysis. *Nature Physics*, 10(9):631–633, July 2014. ISSN 1745-2481. doi: 10.1038/nphys3029.

47 Evan R. MacQuarrie, Christoph Simon, Stephanie Simmons, and Elicia Maine. The emerging commercial landscape of quantum computing. *Nature Reviews Physics*, 2(11):596–598, 2020.

48 Nikolaj Moll, Panagiotis Barkoutsos, Lev S. Bishop, Jerry M. Chow, Andrew Cross, Daniel J. Egger, Stefan Filipp, Andreas Fuhrer, Jay M. Gambetta, Marc Ganzhorn, et al. Quantum optimization using variational algorithms on near-term quantum devices. *Quantum Science and Technology*, 3(3):030503, 2018.

49 Stylianos Mystakidis. Metaverse. *Encyclopedia*, 2(1):486–497, 2022.

50 Michael A. Nielsen and Isaac L. Chuang. *Quantum Computation and Quantum Information: 10th Anniversary Edition*. Cambridge University Press, December 2010. doi: 10.1017/CBO9780511976667.

51 M. Niemiec and A. R. Pach. Management of security in quantum cryptography. *IEEE Communications Magazine*, 51(8):36–41, 2013. doi: 10.1109/MCOM.2013.6576336.

52 Huansheng Ning, Hang Wang, Yujia Lin, Wenxi Wang, Sahraoui Dhelim, Fadi Farha, Jianguo Ding, and Mahmoud Daneshmand. A survey on metaverse: The state-of-the-art, technologies, applications, and challenges. *arXiv preprint arXiv:2111.09673*, 2021.

53 Michael G. Raymer and Christopher Monroe. The us national quantum initiative. *Quantum Science and Technology*, 4(2):020504, 2019.

54 Patrick Rebentrost, Masoud Mohseni, and Seth Lloyd. Quantum support vector machine for big data classification. *Physical Review Letters*, 113:130503, September 2014. doi: 10.1103/PhysRevLett.113.130503.

55 Anton Robert, Panagiotis Kl Barkoutsos, Stefan Woerner, and Ivano Tavernelli. Resource-efficient quantum algorithm for protein folding. *npj Quantum Information*, 7(1):1–5, February 2021.

56 Walid Saad. 6G wireless systems: Challenges and opportunities. In *5G and Beyond: Fundamentals and Standards*, pages 201–229. Springer, 2021.

57 Walid Saad, Mehdi Bennis, and Mingzhe Chen. A vision of 6G wireless systems: Applications, trends, technologies, and open research problems. *IEEE Network*, 34(3):134–142, July 2019.

58 Hojjat Salehinejad, Sharan Sankar, Joseph Barfett, Errol Colak, and Shahrokh Valaee. Recent advances in recurrent neural networks. *arXiv preprint arXiv:1801.01078*, 2017.

59 Valerio Scarani, Helle Bechmann-Pasquinucci, Nicolas J. Cerf, Miloslav Dušek, Norbert Lütkenhaus, and Momtchil Peev. The security of practical quantum key distribution. *Reviews of modern physics*, 81(3):1301, 2009.

60 Philip Tetlow, Dinesh Garg, Leigh Chase, Mark Mattingley-Scott, Nicholas Bronn, Kugendran Naidoo, and Emil Reinert. Towards a semantic information theory (introducing quantum corollas). *arXiv preprint arXiv:2201.05478*, January 2022.

61 Minghao Wang, Tianqing Zhu, Tao Zhang, Jun Zhang, Shui Yu, and Wanlei Zhou. Security and privacy in 6G networks: New areas and new challenges. *Digital Communications and Networks*, 6(3):281–291, 2020.

62 Marvin Weinstein and David Horn. Dynamic quantum clustering: A method for visual exploration of structures in data. *Physical Review E*, 80(6):066117, 2009.

63 Leo Zhou, Sheng-Tao Wang, Soonwon Choi, Hannes Pichler, and Mikhail D. Lukin. Quantum approximate optimization algorithm: Performance, mechanism, and implementation on near-term devices. *Physical Review X*, 10(2):021067, 2020.

12

The Metaverse with Life and Everything: An Overview of Privacy, Ethics, and Governance*

Lik-Hang Lee[1], *Carlos Bermejo*[2], *and Pan Hui*[2]

[1] *Department of Industrial and Systems Engineering (ISE), The Hong Kong Polytechnic University, Hong Kong SAR, China*
[2] *Department of Computer Science and Engineering, School of Engineering, The Hong Kong University of Science and Technology, Hong Kong SAR, China*

After reading this chapter, you should be able to:

- Understand the current trends and challenges that building such a virtual environment will face.
- Focus on three major: privacy, governance, and ethical design, to guide the sustainable yet acceptable development of the Metaverse.
- Illustrate a preliminary modular-based framework for an ethical design of the Metaverse.

12.1 Introduction

Recently, there has been a rise in the significance placed on the Metaverse inside the online space [16]. The Metaverse is expected to be the next major evolution phase of the Internet. The Metaverse will impact human society, industrial production, services, and life. The promise of the Metaverse is shown by online platforms, such as Decentraland and the Sandbox,[1] which are among the first virtual worlds to be implemented that makes use of decentralized capabilities (such as Blockchain; see Figure 12.1). Along this road, there are also various platforms,

* Lik-Hang Lee and Carlos Bermejo equally contributed to this chapter.
1 https://decentraland.org.

Metaverse Communication and Computing Networks: Applications, Technologies, and Approaches, First Edition.
Edited by Dinh Thai Hoang, Diep N. Nguyen, Cong T. Nguyen, Ekram Hossain, and Dusit Niyato.
© 2024 The Institute of Electrical and Electronics Engineers, Inc. Published 2024 by John Wiley & Sons, Inc.

Figure 12.1 Metaverse festival. Source: Decentraland.

such as Second Life,[2] Minecraft,[3] and Roblox,[4] that have contributed to the current interest in the Metaverse. There are also several firms, such as Niantic,[5] Microsoft (with Mesh),[6] and lately Meta, that have contributed to this interest (before formerly known as Facebook).

The Metaverse will have profound repercussions for both our culture and our individual lives. The trading of virtual assets online and in online games, in which players may produce and exchange digital goods such as accessories for avatars, is already undergoing significant transformation. We are already seeing this transition. However, the creation of the Metaverse still faces a number of obstacles, including privacy concerns, ethical considerations, and issues of governance. To begin, the technologies that were used in the production of the Metaverse gave rise to new ethical and privacy concerns. The continual sensing of users' devices in order to give them with more authentic and immersive experiences poses a potential risk to users' privacy, security, and even physical well-being [21]. The information obtained from the users' biometrics, such as their gaze, gait, and heart rate, reveals significant parts of the users' mental makeup [20].

Second, the Metaverse might be seen as a miniature version of our civilization [16]. The administration of such huge virtual worlds involves a number of issues when it comes to the regulation of user behavior [19]. As a result, the Metaverse needs rules and guidelines to follow in order to effectively administer the platform and its users. For instance, some of the problems listed above are

---

2 https://secondlife.com.
3 https://www.minecraft.net/en-us.
4 www.roblox.com.
5 https://nianticlabs.com/en/.
6 https://www.microsoft.com/en-us/mesh.

experienced by users of online virtual environments such as Horizons by Meta. The unethical behavior of which depicts avatars using the virtual environment of the Metaverse as a means to engage in sexual harassment of other avatars.[7]

These behaviors bring up questions about the appropriate penalties for bad behavior as well as the rules and policies that the Metaverse ought to have. The murdering of an avatar by another member of the platform raises the question of how the Metaverse will control this behavior. Will the Metaverse adhere to the policies that are in place according to the various local governments, knowing that governments own different norms and regulations?

In this chapter, we highlight the considerable issues that the Metaverse will face in the future in terms of ethics, governance, and security and privacy concerns. In addition, we take into account the tendencies and strategies that are now being implemented by existing online worlds as part of our solutions and research routes to allow a more sustainable Metaverse. We pose questions to the community about important facets of the administration and design of an accessible and inclusive Metaverse, but we do not provide answers to those concerns. In conclusion, we discuss a basic concept for an ethically designed Metaverse that is built on a modular structure.

12.2 Privacy and Security

For the purpose of creating immersive experiences, the Metaverse utilizes data obtained from the "actual world." A user's avatar may be realistically controlled by sensors that are connected to the user (for example, a gyroscope that tracks the user's head motions). Additionally, the Metaverse presents additional difficulties in the shape of huge virtual worlds, which might expose users to privacy invasions in the form of activities such as listening in on conversations taking place on other platforms. In this part, we will discuss the primary obstacles that will need to be overcome by creators, designers, practitioners, regulators, and users in order to realize the potential of the Metaverse.

12.2.1 Confidentiality with Regard to the Senses

Extended reality (XR) technologies make it possible to have a Metaverse experience that is more immersive, realistic, and satisfying. XR devices are able to collect a great quantity of information, ranging from the biometric data of users to spatial data, including surrounds such as bystanders' physical space (rooms) [4, 8]. This information may include both the user's own biometric data and

---

7 https://www.bbc.com/news/technology-60247542.

geographical data. Previous research [4, 8, 21], and published articles have shown that XR technologies pose a number of privacy and security risks to users and bystanders alike.

In general, these technologies make use of sensors in order to scan and monitor the environment around its users [8]. These scans have the potential to gather information that might be useful to users and bystanders who are located within the monitoring system's coverage zone [4, 21]. Head-mounted displays, or HMDs, are often used to show the Metaverse. These displays have the ability to capture biometric data (such as head movement and eye tracking) that is not immediately apparent to users. For instance, a user's sexual inclinations may be discerned from their gaze data [20]. The most private and sensitive portions of our psyche are placed at danger when we have our biometric data gathered. As a result, these gadgets have to handle the data in accordance with a few guiding principles that safeguard the privacy of the users.

Solutions

Different approaches to protecting the privacy of users in these kinds of situations have been proposed in a few different papers [4, 10, 15]. These works suggest frameworks for controlling the privacy of data by safeguarding the input in the data collection process. These frameworks provide users and developers the ability to exercise granular control over the level of privacy afforded to input data (from different sensors). This granular control over the data that are gathered may be handled by privacy-enhancing technologies (PETs), which scramble any sensitive information that is captured by sensors before it is sent to cloud services (e.g. online games). Despite the fact that these solutions govern the data gathering and exchange with other organizations (for example, cloud servers), the creation of the Metaverse still requires participation from all involved parties (e.g., device manufacturers).

12.2.2 Protection of One's Actions and Conversations in Private

It is possible to think of the Metaverse as a miniature version of our actual world. Users are able to engage in conversation with one another's avatars and other virtual assets. It is possible to deduce the users' routines, activities, and preferences in the Metaverse based on the relationships and social interactions between users. In a manner similar to that of the biometric data, this information may reveal the psyches of the users. In addition, users run the danger of having their privacy compromised by the metadata that is produced during any social engagement with other avatars (such as discussions or responses). The behavior of users might be monitored and controlled more effectively with the aid of this information. Who is in charge of all of this information and how is it being used?

Solutions

When it comes to more particular situations, such as going shopping in the virtual world, we may see people using backup avatars, sometimes known as clones, in order to conceal their true behavior from other avatars that are listening or attacking them [6]. Users of the Metaverse should also have certain choices to govern their personal space in the virtual world that may be configured according to their preferences. For instance, privacy bubbles limit the user's ability to see other avatars that are located outside the bubble. Facebook, which is now known as "Meta," had features of a similar kind on its social platform Horizons.

Artificial intelligence (AI)-driven technologies have the potential to both improve services offered in the Metaverse (for example, conversational AI in the Metaverse) and influence the actions taken by users. In this scenario, privacy protection strategies, such as differential privacy, homomorphic encryption, and federate learning, should be used in order to minimize the amount of users' personally identifiable information that is exposed [12]. In addition, these AI-driven technologies need to give transparency, explainability, and resilience in order for users of the Metaverse to have a complete comprehension of the decision-making process that such methods entail.

12.2.3 The Protection of Participants and Bystanders

Other security issues, such as safety [3, 26], may also have an effect on users in addition to the security difficulties that are involved in the protection of information acquired while it is being shared (for example, sensors, users' actions in the Metaverse), and against manipulation. The existing head-mounted displays (HMDs) that are used to show the Metaverse might obscure the user's view of the real world as well as the user's ability to identify things in the immediate area, which increases the danger of falling [3, 26]. The Metaverse may also be conceptualized as an enhancement of the physical world via the use of mixed reality (MR), as was covered in earlier part of this chapter. Consequently, the user runs the risk of obstructing potential threats such as autos when they are strolling on the streets.

Solutions

The authors of [14] suggest the representation of actual users (those who are situated in the same room as the user) as virtual ("shadow") avatars as a means of preventing collisions during multiuser virtual reality (VR) experiences. It is possible to limit the number of times users collide with real-world objects by rerouting their movement while also breaking their immersion in the virtual environment [1, 24]. Other potential solutions include using sensors that are built into the headsets to do a spatial scan of the room and show the real-world items in the virtual environment if there is a risk of a collision occurring. These solutions shouldn't need the gathering of more sensitive data and should require the same privacy precautions that existing sensors do at the very least.

12.2.4 Open Challenges

There is still a need for XR platforms (device makers, systems, and frameworks) to provide a uniform privacy protection solution across all entities that comply to the Metaverse, despite all of the solutions that have been presented so far. The Future of Privacy Forum[8] is an organization that promotes the idea that people should be the ones to process their own private data (e.g. XR device). XR devices that collect sensitive data should provide granular control (switches) to manage the input data flows from sensors and provide visual cues (for example, an LED in the device) when personal data are collected or transmitted. This is to ensure that users are aware when their data are being collected or transmitted.

Even when key privacy protection mechanisms are included in virtual worlds (such as the privacy bubble in Horizon Worlds), users either are not completely aware of them or do not know how to utilize them, as we have seen. This is despite the fact that these tools are there. Everyone who participates in the Metaverse should be able to easily understand the privacy policies and practices that are in place. In accordance with the recommendations made at the XR Privacy Summit, all of the parties participating in the process of developing and maintaining the Metaverse should implement an institutional review board (IRB) model of some kind inside their respective companies. In order for businesses to win broad customer trust and acceptance, they need to implement a variety of technological solutions and policies while remaining in compliance with the law.

As an example, we may look at local regulators like the California Consumer Privacy Act (CCPA) and the General Data Regulation Protection (GDPR), both of which are geared at the protection of persons in contexts including surveillance. Despite the fact that they are derived from various local laws, these standards are quite similar in how they safeguard the data of users. If the privacy regulatory safeguards for the Metaverse are built on a framework that is modular-based, then the Metaverse will be able to adapt to the requirements of the local authorities while still providing a consistent policy to protect the private of its users. The Metaverse ought to support PETs and in-sensor data processing processes and advocate for them. In the Metaverse, any data collecting and processing activity may be registered on a distributed ledger known as a blockchain by any interested entity. In conclusion, the Metaverse ought to ensure that no party has a monopoly on the gathering of data in any activity, and this applies to all parties.

8 https://fpf.org.

12.3 Governance

A study of the existing approaches that online social media and gaming platforms are using to govern themselves and the societal implications that these approaches have on users is something that may be done before an investigation into the possibilities of governance in the Metaverse [7, 9]. The users of these platforms are subject to problems such as inappropriate behavior, harassment from other users, and conflict with other users of the site [22]. When online communities expand to the point that its moderators, who are originally other members of the community, are unable to keep up with the volume of comments and inappropriate behavior of community members, this gives rise to a number of problems.

Automation technologies have been included into social networks like Facebook and Twitter in order to manage undesirable behavior on these platforms (e.g. banning inappropriate posts). These platforms further depend on the reports of other members in order to monitor spam and incorrect postings made by other users. In a similar manner, players of massively multiplayer online games develop communities in order to self-govern the inappropriate behavior of other players [11]. Examples of such inappropriate behavior include stealing the digital assets of other players and spamming. This section demonstrates how existing strategies from online social and multiuser gaming platforms may be adapted for use in regulating the ("limitless") Metaverse by providing some examples.

12.3.1 Statutes and Regulations

The software code of the Metaverse may be compared to the physical laws that govern the natural world, and like those laws, code can impose limits on the structure of the Metaverse. The behavior of individuals and the places they inhabit online are shaped by code [17]. It is up to the firms and developers who will use the internet platform to select which functions will be included in it [11].

The social behavior of users in the Metaverse may also be influenced by the code rules that govern the environment. The decisions that are made by developers when the Metaverse is being built may have an effect on the ways in which users engage with the platform [17]. For instance, developers may build up a privacy bubble mode, which allows users to choose their own personal zone (the bubble) and control who may access it (e.g. interactions such as chat). Because users always have the option to limit their access to other avatars, the privacy mode has the potential to alter the way individuals interact with one another in the Metaverse.

Users often develop novel methods to engage with online platforms that are not governed by the rules provided by the code [11], despite the fact that there are "rules of the code." Hacking is a fantastic illustration of how alterations to the source code may allow new virtual worlds to be created, complete with new tools for creation, interactions, and overall experiences. For instance, modifications have been implemented into the online versions of games like Grand Theft Auto and Skyrim, which dramatically alters the experience of playing the game [11].

12.3.2 Distributed Ledger Technology (DLT) and Decentralized Autonomous Organizations (DAO)

The ability of blockchain to facilitate the generation of information that is trustworthy as well as smart contracts [19] positions it as a transformation driver for online platforms since it enables the transmission of information and transactions that call for authentication and trust [19]. As such, there is a possibility of improving governance in the Metaverse.

By leveraging Blockchain technology and smart contract, decentralized autonomous organizations, or DAOs, can be built. These decentralized groups make it possible for people all around the world to work together and coordinate using Internet platforms like the Metaverse. In most cases, DAOs take the form of flat, completely democratic organizations in which every member is eligible to take part in the voting process. The system is also capable of handling services automatically, such as the sale of a property asset located inside the Metaverse. An example of a DAO operating in the real world may be seen in the online platforms that strive to produce Metaverse virtual worlds, such as Decentraland and the Sandbox. Users are able to provide input into the decision-making process on these sites.

However, DAOs might run into problems with scale and interaction with the Metaverse community. The fact that many DAOs are structured in a flat-based manner makes it difficult for members to participate in the decision-making process. This is because the sheer volume of voting sessions may be exhausting. The algorithms that are inherent in autonomous choices have the potential to have a significant influence on the Metaverse as a whole. The use of open source methodologies in the construction of the Metaverse is strongly encouraged. However, it may be difficult for members of the general public to comprehend these algorithms. As a result, auditing mechanisms should be accessible from inside the platform. In conclusion, it is important to point out that the computational techniques being discussed here cannot capture the whole of human governance practices. However, they may aid noncomputational processes inside the Metaverse, such as voting or the moderation of material.

12.3.3 Ethical Governance Based on Modules

The difficulty of high-stakes governance in online platforms is made more difficult by the fact that they lack several fundamental characteristics of offline (physical) governing systems [22]. The authors [22] of this article suggest a model to governance online platforms, that is modular and bottom-up in nature. This versatility may make it possible to design portable tools that can be customized to work on a variety of systems and serve a variety of purposes. The governance layer needs to include a diverse range of procedures (such as juries and formal discussions), and it ought to interface with many additional governance systems. DAOs are a good illustration of a governance structure that adheres to the aforementioned characteristics (DAOs). Users are given the opportunity to have an active role in the decision-making process via these governance systems (e.g. Decentraland, Sandbox).

12.3.4 The Role of Online Platforms in Contributing to the Common Good

It is possible that the surroundings that exist online have the potential to favorably influence the behaviors of young people [25]. The results of a two-year research conducted "in the wild" on the popular (custom) Minecraft game reveal that young people (aged 8–13) have the potential for ownership and control mechanisms that may be used to regulate the culture and environment of an online community. The research also emphasizes the significance of moderators, or adult involvement in this particular scenario, in the process of resolving various issues that arise in the community. The authors also recommend that online platforms think about implementing tools to deal with inappropriate behavior from players (i.e. punitive techniques), as well as mechanisms to encourage appropriate behavior (i.e. preventive approaches). In addition to this, they suggest the use of incentive systems in order to encourage good behavior and constrain bad actors. These incentive schemes have the potential to inspire cooperation, as well as collaboration and shared planning. In conclusion, the Metaverse ought to think about building tools (like DAOs) that will enable people to participate in conversations, introspection, and decision-making. We are able to see the significance of community governance in terms of its role in shaping and directing behaviors that occur online (including moderation and conflict resolution).

12.3.5 Open Challenges

The question of who will rule in the Metaverse has to be addressed. These governments are going to be based on the existing local laws, right? The modular techniques provide a solution to the problem of adapting the rules and management of the Metaverse to the circumstance at hand. For instance, if it is mandated that the Metaverse adhere to the laws of the local environment, the modules will switch places appropriately. The next concern that arises is how users who are located in different parts of the globe will be handled, as well as how these locally established guidelines may be integrated into the Metaverse in order to realize the vision of a worldwide online community. It is possible that we will end up with a version of the Metaverse that has borders and in which the rules are implemented in a different way. In an ideal world, as we shall see in the last part, the Metaverse should not be restricted by local rules that limit the freedom of its users and be open to all users. We are able to anticipate that the tools and techniques that have been discussed might mold the Metaverse and adapt the essential rules in the appropriate manner.

12.4 Creation, Social Good, and Ethical Design

The Metaverse has the ability to offer new channels through which we may express ourselves and communicate with others without any restrictions, which might lead to a revolution in our existing society (location, time, race, and gender). This part demonstrates the good effects that the Metaverse has had on our society as well as the ethical framework that we have provided.

12.4.1 The Beginning of Everything in the Metaverse

In the Metaverse, the process of creation will prove to be an extremely valuable asset. The production of digital assets has resulted in the opening of a new market, which has led to an increase in the number of work opportunities as well as monetary and other kinds of revenue. This is something that we have already seen with existing platforms such as Decentraland and the Sandbox. The ownership of a specific object, such as digital artwork or collectables, may be represented by a nonfungible token (NFT), which is a one-of-a-kind digital token that is created with the use of blockchain technology. NFTs function as a one-to-one mapping between an owner, who is denoted by a crypto wallet address, and the asset that the NFT references (usually by a uniform resource identifier, URI). The characteristics of real-world things, such as scarcity and one-of-a-kindness, are imitated by NFTs [23]. For instance, the virtual lands and other digital assets in the game Decentraland, such as clothes and accessories for avatars, are managed with the use of NFTs.

A new revolution is taking place in online platforms such as play-to-earn games. NFTs-powered games, such as Axie Infinite,[9] which is a monster-battling game similar to Pokemon, allow players to earn money while they play by allowing them

9 https://axieinfinity.com/.

to sell their improved monsters. In addition to the markets that trade in NFTs, we can see this revolution taking place. The create-to-earn model is another one that virtual worlds are beginning to include into their platforms (for example, Sandbox). In this model, users of the platform may contribute to the creation of the platform while also selling their own developed digital products.

The absence of centralized authority gives NFTs the ability to serve as a tool for the commercialization and democratization of the production of content. However, con artists and other malevolent content makers are able to take advantage of the system to sell copies of low-quality NFTs since these democratization tools, which decrease the barriers to making and distributing material, making it easier for people to do so. A number of the NFT trading platforms have implemented invite-only practices, which restrict access to their platforms to a predetermined set of content providers. The benefits of NFTs as an open-access content generation tool are diminished when policies of this sort are implemented. The use of DAOs and users of the platform to construct a reputation-based system where everyone may vote and enforce rules in order to maintain the quality of NFTs and limit the number of scams is one potential approach to the problem.

Additionally, several research works [2, 13, 23] emphasize the application of moderation methods employing AI technology to minimize toxic behaviors in games, online forums, and content production. These practices aimed at reducing the negative impact of these behaviors. Users of the platform and AI tools are included in these AI-based and cross-modality solutions in order to give a scalable moderation framework that can be deployed in the Metaverse. In order to strike a healthy balance between NFTs and quality control, further research is required.

Digital Twins

Digital twins are virtual things that are built to replicate the look and physical behavior of their physical counterparts. This definition encompasses both aspects of the actual item (of the real world). Users of the Metaverse will be able to view themselves not only in the realms of virtual worlds in VR but also in other paradigms such as mixed reality (MR), in which the virtual and the physical are merged. This will be made possible by digital twins, which are able to synchronize their physical and virtual states.

Furthermore, we are able to anticipate that the activities of users in the real world, such as going on vacation or shooting pictures, will provide content for the Metaverse and the avatars that exist inside it (e.g. displaying the photo in the virtual world). When this occurs, the Metaverse will be a realm that is always changing and is in step with the physical world. There are still a few obstacles to overcome when it comes to the ownership of digital twins. Utilizing a hyperledger such as Blockchain is the solution that provides the highest level of simplicity when it comes to safeguarding the authenticity and origin of digital twins.

12.4.2 The Online Community Space

Accessibility, variety, equality, and humanism are all areas in which the Metaverse has the potential to have a beneficial influence on social good [5]. Accessibility. The Metaverse has the potential to facilitate worldwide cooperation in spite of the physical distances involved. In addition to that, the Internet world has the potential to make traditionally offline activities, such as concerts, more accessible. The Metaverse has the potential to facilitate a broad variety of social gatherings that would be physically impossible to stage, such as concerts attended by millions of people all over the globe. Take, for instance, the graduation ceremony that was conducted at UC Berkeley in the year 2020; it was held in Minecraft.

Diversity

When using the Metaverse, one is able to overcome the constraints that are imposed by the physical reality. The Metaverse may include an infinite number of virtual worlds and places. In addition, the Metaverse places no restrictions on the things that we may do there. We may see the Metaverse as a location where people can exhibit their artwork, engage in social activities, play games, and learn new things, among other things.

Equality

One may consider the Metaverse to be a place of equality, free from the constraints of gender, color, handicap, and socioeconomic standing. Users are able to personalize their avatars, and the only restriction is their own creativity (e.g. they can be a cat). Because of this characteristic, the Metaverse will be able to create a civilization in the virtual world that is more equitable and more sustainable. Design AI systems with Equality will significantly impact the user acceptability to the Metaverse [27], while various stakeholders and their interests should be addressed by offering equal access to the Metaverse [28].

Humanity

The Metaverse has the potential to open up channels of communication across cultures and to safeguard such channels. For instance, the Metaverse may serve as the platform for the preservation and restoration of works of art. On the other hand, this may provide difficulties for the platform in terms of how it will sustain various cultural and creative traditions. The Metaverse has the potential to generate an infinite number of communities, each with their own unique set of values and perspectives.

Trust

On the other hand, as was discussed in the introduction, the anonymity that is afforded to members of online communities may give rise to issues about the manner in which information is transmitted in the Metaverse. We are able to witness rising worries with stories of false news and the repercussions that the "bad" Internet may have in terms of distributing information and, therefore, knowledge. Humans, in the past, have usually gained knowledge about the physical world via rule-based studies. Misinformation, such as false news, has a negative impact on those who are responsible for spreading it as well as those who are foolish enough to do so (e.g. inform, warn).

Testimonies and trust will play an even more significant part in the Metaverse since, in many instances, the individual giving the testimony will not be a real person but rather their avatar. This will make it more difficult to verify the authenticity of the testimony. Several industry professionals investigate several extra response concepts, such as punishment, in order to demonstrate why truth denial is a more serious issue than false belief.

A vital piece of functionality in the fight against the spread of false information will be incentive systems that encourage avatars to trust one another. The Extended Reality Safety Initiative (XRSI)[10] and the XR Association[11] are two organizations that advocate for the responsible development and deployment of XR technologies (including the Metaverse). These organizations have the ability to persuade corporations and establishments to adhere to ethical designs in their implementations of the Metaverse (e.g. XR devices).

12.4.3 Ethical Structure with Modular Designs

Don Norman was one of the many people who contributed to the development of the Human-Centered Design (HCD) ideology [18]. We take into consideration this design strategy for the Metaverse since it needs the active participation of users in the decision-making process as well as the design process. As we have shown, the Metaverse will have a tremendous effect on human civilization as well as production and life in general. An architecture for the Metaverse that is built on modules will make it possible to conform to the prerequisites and needs of a global platform. As a result, the early strategy that we have devised is to include every required member (developers, regulators, users, and content providers) in the creation and execution of the metaverse.

In Figure 12.2, many different instances of modules that will complete a certain job are shown. In the decision-making module, for instance, there will be participation from members, regulators, and software developers. We think that distributed autonomous organizations, or DAOs, have the potential to tackle scalability issues when such issues are distributed among several Metaverse characteristics (modular approach). These modules might be seen as an example of a

10 https://xrsi.org.
11 https://xra.org.

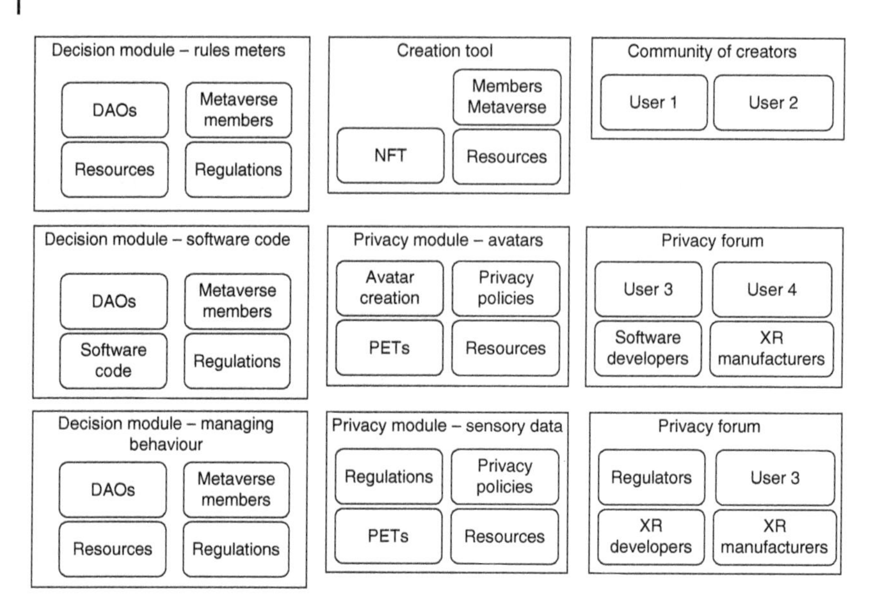

Figure 12.2 An example of a modular-based Metaverse architecture motivated the "Ethical Hierarchy of Needs," where each module is interchangeable.

federated approach. These decision modules are still linked to other decision modules, resources, and regulations, despite the fact that they are capable of making autonomous choices, such as how to respond to inappropriate behavior.

These decision algorithms must to be open and be accessible to any participant in the Metaverse, as was mentioned in earlier section of this chapter. As a result of the potential for code and hardware implementations to have an impact on the users' sense of privacy and safety, the Metaverse will undergo several adjustments. In addition, there should be certain default privacy protection measures created in order to secure the information of users and limit the likelihood of data monopoly, which is something that is already occurring with our existing Internet. These choices will adhere to (if required) the rules set out by the local government and adjust to any changes that may occur in the future.

The Metaverse will need the development of an effective mechanism for the distribution of trust across all levels. A reputation-based system that operates on top of the Blockchain will provide the Metaverse with a tool that can counteract assaults made during decision-making processes and restrict the spread of false information (e.g. fake news). Tools that use machine learning or AI to automate procedures in the Metaverse need to be the topic of discussion and ought to adhere to the regulations that are now in place regarding explainability and transparency in accordance with GDPR. We show the alignments of the Metaverse with the "Ethical Hierarchy of Needs" (licensed under CC BY 4.0) for an ethical design by using the following examples:

Human Rights

Every person should feel welcome in the Metaverse, which is why it should be easily accessible, varied, and inclusive. Users have an inherent right to their privacy, which they may defend using the many instruments that are at their disposal (PETs, avatar creation). All of the modules may be swapped out for one another, and they will employ existing technologies such as Blockchain and DAOs to govern any choice that is made in the Metaverse (for example, a change in the regulations already in place regarding users' inappropriate behavior). Any participant on the platform need to be able to see through and comprehend all of the active aspects of the Metaverse, including the code. This may make voting (using DAOs) in the decision-making processes simpler, reducing the amount of complexity involved. The Metaverse is able to deliver a decentralized, open, sustainable, and interoperable system because to the modular-based architecture that is used. In addition to this, the technologies that are employed in the system should contain the technologies that are required for a Metaverse that is both private and secure (e.g. AI-driven services using PETs).

Human Endeavor

The Metaverse will include a reputation-based system that will be inextricably linked to users and will be administered via Blockchain and DAOs. People will be able to report malevolent users' inappropriate behavior and malpractice when voting utilizing DAOs thanks to this reputation system. Participants in the Metaverse, including developers, content providers, regulators, and users, will all have a voice in the organization's decision-making process. Together, the reputation-based system and the flexibility of the proposed solution have the potential to provide users with an experience that is both dependable and useful inside the Metaverse. For instance, the reputation will offer an additional layer of protection during the voting procedures, which will help minimize the number of assaults and inappropriate behaviors carried out by members of the Metaverse.

Human Experience

The Metaverse will be an immersive experience that will allow new forms of social interaction and content production that go beyond the limitations of our existing physical reality (e.g. regulations, geographical location). It is anticipated that the Metaverse will owing to the fact that it is possible to create a vast number of avatars and groups, the virtual world is open and welcoming to all users. As mentioned in earlier sections, the modules should be self-explanatory and transparent to users. This means that the code, decision-making, resources, and technologies (for example, at the sensor level) should be easy to understand. This gives users the ability to create, share, and "live" in the Metaverse. The Metaverse built on modular components will streamline the currently burdensome one-size-fits-all systems in an effort to handle several facets of the Metaverse (e.g. all decisions using one DAO).

Limitations

Both a modular method for the Metaverse and an approach to allow ethical creation are part of what we propose here. However, we have not shown that the use of such tools would result in the participation of all relevant parties (for example, users of the Metaverse) in the making of just judgments that will lead to the achievement of an ethical objective. However, examples from already available platforms such as Decentraland, the Sandbox, and massively multiplayer online games (MMOGs) have shown the Metaverse's potential as a social good as well as the involvement and cooperation of users in the decision-making process (using DAOs in case of the first two).

12.5 Conclusions and Future Research Directions

In this chapter, an outline of the most significant difficulties that the Metaverse will confront in the future in terms of privacy, governance, and ethics is presented. In addition to this, we demonstrate the first step toward developing an ethical architecture for the Metaverse. However, there is still a lot of work to be done in terms of study and testing before the Metaverse can become the paradise that many people hope it will become.

Regarding future research directions, it is challenging to maintain privacy, governance, and ethics in the Metaverse, especially when significant numbers of users with diversified roles have participated in the immersive environments. It is also important to note that once things go wrong in terms of privacy, governance, and ethics, the mechanism of information sharing among shareholders (e.g. virtual space owners and platform service providers) for crisis management will become an indispensable issue in such decentralized frameworks.

Bibliography

1 Eric R. Bachmann, Eric Hodgson, Cole Hoffbauer, and Justin Messinger. Multi-user redirected walking and resetting using artificial potential fields. *IEEE Transactions on Visualization and Computer Graphics*, 25(5):2022–2031, 2019.
2 Eshwar Chandrasekharan, Chaitrali Gandhi, Matthew Wortley Mustelier, and Eric Gilbert. Crossmod: A cross-community learning-based system to assist reddit moderators. *Proceedings of the ACM on Human-Computer Interaction*, 3(CSCW):1–30, 2019.

3 Emily Dao, Andreea Muresan, Kasper Hornbæk, and Jarrod Knibbe. Bad breakdowns, useful seams, and face slapping: Analysis of VR fails on youtube. In *Proceedings of the 2021 CHI Conference on Human Factors in Computing Systems*, pages 1–14, 2021.

4 Jaybie A. De Guzman, Kanchana Thilakarathna, and Aruna Seneviratne. Security and privacy approaches in mixed reality: A literature survey. *ACM Computing Surveys (CSUR)*, 52(6):1–37, 2019.

5 Haihan Duan, Jiaye Li, Sizheng Fan, Zhonghao Lin, Xiao Wu, and Wei Cai. Metaverse for social good: A university campus prototype. In *Proceedings of the 29th ACM International Conference on Multimedia*, pages 153–161, 2021.

6 Ben Falchuk, Shoshana Loeb, and Ralph Neff. The social metaverse: Battle for privacy. *IEEE Technology and Society Magazine*, 37(2):52–61, 2018.

7 Tarleton Gillespie. Custodians of the internet: Platforms, content moderation, and the hidden decisions that shape social media. 01 2018.

8 Jaybie Agullo de Guzman, Aruna Seneviratne, and Kanchana Thilakarathna. Unravelling spatial privacy risks of mobile mixed reality data. *Proceedings of the ACM on Interactive, Mobile, Wearable and Ubiquitous Technologies*, 5(1):1–26, 2021.

9 Oliver L. Haimson and Anna Lauren Hoffmann. Constructing and enforcing "authentic" identity online: Facebook, real names, and non-normative identities. *First Monday*, 2016.

10 Jinhan Hu, Andrei Iosifescu, and Robert LiKamWa. LensCap: Split-process framework for fine-grained visual privacy control for augmented reality apps. In *Proceedings of the 19th Annual International Conference on Mobile Systems, Applications, and Services*, pages 14–27, 2021.

11 Sal Humphreys. Ruling the virtual world: Governance in massively multiplayer online games. *European Journal of Cultural Studies*, 11(2):149–171, 2008.

12 Thien Huynh-The, Quoc-Viet Pham, Xuan-Qui Pham, Thanh Thi Nguyen, Zhu Han, and Dong-Seong Kim. Artificial intelligence for the metaverse: A survey. *arXiv preprint arXiv:2202.10336*, 2022.

13 Yubo Kou and Xinning Gui. Mediating community-AI interaction through situated explanation: The case of AI-led moderation. *Proceedings of the ACM on Human-Computer Interaction*, 4(CSCW2):1–27, 2020.

14 Eike Langbehn, Eva Harting, and Frank Steinicke. Shadow-avatars: A visualization method to avoid collisions of physically co-located users in room-scale VR. In *IEEE Workshop on Everyday Virtual Reality*, 2018.

15 Kiron Lebeck, Kimberly Ruth, Tadayoshi Kohno, and Franziska Roesner. Securing augmented reality output. In *2017 IEEE Symposium on Security and Privacy (SP)*, pages 320–337. IEEE, 2017.

16 Lik-Hang Lee, Tristan Braud, Pengyuan Zhou, Lin Wang, Dianlei Xu, Zijun Lin, Abhishek Kumar, Carlos Bermejo, and Pan Hui. All one needs to know

about metaverse: A complete survey on technological singularity, virtual ecosystem, and research agenda. *arXiv preprint arXiv:2110.05352*, 2021.

17 Lawrence Lessig. *Code and Other Laws of Cyberspace*. Basic Books, Inc., USA, 1999. ISBN 046503912X.

18 Donald A. Norman. Human-centered design considered harmful. *Interactions*, 12(4):14–19, 2005.

19 Svein Ølnes, Jolien Ubacht, and Marijn Janssen. Blockchain in government: Benefits and implications of distributed ledger technology for information sharing. *Government Information Quarterly*, 34(3):355–364, 2017.

20 Patrice Renaud, Joanne L. Rouleau, Luc Granger, Ian Barsetti, and Stéphane Bouchard. Measuring sexual preferences in virtual reality: A pilot study. *CyberPsychology & Behavior*, 5(1):1–9, 2002.

21 Franziska Roesner, Tadayoshi Kohno, and David Molnar. Security and privacy for augmented reality systems. *Communications of the ACM*, 57(4):88–96, 2014.

22 Nathan Schneider, Primavera De Filippi, Seth Frey, Joshua Z. Tan, and Amy X. Zhang. Modular politics: Toward a governance layer for online communities. *Proceedings of the ACM on Human-Computer Interaction*, 5(CSCW1):1–26, 2021.

23 Tanusree Sharma, Zhixuan Zhou, Yun Huang, and Yang Wang. " It's a blessing and a curse": Unpacking creators' practices with non-fungible tokens (NFTs) and their communities. *arXiv preprint arXiv:2201.13233*, 2022.

24 Qi Sun, Anjul Patney, Li-Yi Wei, Omer Shapira, Jingwan Lu, Paul Asente, Suwen Zhu, Morgan McGuire, David Luebke, and Arie Kaufman. Towards virtual reality infinite walking: Dynamic saccadic redirection. *ACM Transactions on Graphics (TOG)*, 37(4):1–13, 2018.

25 Katie Salen Tekinbaş, Krithika Jagannath, Ulrik Lyngs, and Petr Slovák. Designing for youth-centered moderation and community governance in minecraft. *ACM Transactions on Computer-Human Interaction (TOCHI)*, 28(4):1–41, 2021.

26 Wen-Jie Tseng, Elise Bonnail, Mark Mcgill, Mohamed Khamis, Eric Lecolinet, Samuel Huron, and Jan Gugenheimer. The dark side of perceptual manipulations in virtual reality. *arXiv preprint arXiv:2202.13200*, 2022.

27 Pengyuan Zhou, Benjamin Finley, Lik-Hang Lee, Yong Liao, Haiyong Xie, and Pan Hui. Towards user-centered metrics for trustworthy AI in immersive cyberspace. *ArXiv*, abs/2203.03718, 2022.

28 Pengyuan Zhou, Hengwei Xu, Lik Hang Lee, Pei Fang, and Pan Hui. Are you left out? An efficient and fair federated learning for personalized profiles on wearable devices of inferior networking conditions. *Proceedings of the ACM Interactive, Mobile, Wearable and Ubiquitous Technologies*, 6(2), July 2022. doi: 10.1145/3534585.

Index

Metaverse Communication and Computing Networks: Applications, Technologies, and Approaches, First Edition.
Edited by Dinh Thai Hoang, Diep N. Nguyen, Cong T. Nguyen, Ekram Hossain, and Dusit Niyato.
© 2024 The Institute of Electrical and Electronics Engineers, Inc. Published 2024 by John Wiley & Sons, Inc.

b

d

h

i

j

k

p

u

v

W

X

www.ingramcontent.com/pod-product-compliance
Lightning Source LLC
Chambersburg PA
CBHW072110250125
20788CB00003B/20

* 9 7 8 1 3 9 4 1 5 9 9 8 7 *